T. R. OKE

Boundary Layer Climates

LONDON
METHUEN & CO LTD
A HALSTED PRESS BOOK
JOHN WILEY & SONS
NEW YORK

First published 1978
by Methuen & Co Ltd
11 *New Fetter Lane London* EC4P 4EE

Printed in Great Britain by
William Clowes and Sons Limited
London, Beccles and Colchester

ISBN (*hardbound*) 0 416 70520 0
ISBN (*paperback*) 0 416 70530 8

Library of Congress Cataloging in Publication Data

Oke, T R
Boundary layer climates.

'A Halsted Press book.'
Bibliography: p.
Includes indexes.
1. Microclimatology. 2. Planetary boundary layer.
I. Title.
QC981.7.M5034 1978 551.6'6 77-25266
ISBN 0-470-99364-2 (Cloth)
0-470-99381-2 (Paper)

To my mother Kathleen
and my wife Midge

Contents

Acknowledgements

The author and publishers would like to thank the following for permission to reproduce copyright figures:

Academic Press, London for 2.9b, 4.2a, 4.10, 4.12 and A2.6; Academic Press, New York for 1.9; Aerospace Medical Association, Washington for 6.15; American Association for the Advancement of Science, Washington for 6.6, 6.7 and 9.16; American Geophysical Union, Washington for 3.1, 4.21b and 4.22; American Meteorological Society, Boston for 5.8, 7.11, 8.10, 8.17, 8.18, 8.20, 9.6, 9.9 and 9.17; Edward Arnold, London for 5.9 and 6.4; *Atmosphere* for 8.6 and 8.14; Blackwell Scientific Publications, Oxford for 4.13; Butterworths, London for 6.8; Cahiers de Géographie de Québec, Quebec for 9.13a; California Agricultural Experimental Station for 7.5; Cambridge University Press, Cambridge for 4.4 and 4.16; Cambridge University Press, New York for 6.11; Clarendon Press, Oxford for 4.6; Colorado State University for 5.11; Connecticut Agricultural Experiment Station, New Haven for 7.3; John Davies, McMaster University for 3.11; Elsevier, Amsterdam for 2.8, 2.16, 4.18 and 5.12; *Ergonomics*, London for 6.16; W. H. Freeman, San Francisco for 6.5; Harper & Row, New York for 6.13; Harvard University Press, Cambridge for 3.3; Hemisphere, Washington for 4.8; Her Majesty's Stationery Office, London for 2.9 and 8.3; Institute of Hydrology, Wallingford for 4.20; Johns Hopkins University Press, Baltimore for 8.8; *The Lancet*, London for 6.12; Macmillan, New York for 6.2; Marine Biological Laboratory, Massachusetts for 6.9; Meteorological Institute, Uppsala University for

3.7; National Research Council, Ottawa for 6.14; New Science Publications London for 8.4; Pergamon Press, Oxford for 3.4, 4.9, 4.11, 8.15 and 8.16; D. Reidel, Dordrecht for 3.10 and 8.19; Royal Meteorological Society, Bracknell for 3.9, 5.13b, 7.1 and 7.4; Springer Verlag, Berlin for 4.13 and 4.17; Steno Memorial Hospital, Copenhagen for 6.3; Taft Sanitary Engineering Centre, Cincinnati for 9.13b; United States Department of Commerce, Washington for 3.13, 3.14 and 3.15; United States Government Printing Office, Washington for 9.5, 9.7 and 9.8; University of Chicago Press, Chicago for 2.10, 2.11 and 2.15; University of Oregon, Corvallis for 3.2; University of Texas Press, Austin for 8.9; Volcani Institute for Agricultural Research, Israel for 4.19; World Meteorological Organization, Geneva for 8.13.

Preface

This book is designed to introduce students to the nature of the atmosphere near the ground. It is especially aimed at those whose curiosity about the atmosphere has been raised by an introductory course or general interest, but who are daunted by the technical nature of most micro- or biometeorological texts which assume a reasonably advanced ability in physics and mathematics. I believe that the frustration felt by such students can be alleviated by the more qualitative approach to the physics of the lower atmosphere used here. The book is based on the application of simple physical principles, and an exposition that is *explanatory* rather than *descriptive*. It is expected to appeal to two groups of students: first, those who want to know about the role of the atmosphere in environmental science and its applications in geography, agriculture, forestry, ecology, engineering and planning; second, those embarked on a more specialist course in small scale meteorology who wish to supplement their technical material with examples of atmospheric systems from many real world environments.

The book is organized in three sections. Part I – Atmospheric Systems – provides a simple scientific introduction to atmospheric processes operating in the planetary boundary layer. The idea of the flows and transformations of energy and mass through systems is developed. This supplies the basic concepts embodied in modern energy and water balance climatology and sets the framework for the subsequent analysis of the climates of a wide range of surface environments. Thus in Part II – Natural

Atmospheric Environments – climates are presented by first considering the special mix of physical properties (radiative, thermal, moisture and aerodynamic) characterizing a particular system, then interpreting the effects they have upon the exchanges of energy, mass and momentum, before describing the resulting climatic qualities. Boundary layer climates are therefore viewed as the outcome of the unique way each surface responds to external forcing functions such as solar heating, precipitation and airflow. This provides a process-response (cause and effect) framework rather than the more traditional case study approach to understanding climates, and a rational basis for the analysis of the climatic impact of human activities in Part III – Man-Modified Atmospheric Environments.

The text attempts to be illustrative rather than comprehensive. The reader who requires further examples and case studies is recommended to consult Geiger's classic study *The Climate Near the Ground* (1965). For those with the appropriate scientific background who wish to extend their knowledge into the more technical fields of micro- and biometeorology Appendix A2 provides an introduction to some of the approaches used to evaluate quantitatively the flows of energy, mass and momentum. Throughout the book Système International (SI) units are used. Thus the basic units of length, mass and time are the metre, kilogram and second respectively. A fuller outline of SI units and their equivalents in other systems is given in Appendix A4. Mathematical equations have been kept to a minimum and the symbols used conform closely to those suggested by the World Meteorological Organization.

I am very grateful to the following who greatly helped the preparation of this book by reviewing individual chapters: Dr T. A. Black, University of British Columbia (Ch. 4); Professor T. J. Chandler, Manchester University (Chs 8 and 9); Professor J. A. Davies, McMaster University (Chs 1 and 2); Professor H. Flohn, University of Bonn (Ch. 5); Professor B. J. Garnier, McGill University (Ch. 6); Professor F. K. Hare, University of Toronto (Ch. 6); Dr J. D. Kalma, CSIRO, Canberra (Ch. 7); Professor H. E. Landsberg, University of Maryland (Ch. 8); Professor D. H. Miller, University of Wisconsin-Milwaukee (Ch. 3); Dr R. E. Munn, Canadian Atmospheric Environment Service (Ch. 9); and Dr D. S. Munro, Erindale College, University of Toronto (Ch. 3). Special thanks are due to my colleague Dr J. E. Hay who commented on a large portion of the manuscript. Naturally I remain responsible for any remaining errors. Thanks are also due to those authors and publishers who kindly gave permission to reproduce photographs and diagrams.

Finally I wish to thank my wife and children for affording me the time and the happy supportive environment necessary for writing.

Vancouver T.R.O.
1977

Symbols

Symbol	*Quantity*	*SI Units*
Roman capital letters		
A	horizontal moisture transport in the air per unit horizontal area	$kg\ m^{-2}\ s^{-1}$
A'	lot area	m^2
A^*	silhouette area	m^2
B	water intake by an animal	$kg,\ kg\ m^{-2}\ s^{-1}$
C	volumetric heat capacity of a substance	$J\ m^{-3}\ K^{-1}$
	carbon dioxide flux density (Chapter 4)	$kg\ m^{-2}\ s^{-1}$
	Dalton Number (Appendix A2)	
D	diffuse-beam short-wave radiation	$W\ m^{-2}$
E	evapotranspiration	$mm,\ kg\ m^{-2}\ s^{-1}$
ELR	environmental lapse rate	$K\ m^{-1}$
F	anthropogenic water release due to combustion (Chapter 8)	$mm,\ kg\ m^{-2}\ s^{-1}$
	vertical flux of pollution (Chapter 9)	$kg\ m^{-2}\ s^{-1}$
H	effective stack height	m
I	piped water supply per unit horizontal area (Chapter 8)	$mm,\ kg\ m^{-2}\ s^{-1}$
K^*	net short-wave radiation	$W\ m^{-2}$
$K\downarrow$	incoming short-wave radiation	$W\ m^{-2}$

Symbol	*Quantity*	*SI Units*
$K\uparrow$	reflected short-wave radiation	$W\ m^{-2}$
K_C	eddy diffusion coefficient for carbon dioxide	$m^2\ s^{-1}$
K_{Ex}	extraterrestrial solar radiation	$W\ m^{-2}$
K_F	eddy diffusion coefficient for pollution	$m^2\ s^{-1}$
K_H	eddy conductivity	$m^2\ s^{-1}$
K_M	eddy viscosity	$m^2\ s^{-1}$
K_W	eddy diffusivity for water vapour	$m^2\ s^{-1}$
L	latent heat, of fusion (L_f), of vaporization (L_v), of sublimation (L_s)	$J\ kg^{-1}$
L^*	net long-wave radiation	$W\ m^{-2}$
$L\downarrow$	incoming long-wave radiation from the atmosphere	$W\ m^{-2}$
$L\uparrow$	long-wave radiation emitted by a surface	$W\ m^{-2}$
M	vertical flux of soil water	$kg\ m^{-2}\ s^{-1}$
P	total atmospheric pressure	Pa
	rate of gross photosynthesis (Chapter 4)	$kg\ m^{-2}\ s^{-1}$
	wave period (Chapter 2)	s
	population of a settlement (Chapter 8)	
Q	heat energy	J
Q^*	net all-wave radiation flux density	$W\ m^{-2}$
$Q\downarrow$	total incoming short- and long-wave radiation	$W\ m^{-2}$
$Q\uparrow$	total outgoing short- and long-wave radiation	$W\ m^{-2}$
Q_A	horizontal energy transport in the air per unit horizontal area	$W\ m^{-2}$
Q_E	turbulent latent heat flux density	$W\ m^{-2}$
Q_F	anthropogenic heat flux density due to combustion	$W\ m^{-2}$
Q_G	sub-surface heat flux density	$W\ m^{-2}$
Q_H	turbulent sensible heat flux density	$W\ m^{-2}$
Q_M	metabolic heat production by animals	$W\ m^{-2}$
Q_P	rate of energy storage in photosynthesis	$W\ m^{-2}$
Q_R	rate of heat supply by rainfall	$W\ m^{-2}$
Q_S	rate of energy storage	$W\ m^{-2}$
R	rate of CO_2 respiration by plants	$kg\ m^{-2}\ s^{-1}$
Ri	Richardson's Number	
S	direct-beam short-wave radiation	$W\ m^{-2}$
	soil moisture content or water storage	
SVF	sky view factor	
T	temperature	K, (°C)
T_b	animal body (or core) temperature	K, (°C)

Symbol	*Quantity*	*SI Units*
U	urinary water loss by an animal	kg, kg m^{-2} s^{-1}
V	voltage	V

Roman small letters

a	absorptivity	
c	specific heat of a substance	J kg^{-1} K^{-1}
	concentration of carbon dioxide	kg m^{-3}, ppm
c_p	specific heat of air at constant pressure	J kg^{-1} K^{-1}
d	zero plane displacement	m
e	vapour pressure	Pa
	base of Naperian logarithms	
$e^*_{(T)}$	saturation vapour pressure	Pa
f	moisture infiltration	mm, kg m^{-2} s^{-1}
g	acceleration due to gravity	m s^{-2}
h	height of an object (e.g. crop or shelterbelt)	m
h^*	depth of the surface mixed layer	m
h_s	stack height	m
k	thermal conductivity	W m^{-1} K^{-1}
	von Karman's constant	
m	soil moisture tension	Pa
p	precipitation	mm
q	specific humidity	kg kg^{-1}
r	resistance to flow	s m^{-1}
r_a	aerodynamic resistance	s m^{-1}
r_b	laminar boundary layer resistance	s m^{-1}
r_c	canopy (or surface) resistance	s m^{-1}
r_m	mesophyll resistance	s m^{-1}
r_{st}	stomatal resistance	s m^{-1}
t	time	s
	transmissivity	
u	horizontal wind speed	m s^{-1}
u_*	friction velocity	m s^{-1}
vpd	vapour pressure deficit	Pa
w	vertical wind speed	m s^{-1}
x	horizontal (along-wind) distance	m
y	horizontal (across-wind) distance	m
z	vertical distance	m
z_0	roughness length	m

Greek capital letters

Γ	dry adiabatic lapse rate	K m^{-1}

Symbol	*Quantity*	*SI Units*
Δ	finite difference approximation (i.e. difference or net change in a quantity)	
ΔA	net moisture advection; rate per unit volume (per unit horizontal area)	kg; $kg\,m^{-3}\,s^{-1}$ ($kg\,m^{-2}\,s^{-1}$)
ΔP	net rate of photosynthesis, rate of net CO_2 assimilation	$kg\,m^{-2}\,s^{-1}$
ΔQ_A	net energy (sensible and latent) advection; rate per unit volume (per unit horizontal area)	J; $W\,m^{-3}$ ($W\,m^{-2}$)
ΔQ_P	net energy storage due to photosynthesis; rate per unit volume (per unit horizontal area)	J; $W\,m^{-3}$ ($W\,m^{-2}$)
ΔQ_S	net energy storage; rate per unit volume (per unit horizontal area)	J; $W\,m^{-3}$ ($W\,m^{-2}$)
ΔS	net moisture storage; rate per unit volume (per unit horizontal area)	kg; $kg\,m^{-3}\,s^{-1}$; ($kg\,m^{-2}\,s^{-1}$)
Δh	height of plume rise	m
Δr	net run-off	mm, $kg\,m^{-2}\,s^{-1}$
Θ	angle	
$\Phi_C, \Phi_H, \Phi_M, \Phi_W$	dimensionless stability functions for carbon dioxide, heat, momentum and water vapour respectively.	
X	rate of pollution emission	$kg\,m^{-2}\,s^{-1}$, $m^3\,s^{-1}$

Greek small letters

α	surface albedo	
β	Bowen's ratio (Q_H/Q_E)	
ε	surface emissivity	
θ	potential temperature	K
κ	thermal diffusivity of a substance	$m^2\,s^{-1}$
$\kappa_H, \kappa_M, \kappa_W$	molecular diffusion coefficients for heat, momentum (viscosity) and water vapour in air.	$m^2\,s^{-1}$
λ	wavelength	m
ρ	density of a substance	$kg\,m^{-3}$
σ	Stefan-Boltzmann constant	$W\,m^{-2}\,K^{-4}$
σ_y, σ_z	standard deviations of horizontal and vertical pollutant distribution	m
τ	momentum flux per unit surface area	Pa
ϕ	heat of assimilation of carbon	$J\,kg^{-1}$
χ	concentration of an air pollutant	$kg\,m^{-3}$, ppm
ψ	hydraulic conductivity	

Symbol Quantity

Subscripts

(A)	Atmosphere
(E)	Earth
a	air
(b)	bottom (abaxial) surface of a leaf
c	cloud
	canopy
f	forest floor
l	liquid
o	surface
(o)	cloudless
r	rural
s	soil
(t)	top (adaxial) surface of a leaf
u	urban
v	vapour

PART I

Atmospheric systems

CHAPTER 1

Energy and mass exchanges

1 Atmospheric scales

The Atmosphere is characterized by phenomena whose space and time scales cover a very wide range. The space scales of these features are determined by their typical size or wavelength, and the time scales by their typical lifetime or period. Figure 1.1 is an attempt to place various atmospheric phenomena (mainly associated with motion) within a grid of their probable space and time limits. The features range from small scale turbulence (tiny swirling eddies with very short life spans) in the lower left-hand corner, all the way up to jet streams (giant waves of wind encircling the whole Earth) in the upper right-hand corner.

In reality none of these phenomena is discrete but part of a continuum, therefore it is not surprising that attempts to divide atmospheric phenomena into distinct classes have resulted in disagreement with regard to the scale limits. Most classification schemes use the characteristic horizontal distance scale as the sole criterion. A reasonable consensus of these schemes gives the following scales and their limits (see the top of Figure 1.1):

Micro-scale	10^{-2} to 10^{3} m
Local scale	10^{2} to 5×10^{4} m
Meso-scale	10^{4} to 2×10^{5} m
Macro-scale	10^{5} to 10^{8} m

In these terms this book is mainly restricted to atmospheric features

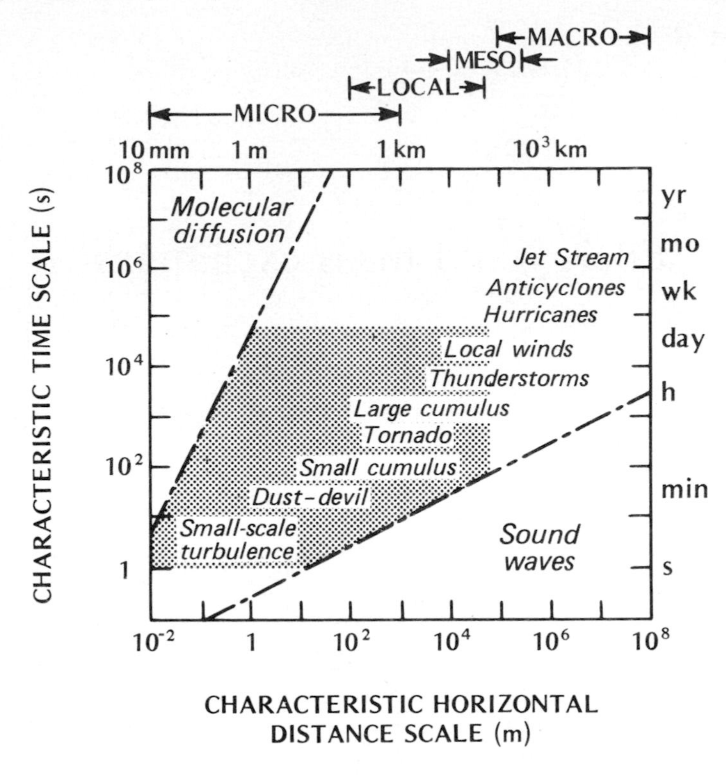

Figure 1.1 Time and space scales of various atmospheric phenomena. The shaded area represents the characteristic domain of boundary layer features (modified after Smagorinsky, 1974).

whose horizontal extent falls within the micro- and local scale categories. A fuller description of its scope is given by also including characteristic vertical distance, and time scales.

This book is concerned with the interaction between the Atmosphere and the Earth's surface. The influence of the surface is effectively limited to the lowest 10 km of the Atmosphere in a layer called the *troposphere* (Figure 1.2). (*Note*—terms introduced for the first time are italicized and the meaning of those not fully explained in the text is given in Appendix A5.) Over time periods of about one day this influence is restricted to a very much shallower zone known as the *planetary or atmospheric boundary layer*, hereinafter referred to simply as the *boundary layer*. This layer is particularly characterized by well developed mixing (turbulence) generated by frictional drag as the Atmosphere moves across the rough and rigid surface of the Earth, and by the 'bubbling-up' of air parcels from the heated surface. The boundary layer receives much of its heat and all of its water through this process of turbulence. The height of the boundary layer (i.e.

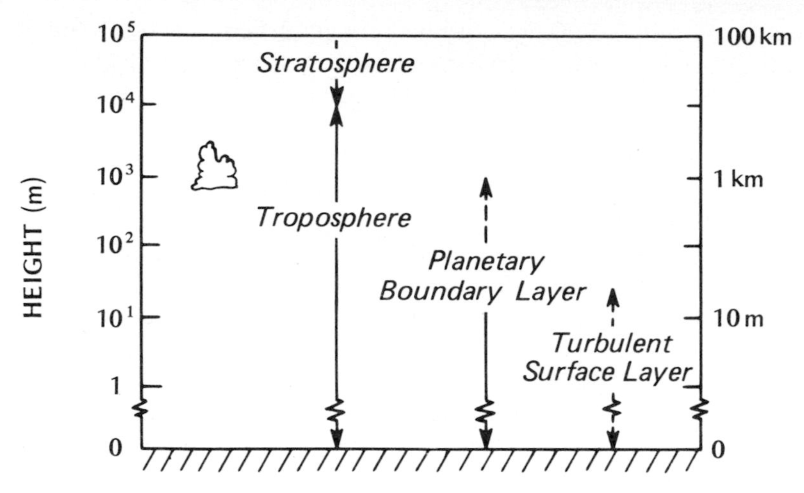

Figure 1.2 The vertical structure of the atmosphere (modified after Tennekes, 1974).

the depth of surface-related influence) is not constant with time, it depends upon the strength of the surface-generated mixing. By day, when the Earth's surface is heated by the Sun, there is an upward transfer of heat into the cooler Atmosphere. This vigorous thermal mixing (convection) enables the boundary layer depth to extend to about 1 to 2 km. Conversely by night, when the Earth's surface cools more rapidly than the Atmosphere, there is a downward transfer of heat. This tends to suppress mixing and the boundary layer depth may shrink to less than 100 m. Thus in the simple case we envisage a layer of influence which waxes and wanes in a rhythmic fashion in response to the daily solar cycle. Naturally this ideal picture can be considerably disrupted by large-scale weather systems whose wind and cloud patterns are not tied to surface features, or to the daily heating cycle. For our purposes the characteristic horizontal distance scale for the boundary layer can be related to the distance air can travel during a heating or cooling portion of the daily cycle. Since significant thermal differences only develop if the wind speed is light (say less than 5 m s^{-1}) this places an upper horizontal scale limit of about 50 to 100 km. With strong winds mixing is so effective that small-scale surface differences are obliterated. Thus, except for the dynamic interaction between airflow and the terrain, the boundary layer characteristics are dominated by tropospheric controls. In summary the upper scale limits of boundary layer phenomena (and the subject matter of this book) are vertical and horizontal distances of ~1 km and ~50 km respectively, and a time period of ~1 day.

Within the boundary layer there are two other layers controlled by surface features. In immediate contact with the surface is the *laminar boundary layer* whose depth is at most a few millimetres (Figure 2.3). This is a layer of non-turbulent air which adheres to all surfaces thus establishing an efficient buffer from the *turbulent surface layer* above, and providing the lower vertical scale size for this book. The turbulent surface layer (Figure 1.2) is characterized by intense small-scale turbulence generated by the surface roughness and convection; by day it may extend to a height of about 50 m, but at night when the boundary layer shrinks it may be at most only a few metres in depth.

The lower horizontal scale limit is dictated by the dimensions of relevant surface units and since the smallest climates covered are those of insects and leaves, this limit is of the order of 10^{-2} to 10^{-3} m. It is difficult to set an objective lower cut-off for the time scale. An arbitrary period of approximately 1 s is suggested.

The shaded area in Figure 1.1 gives some notion of the space and time bounds to boundary layer climates as discussed in this book (except that it requires a third co-ordinate to show the vertical space scale). Two aberrations from this format should be noted. First, it should be pointed out that convective cloud and precipitation phenomena, and violent weather events (such as tornados), which might be classed as boundary layer phenomena, have been omitted. The former, although deriving their initial impetus near the surface, owe their internal dynamics to condensation which often occurs at the top or above the boundary layer. The latter are dominated by weather dynamics occurring at much larger scales than outlined above. Second, the boundary layer treated herein includes the uppermost layer of the underlying material (soil, water, snow etc.) extending to a depth where diurnal exchanges of water and heat become negligible.

2 Energy, water and climate

The classical climatology practised in the first half of the twentieth century was almost entirely concerned with the distribution of the principal climatological parameters (e.g. air temperature and humidity) in time and space. While this information conveys a useful impression of the state of the atmosphere at a location it does little to explain how this came about. Such parameters are really only indirect measures of more fundamental quantities. Air temperature and humidity are really a gauge of the thermal energy and water status of the atmosphere respectively, and these are tied to the fundamental energy and water cycles of the *Earth-Atmosphere system.* Study of these cycles, involving the processes by which energy and mass are transferred, converted and stored, forms the basis of modern physical climatology.

The relationship between energy flow and the climate can be illustrated in the following simple manner. The First Law of Thermodynamics (conservation of energy) states that energy can be neither created nor

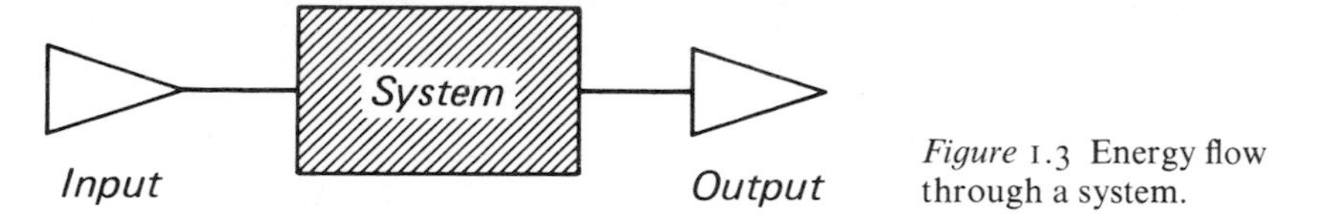

Figure 1.3 Energy flow through a system.

destroyed, only converted from one form to another. This means that for a simple system such as that in Figure 1.3, two possibilities exist. Firstly:

$$\text{Energy Input} = \text{Energy Output} \tag{1.1}$$

in this case there is no change in the net energy status of the system through which the energy has passed. It should however be realized that this does not mean that the system has no energy, merely that *no change* has taken place. Neither does it mean that the Output energy is necessarily in the same *form* as it was when it entered. Energy of importance to climatology exists in the Earth-Atmosphere system in four different forms (*radiant*, *thermal*, *kinetic* and *potential*) and is continually being transformed from one to another. Hence, for example, the Input energy might be entirely radiant but the Output might be a mixture of all four forms. Equally the Input and Output *modes* of energy transport may be very different. The transmission of energy in the Earth-Atmosphere system is possible in three modes (*conduction*, *convection* and *radiation* (see Section 3 for explanation)). The second possibility in Figure 1.3 is:

$$\text{Energy Input} = \text{Energy Output} + \text{Energy Storage Change}$$

For most natural systems the equality, Input = Output, is only valid if values are integrated over a long period of time (e.g., a year). Over shorter periods the energy balance of the system differs significantly from equality. The difference is accounted for by energy accumulation or depletion in the system's energy store. (The energy storage term may have a positive or negative sign. By convention a positive storage indicates the addition of energy.) In climatic terms, for example, if energy is being accumulated in a soil-atmosphere system it probably means an increase in soil and/or air temperature.

Hence we see the link between process (energy flow) and response (temperature change). The whole relationship is then referred to as a *process-response system*, which in essence describes the connection between cause and effect. The degree of detailed understanding of the system depends on how well the internal workings of the 'box' in Figure 1.3 are known. Inside the box the energy is likely to be partitioned into different flow channels, and converted into different combinations of energy forms and modes of transport. Some of the channels will lead to energy storage change and others to energy output from the system. This partitioning is not haphazard, it is a function of the system's physical properties. In the case of energy these properties include its ability to absorb, transmit, reflect and emit radiation, its ability to conduct and convect heat, and its capacity to store energy.

The process of energy channelling is referred to as a *cascade* because it is such a good analogue of a cascade of water in a river. Up to the point of the cascade the river (input) is concentrated in a single channel. Thereafter it disperses into a multiplicity of flows around rocks, and some is temporarily held in pools (stores), before it emerges again as a more or less single flow (output) to continue downstream. At all times the mass of water in the system is the same (conserved), but it follows different routes and at different rates as a direct result of the river channel form and the arrangement of rocks etc. (physical properties).

In the case of water flow in a soil-atmosphere system the mass of water is conserved at all times but it may be found in three different *states* (vapour, liquid and solid); be transported in a number of modes (including convection, precipitation, percolation, and run-off); and its accumulation or depletion in stores is measured as changes of water content (atmospheric humidity, soil moisture or the water equivalent of a snow or ice mass). Similar analogues can be extended to the mass balances of other substances cycled through systems as a result of natural or human (anthropogenic) activities including sulphur, carbon, nitrogen, and particulates. In the case of atmospheric systems the accumulation of these substances beyond certain levels constitutes atmospheric pollution. This occurs when the natural cycling of substances is upset by human activities. For example, in urban areas, if the emission (input) of these materials exceeds the physical capability of the local atmospheric system to flush itself (output) in a short period of time, the result is an increase in the local concentration of that substance (i.e., increased storage). Therefore in the most general form we may write the following energy or mass balance equation for a system:

$$\text{Input} - \text{Output} - \text{Storage Change} = 0 \qquad (1.2)$$

There are two fundamental cascades of importance in understanding atmospheric systems. These are the solar cascade of energy (heat), and the hydrologic cascade of mass (water). The remainder of this chapter is concerned with a description of the workings of these two cascades. This is followed in Chapter 2 by an explanation of the way these exchange processes are linked to the vertical distributions of such climatological elements as temperature, humidity and wind speed in the boundary layer.

3 Energy cascades

The total Earth-Atmosphere (E-A) system is a *closed* system. This means that it is closed to the import or export of mass, but it does allow exchange of energy with the exterior (Space). The sole energy input to the E-A system is *radiation* emitted by the Sun. The energy cascade of the E-A system is therefore driven by solar radiation.

(a) RADIATION CHARACTERISTICS

Radiation is the transfer of energy by the rapid oscillations of electromagnetic fields. These oscillations may be considered as travelling waves characterized by their wavelength λ (distance between successive wave-crests). In most atmospheric applications we are concerned with wavelengths in the approximate range 0·1 to 100 μm (1 μm $= 10^{-6}$ m), representing only a very small portion of the total electromagnetic spectrum (Figure 1.4). The visible portion of the spectrum, to which the human eye is sensitive, is an even smaller fraction extending from 0·36 μm (violet) to 0·75 μm (red). Radiation is able to travel in a vacuum, and all radiation moves at the speed of light (3×10^8 m s^{-1}), and in a straight path.

All bodies possessing energy (i.e. whose temperatures are above absolute zero, 0 K = −273·2°C) emit radiation. If a body at a given temperature emits the maximum possible amount of radiation per unit of its surface area in unit time then it is called a *black body* or *full radiator*. Such a body has a surface *emissivity* (ε) equal to unity. Less efficient radiators have emissivities between zero and unity. The relation between the amount of radiation emitted by a black body, and the wavelength of that radiation at a given temperature is given by Planck's Law. In graphical form this law shows the spectral distribution of radiation from a full radiator to be a characteristic curve (Figure 1.5). The shape consists of a single peak of emission at one wavelength (λ_{max}), and a tailing-off at increasingly longer wavelengths. The form is so characteristic that in Figure 1.5 the same Planck curve on different scales, describes the emission spectra from full radiators at 300 and 6000 K. However, the total amount of radiation given out and its spectral composition are very different for the two cases.

The total energy emitted by each body in Figure 1.5 is proportional to the area under the curve (including the tail at longer wavelengths that has been truncated). This is the basis of the Stefan–Boltzmann Law:

$$\text{Energy emitted} = \sigma T_0^4 \qquad (1.3)$$

where, σ – Stefan–Boltzmann proportionality constant $= 5{\cdot}67 \times 10^{-8}$ W m^{-2} K^{-4}, and T_0 – surface temperature of the body (K). If the body is not a full radiator, equation 1·3 can be re-written to include the value of the surface emissivity:

$$\text{Energy emitted} = \varepsilon\sigma T_0^4 \qquad (1.4)$$

The energy emission given by these equations is the *flux* (rate of flow) of energy (J s^{-1} = W) from unit area (m^2) of a plane surface into the

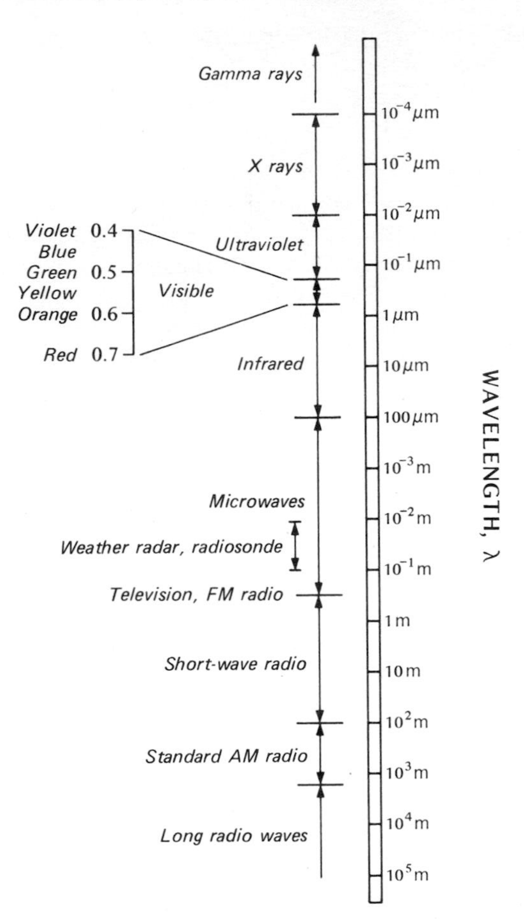

Figure 1.4 The electromagnetic spectrum.

overlying hemisphere. The flux per unit area is termed the *flux density* (W m^{-2}). In the typical range of temperatures found in the E-A system (−15 to 45°C) a difference of 1 K results in a change of the emitted radiation flux density of 4 to 7 W m^{-2}.

The effect of temperature change on the wavelength composition of the emitted radiation is embodied in Wien's Displacement Law. It states that a rise in the temperature of a body not only increases the total radiant output, but also increases the proportion of shorter wavelengths of which it is composed. Thus as the temperature of a full radiator increases, the Planck curve is progressively shifted to the left, and the wavelength of peak emission moves with it so that:

$$\lambda_{max} T_0 = 2{\cdot}88 \times 10^{-3} \qquad (1.5)$$

with λ_{max} in metres and T_0 on the Kelvin scale.

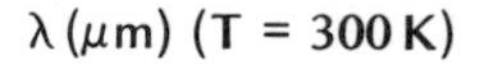

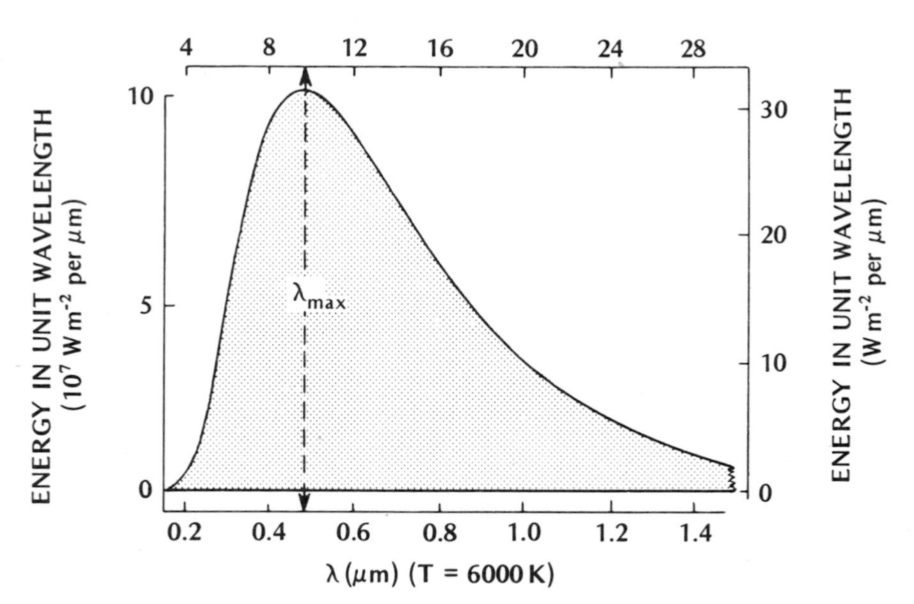

Figure 1.5 Spectral distribution of radiant energy from a full radiator at a temperature of (a) 6000 K, left-hand vertical and lower horizontal axis, and (b) 300 K, right-hand vertical and upper horizontal axis. λ_{max} is the wavelength at which energy output per unit wavelength is maximal (after Monteith, 1973).

The temperatures in Figure 1.5 were chosen because they approximately represent the average surface temperatures of the Sun and the E-A system, and thus illustrate the nature of the radiation each emits. Obviously the E-A system emits smaller total amounts of energy than the Sun, but also the wavelength composition is very different. From equation 1.5 it can be seen that the Sun's peak wavelength is about 0·48 μm (in the middle of the visible spectrum), whereas for the E-A system it is about 10 μm. Typical wavelengths for solar radiation extend from 0·15 μm (ultra-violet) to about 3·0 μm (near infra-red), whereas E-A system wavelengths extend from 3·0 μm to about 100 μm, well into the infra-red. In fact the difference between the two radiation regimes is conveniently distinct because about 99% of the total energy emitted by the two planets lies within these limits. On this basis atmospheric scientists have designated the radiation observed in the range 0·15–3·0 μm to be *short-wave* or *solar radiation*, and that in the range 3·0–100 μm to be *long-wave* radiation.

Radiation incident upon a substance must either be transmitted through the substance, or reflected from its surface, or be absorbed. Only the portion absorbed results in energy retention. It follows that:

$$t + \alpha + a = 1 \qquad (1.6)$$

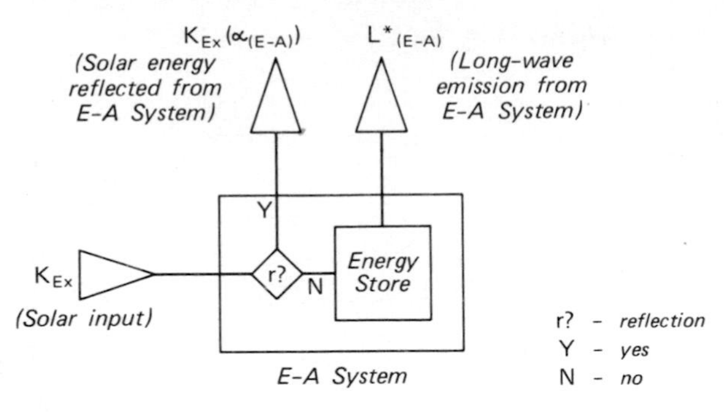

Figure 1.6 Simple representation of the solar energy cascade for the total Earth-Atmosphere system.

where t-transmissivity, α-reflectivity, and a-absorptivity of the substance, and are dimensionless numbers between zero and unity. Strictly equation 1.6 is only valid for individual wavelengths, but especially with regard to short-wave radiation it is usually acceptable to apply it to the complete range of wavelengths, whereupon the shortwave reflectivity of a surface is referred to as its *albedo* (α).

(b) ENERGY CASCADE OF THE TOTAL EARTH-ATMOSPHERE SYSTEM

The next step in elucidating the energy cascade is to investigate the internal workings of the E-A system, as illustrated in Figure 1.6. This is a system diagram similar to Figure 1.3 but embodies not only input and output but also an energy store and a decision regulator inside the system. The role of the energy store is self-evident but the decision regulator perhaps requires elaboration. Decision regulators act like 'valves' in the flow system. These are points in a system where energy (or mass) flows are either diverted to a system store, or allowed as through-put to other subsystems. They usually relate to a physical property of the system determining flow apportionment. For example each of the terms in equation 1.6 act as decision regulators in channelling radiation through a system.

The only energy input to the E-A system is solar radiation, shown in Figure 1.6 as the *extra-terrestrial short-wave radiation* flux density (K_{Ex}). On a mean annual basis exactly the same amount of energy must be lost from the E-A system to Space. This is necessary if the E-A system is not to experience a net energy gain or loss, since such a storage change would mean the average E-A system temperature would rise or fall (i.e. a climatic shift). The loss is accomplished via two channels of radiation. Firstly, some of the solar input is directly lost by scattering and reflection from atmospheric particles, clouds and the Earth's surface. The amount reflected is directly related to the value of the E-A system albedo ($\alpha_{E\text{-}A}$), which is, therefore, a very important system property. The total reflected energy is $K_{Ex}(\alpha_{(E\text{-}A)})$, and since the E-A system transmissivity is zero the remaining extra-terrestrial input must be absorbed (i.e. $K_{Ex}(1 - \alpha_{(E\text{-}A)})$). Clearly for balance to be maintained this same amount must be re-radiated from the E-A system, and in accordance with Figure 1.5 this will be a net long-wave radiative loss ($L^*_{(E\text{-}A)}$).

Further consideration of the role of the albedo as a decision regulator is instructive. Imagine that the value of $\alpha_{(E\text{-}A)}$ is changed as a result of an increased global cloud cover produced by the contrails of high-altitude jet aircraft. In moving towards a new equilibrium position a chain of interactions may be set in motion such as that shown in Figure 1.7. The

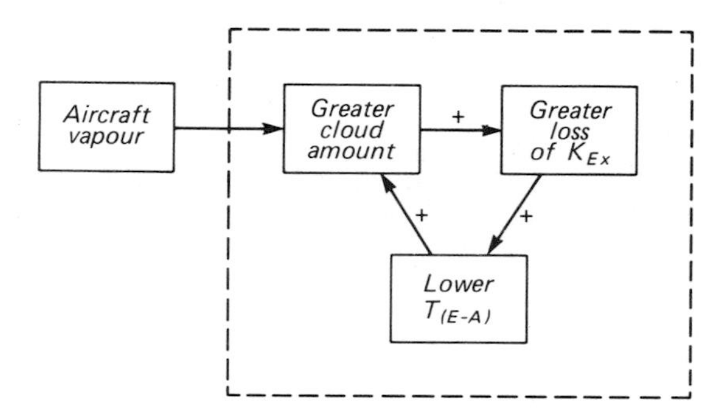

Figure 1.7 A positive feedback loop triggered by an increase in cloud cover produced by jet aircraft contrails.

aircraft activity is shown to increase cloud cover; this decreases the short-wave radiant energy absorbed. On an annual basis this loss of energy status for the system would lower the E-A system temperature. This may cause yet more cloud, due to the condensation of atmospheric moisture, and effectively 'close the circle'. Such a chain of events is termed a *feedback loop*, defined as the situation where a change in one system variable is transmitted through the system structure until it returns to re-affect the original variable. If, as in the case described, it returns to enhance the change it is a *positive feedback*; if on the other hand it dampens the change it is a *negative feedback*. Positive feedback is a 'snow-balling' effect, and potentially destructive; negative feedback is equilibrating. In full reality the situation described in Figure 1.7 may be very much more complex because these effects would certainly trigger other activity in the system such as long-wave exchanges beneath cloud base, changes in surface evaporation etc. In summary, however, this example demonstrates two significant general points. First, it shows that decision regulators are important and sensitive points in cascading systems; and second, that a small change in one system variable may produce a multiplicity of ramifications throughout the system. It should also be noted that human operations are capable of altering the setting of these 'valves' and upsetting the equilibrium of natural systems. This is investigated further in Part III.

Further elaboration of the annual energy cascade of the E-A system is presented in Figure 1.8. This recognizes the Earth and Atmosphere as separate subsystems, and places magnitudes on the energy flows. On an annual basis, spread over the outer surface of the Atmosphere the average input of solar radiation ($\overline{K}_{Ex}$) is approximately 338 W m^{-2} (29·2 MJ m^{-2} day^{-1}), and in Figure 1.8 all fluxes are represented as percentages of this value.

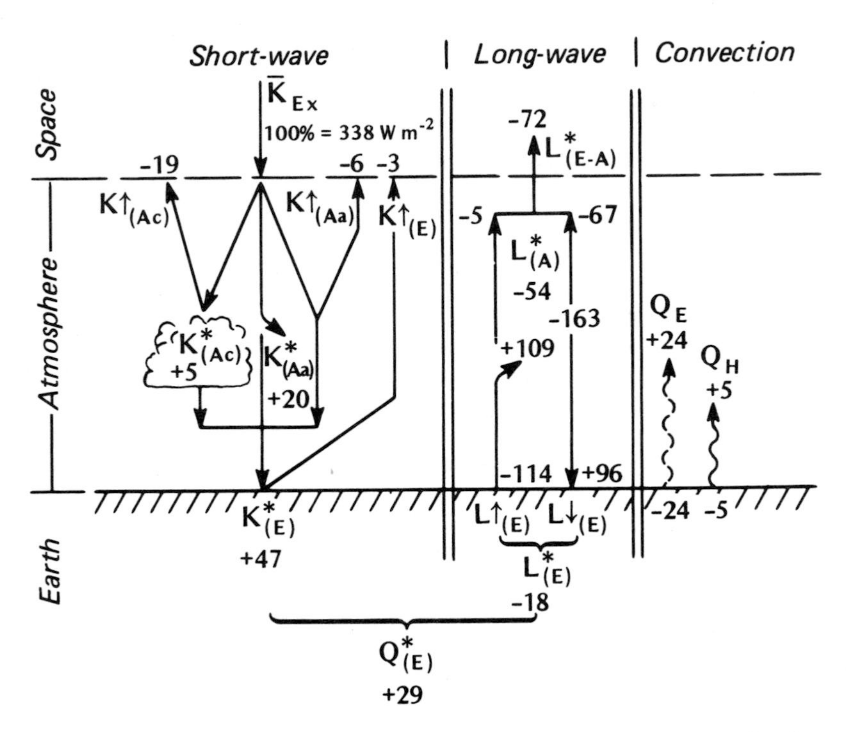

Figure 1.8 Schematic diagram of the average annual solar energy cascade of the Earth-Atmosphere system. Values are expressed as percentages of the average annual extra-terrestrial solar radiation ($\overline{K}_{Ex}$ = 338 W m^{-2}) (data from Rotty and Mitchell, 1974).

Incoming short-wave radiation encounters clouds and other atmospheric constituents such as water vapour, salt crystals, dust particles and various gases. In accord with equation 1.6, a part of the beam is scattered or reflected, a part is absorbed, and the remainder is transmitted on down to the Earth's surface. Clouds reflect about 19% of $\overline{K}_{Ex}$ back to Space ($K\uparrow_{(Ac)}$), and absorb about 5% ($K^*_{(Ac)}$). Atmospheric constituents scatter and reflect about 6% to Space ($K\uparrow_{(Aa)}$) and absorb about 20% ($K^*_{(Aa)}$). The remainder of the original beam is transmitted to the Earth's surface where approximately 3% is reflected to Space ($K\uparrow_{(E)}$), and the remaining 47% is absorbed ($K^*_{(E)}$). Thus, in summary, the solar radiation input is disposed of in the following manner:

$$\begin{aligned} \overline{K}_{Ex} &= K\uparrow_{(Ac)} + K\uparrow_{Aa)} + K^*_{(Ac)} + K^*_{(Aa)} + K\uparrow_{(E)} + K^*_{(E)} \\ 100 &= \underbrace{19 + 6}_{\text{Atmospheric reflection}} + \underbrace{5 + 20}_{\text{Atmospheric absorption}} + \underset{\text{Earth reflection}}{3} + \underset{\text{Earth absorption}}{47} \end{aligned} \qquad (1.7)$$

Three basic features of the short-wave radiation cascade emerge. First, 28% of the E-A input is reflected to Space and does not participate further in the E-A system energy cascade. Second, only 25% of the input is

TABLE 1.1 Radiative properties of natural materials.

Surface	Remarks	Albedo α	Emissivity ε
Soils	Dark, wet	0·05–0·40	0·90–0·98
	Light, dry		
Desert		0·20–0·45	0·84–0·91
Grass	Long (1·0 m)	0·16–	0·90–
	Short (0·02 m)	0·26	0·95
Agricultural crops, tundra		0·18–0·25	0·90–0·99
Orchards		0·15–0·20	
Forests			
Deciduous	Bare	0·15–	0·97–
	Leaved	0·20	0·98
Coniferous		0·05–0·15	0·97–0·99
Water	Small zenith angle	0·03–0·10	0·92–0·97
	Large zenith angle	0·10–1·00	0·92–0·97
Snow	Old	0·40–	0·82–
	Fresh	0·95	0·99
Ice	Sea	0·30–0·45	0·92–0·97
	Glacier	0·20–0·40	

Sources: Sellers (1965), List (1966), Paterson (1969) and Monteith (1973).

absorbed by the Atmosphere. Thus the Atmosphere is semi-transparent to short-wave radiation and consequently is not greatly heated by it. Third, almost one half (47%) of the input is absorbed at the Earth's surface. This considerable amount of energy is converted from radiation into thermal energy which warms the surface.

The Earth's surface emits long-wave radiation in accord with equation 1.4. Most natural surfaces have emissivities close to unity (Table 1.1), and with the Earth's mean annual temperature of approximately 288 K this results in an upward emission ($L\uparrow_{(E)}$) of 114% of $\overline{K}_{Ex}$. This apparent anomaly is possible because the Atmosphere blocks the loss of $L\uparrow_{(E)}$ and forces the surface temperature above the value it would have with no Atmosphere. In fact only 5% is lost directly to Space, the remaining 109% being absorbed by the Atmosphere. It should also be noted that the Earth emits long-wave over its entire surface area, but only receives short-wave over the sunlit hemisphere.

The absorption (a) spectrum of the Atmosphere and some of its most

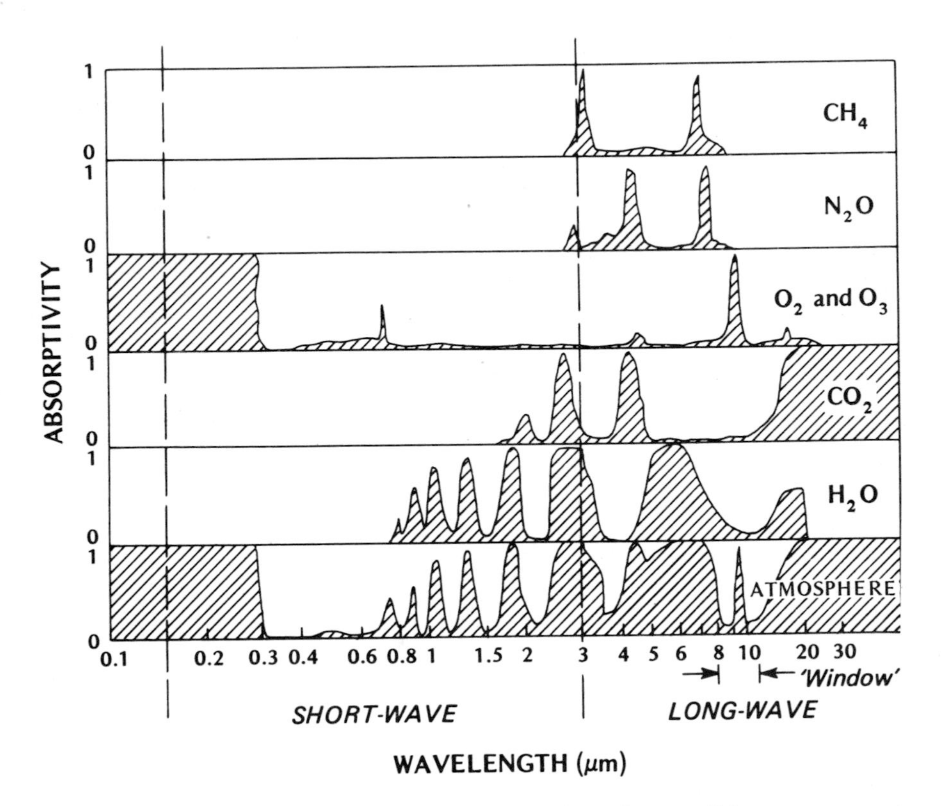

Figure 1.9 Absorption at various wavelengths by constituents of the Atmosphere, and by the Atmosphere as a whole (after Fleagle and Businger, 1963).

influential constituent gases is shown in Figure 1.9. This graph explains why the Atmosphere is such a poor absorber of short-wave, but such a good absorber of long-wave radiation. In the short-wave band (0·15 to 3·0 μm) there are few major absorbing agents. Ozone (O_3) is very effective in filtering out ultra-violet radiation at less than 0·3 μm, and water vapour becomes increasingly important at wavelengths greater than 0·8 μm. But in the band 0·3 to 0·80 μm where the intensity of solar radiation is greatest (Figure 1.5), the Atmosphere is relatively transparent. On the other hand it is almost opaque to long-wave radiation (3·0 to 100 μm). This is particularly due to the absorptivities of water vapour (H_2O), carbon dioxide (CO_2) and ozone. Of these water vapour is by far the most important. If liquid water is present, as in cloud droplets, the absorptivity is even greater still. There is, however, one important gap in a cloudless Atmosphere's absorption spectrum for long-wave radiation. Except for a narrow band of ozone absorption (9·6 to 9·8 μm), the Atmosphere is open to the transmission of radiation in the 8 to 11 μm band. This gap is called the *atmospheric 'window'*. It is through this 'window' that most of the E-A system long-wave loss to Space occurs. However, even this 'window' can be partially closed by clouds, or atmospheric pollutants. Thus in general the Atmosphere allows most incoming solar energy to penetrate easily to the Earth's surface, but severely hinders the loss of radiation back to space. This arrangement serves to keep the E-A system warmer than it would otherwise be (e.g. by comparison with a similar planet with no atmosphere).

The actual process of long-wave radiative exchange in the Atmosphere is very complex. At all levels the Atmosphere absorbs long-wave radiation depending upon the amounts of absorbing constituents present, and re-emits long-wave consistent with its temperature and emissivity (equation 1.4). The emission is directed both upwards and downwards. These processes of absorption and re-emission take place on a continuous basis throughout the Atmosphere, although quantitatively they are most important in the lowest layers where the concentrations of water vapour and carbon dioxide are greatest. Eventually a net portion emerges from the top of the Atmosphere and is lost to Space. On a mean annual global basis (Figure 1.8) this loss ($L^*_{(E\text{-}A)}$) amounts to 72% of $\bar{K}_{Ex}$, and is the sum of the net upward emission from the Atmosphere (67%), and that portion of the surface loss ($L\uparrow_{(E)}$) which passes directly out through the atmospheric 'window' (5%). The net downward emission by the Atmosphere (96%) is often called *counter radiation* ($L\downarrow_{(E)}$) and contributes to a net warming of the Earth's surface. The combined upward and downward emission from the Atmosphere amounts to a total output of 163% (Figure 1.8).

The net annual radiative budgets of the Earth and Atmosphere subsystem can now be summarized from Figure 1.8. The net radiative budget (Q^*) for a system is composed of the algebraic sum of its net short-wave (K^*), and net long-wave (L^*) radiation exchanges. Hence the net all-wave radiation balance is written:

$$Q^* = K^* + L^* \tag{1.8}$$

where both Q^* and L^* may be positive (net gain) or negative (net loss), but since no system emits short-wave radiation *inside* the E-A system, K^* can only be positive (net gain) or zero (at night). Equation 1.8 can be evaluated for the Earth using the values in Figure 1.8. The Earth's surface receives a net input of short-wave radiation ($K^*_{(E)}$) equivalent to 47% of $\overline{K}_{Ex}$, but experiences a net loss of long-wave radiation ($L^*_{(E)}$) of 18% (because it emits 114% to the Atmosphere, but only receives 96% in return). Thus the net all-wave, radiation budget for the Earth ($Q^*_{(E)}$) is positive, and represents 29% (47–18) of the original extra-terrestrial input. In the case of the Atmosphere it gains 25% as $K^*_{(A)}$ due to absorption by clouds and atmospheric constituents (equation 1.7), but loses 54% as $L^*_{(A))}$ (because although it absorbs 109% from the surface it emits 163% to space and back to the surface). Thus the net all-wave radiation budget of the Atmosphere ($Q^*_{(A)}$) is −29% (25–54).

Now we see that the Earth has an annual radiant energy surplus of 29% and the Atmosphere has an annual radiant energy deficit of the same amount. This does not mean there is a balance. Considering their respective physical and thermal properties this situation would result in the Earth warming up at the rate of approximately 250°C day^{-1} and the Atmosphere cooling at approximately 1°C day^{-1}. Since such heating and cooling rates are not observed on a mean annual basis there must be a means of transferring the Earth's surplus into the Atmosphere to balance its deficit. The E-A system, and both of its component subsystems, would then exist in thermal equilibrium.

Since all the radiative terms have been considered, this leaves conduction and convection as the two potential processes capable of accomplishing the required transfer (p. 7). Thermal *conduction* is the process whereby heat is transmitted within a substance by the collision of rapidly moving molecules. It is usually an effective mode of transfer in solids, less so in liquids and least important in gases. In general pure molecular conduction is negligible in atmospheric applications, except within the very thin laminar boundary layer (p. 33) where it is the *only* mode of heat transport On the scale considered in Figure 1.8 it may be ignored as a means of transferring the surface radiative surplus into the bulk of the Atmosphere. On the other hand, conduction of heat down into the Earth certainly occurs, but it does not enter Figure 1.8 because on a mean annual basis net sub-surface storage must be zero.

The process of *convection* involves the vertical interchange of air masses, and can only occur in liquids and gases. In the Atmosphere the parcels of air (or *eddies*) transport energy and mass from one location to another. The eddies may be set into turbulent motion by free or forced convection. *Free convection* is due to the parcel of air being at a different density than the surrounding fluid. If for example a parcel is warmer than its surroundings it will be at a lower density and will tend to rise. Conversely if it is cooler it will be denser and tend to sink. The motion of water in a heated kettle is free convection, and a similar 'bubbling-up' of air parcels occurs when the Earth's surface is strongly heated by solar radiation. If the state of the Atmosphere is conducive to free convection it is said to be *unstable*, if it inhibits such motion it is *stable*. (See Appendix A1.) The atmosphere near the Earth's surface may also be physically thrown into motion when it flows over obstacles. This is *forced*, or *mechanical convection* and depends upon the roughness of the surface, and the speed of the horizontal flow over them. Often free and forced convection coexist giving *mixed convection* (p. 49).

Convection transports energy as both *sensible* (Q_H) and *latent heat* (Q_E) in the Atmosphere (p. 27). Sensible and latent heat imparted to the air near the surface is carried upwards by eddy motion. The sensible heat is released upon mixing with cooler air, and the latent heat is liberated when the water vapour condenses into cloud droplets. Thus combining these terms we may write for the mean annual energy balance:

$$Q^* - Q_H - Q_E = 0 \qquad (1.9)$$

for both the Earth and Atmosphere subsystems. The 29% radiative surplus of the Earth is balanced by a similar convective loss, of which 24% is by Q_E and 5% is by Q_H (Figure 1.8). Similarly, the Atmosphere's deficit is alleviated by this convective warming. With these exchanges the E-A system, and both of its subsystems are seen to conform to the simple energy balance of equation 1.1.

The case of the annual energy cascade of the E-A system has been used to illustrate the nature of energy, energy flow and energy balance. It also serves the purpose of providing the total energy context within which all E-A system climates (macro-, meso- local and micro-) operate. This approach also provides a convenient setting for the cascade of water through the E-A system, outlined in section 4.

(c) DIURNAL ENERGY CASCADE AT AN 'IDEAL' SITE

It was established in Section 1 that boundary layer climates respond to processes operating on time scales of less than one day. This section therefore outlines the most important general features of the diurnal energy cascade at a given site. This is more easily accomplished by considering the case of an 'ideal' site. Such a location presents the minimum complication being horizontal, homogeneous, and extensive. These constraints ensure that surface/atmosphere fluxes are spatially uniform and confined to the vertical direction. To minimize fluctuations in the time domain only cloudless conditions are considered so that the solar input is a smooth wave. The surface itself consists of a moist soil with a short (less than 20 mm) grass cover. Finally we restrict consideration to a mid-latitude area in the warm season. In Part II these constraints will be relaxed, and the emphasis is placed upon differences in surface character rather than similarities.

TABLE 1.2 Radiant energy cascade for a prairie grassland site. Data are daily totals (MJ m^{-2} day^{-1}) for 30 July 1971, at Matador, Saskatchewan (50°N) (after Ripley and Redmann, 1976).

$K\downarrow$	27·3	$L\downarrow$	27·5
$K\uparrow$	4·5	$L\uparrow$	36·8
K^*	22·7	L^*	−9·3
α†	0·16	Q^*	13·4

† Dimensionless

Figure 1.10 shows the diurnal variation of most of the important radiation budget components at a site reasonably conforming to our 'ideal' requirements (at least radiatively), and Table 1.2 summarizes the daily energy totals. The short-wave input to the surface ($K\downarrow$) consists of both *direct-beam* (S), and *diffuse-beam* (D) radiation, so that:

$$K\downarrow = S + D \tag{1.10}$$

On a cloudless day D is about 15–25% of $K\downarrow$ at the surface, and of the total short-wave received approximately 50% is in the visible portion of the electromagnetic spectrum. The pattern of $K\downarrow$ is of course controlled by the *azimuth* and *altitude* of the Sun relative to the horizon, with a single peak at local solar noon. The azimuth and altitude are controlled by Earth–Sun geometrical relationships and vary with latitude, time of day, and date. The peak value of $K\downarrow$ in Figure 1.10 is approximately 860 W m^{-2}, this is less than the maximum possible at this location (50°N) because the date is past the summer solstice (June 21). The maximum possible short-wave receipt anywhere in the E-A system with cloudless skies is 1353 ± 10 W m^{-2} (Thekaekara and Drummond, 1971). This value, known as the *solar constant*, is that observed outside the Atmosphere on a plane surface placed normal to the solar beam. It is approached at the surface at some tropical locations when the Sun is in the zenith (directly overhead), and the air is very clean, as at high altitude.

The reflected short-wave radiation ($K\uparrow$) depends on the value of $K\downarrow$ and the surface albedo (α, see Table 1.1 for typical values), so that:

$$K\uparrow = K\downarrow(\alpha) \tag{1.11}$$

Although α for most surfaces is not perfectly constant through the day (see

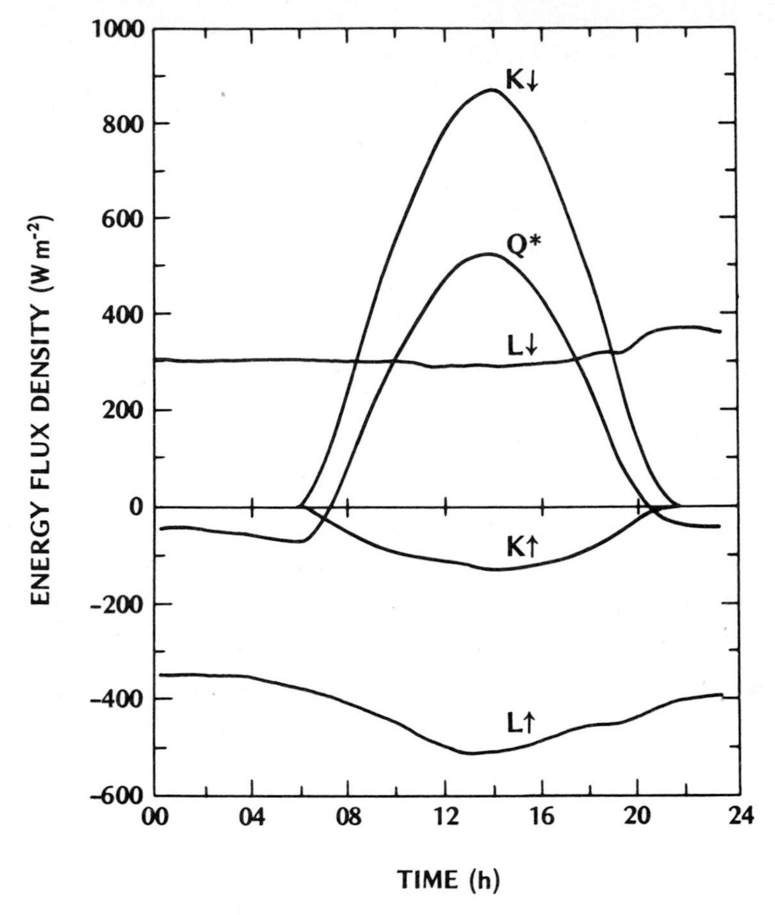

Figure 1.10 Radiation budget components for 30 July 1971, at Matador, Saskatchewan (50°N) over a 0·2 m stand of native grass. Cloudless skies in the morning, increasing cloud in the late afternoon and evening (after Ripley and Redmann, 1976). (Note – In the text no signs have been given to individual radiation fluxes, only to net fluxes (K^*, L^* and Q^*.) However, in figures such as this radiative inputs to the surface ($K\downarrow$, $L\downarrow$) have been plotted as positive, and outputs ($K\uparrow$, $L\uparrow$) as negative to aid interpretation.)

Figures 3.10 and 4.12), as a first approximation $K\uparrow$ is a reduced mirror-image of $K\downarrow$. At the site in Figure 1.10, $\alpha = 0{\cdot}16$, so that approximately one sixth of $K\downarrow$ is lost. Since the surface under consideration is opaque to short-wave radiation (i.e. $t = 0$), that portion of $K\downarrow$ not reflected is absorbed. Thus the net short-wave radiation (K^*) is given by:

$$K^* = K\downarrow - K\uparrow$$

or $$= K\downarrow(1 - \alpha) \qquad (1.12)$$

Although K^* is not plotted in Figure 1.10 it would obviously describe a curve with positive values of about $0{\cdot}84K\downarrow$ in magnitude.

The incoming long-wave radiation emitted by the atmosphere ($L\downarrow$) in the absence of cloud depends upon the bulk atmospheric temperature, and emissivity (which itself depends on the distributions of temperature, water vapour and carbon dioxide) in accord with the Stefan–Boltzmann Law (equation 1.4). Neither of these properties fluctuates rapidly and hence $L\downarrow$ is almost constant through the day (Figure 1.10; note the increase of $L\downarrow$ between 18 and 24 h is discussed later p. 25). The long-wave emitted by the surface ($L\uparrow$) is similarly governed by its temperature and emissivity (Table 1.1). Both are greater than their bulk atmospheric counterparts, and the temperature varies markedly through the day. As a result $L\uparrow$ is both greater in magnitude, and more variable than $L\downarrow$. The difference between these two long-wave fluxes is the surface net long-wave radiation budget (L^*):

$$L^* = L\downarrow - L\uparrow \tag{1.13}$$

The value of L^* (not plotted in Figure 1.10) is usually negative, and relatively small (75 to 125 W m^{-2}) if the surface and air temperatures are not significantly different. If the surface is considerably warmer than the air (e.g. as in Figure 7.3) L^* may be much larger. The diurnal course of L^* is usually in phase with $L\uparrow$.

The net all-wave radiation (Q^*) is the most important energy flux, because for most systems it represents the limit to the available energy source or sink. The daytime budget may be written:

$$\begin{aligned} Q^* &= K\downarrow - K\uparrow + L\downarrow - L\uparrow \\ &= K^* + L^* \end{aligned} \tag{1.14}$$

and, at night solar radiation is absent so that:

$$\begin{aligned} Q^* &= L\downarrow - L\uparrow \\ &= L^* \end{aligned} \tag{1.15}$$

Thus the typical diurnal course of Q^* (Figure 1.10) involves a daytime surface radiant surplus when the net short-wave gain exceeds the net long-wave loss; and a nocturnal surface deficit when the net long-wave loss is unopposed by solar input. At a given location the terms $K\downarrow$ and $L\downarrow$ are unlikely to show significant spatial variability because they are governed by large-scale atmospheric, or Earth–Sun geometric relationships. On the other hand $K\uparrow$ and $L\uparrow$ are governed by sensitive site specific factors (i.e. $K\uparrow$ by α; $L\uparrow$ by T_0 and ε). Thus it is these terms which govern the differences in radiation budget (Q^*) between surfaces in the same local region. In conclusion it should be noted that the range of Q^* values over different surfaces is damped somewhat by a built-in negative feedback mechanism. The range of natural surface ε values is small (Table 1.1) and hence differences in Q^* effectively depend upon the values of α and T_0. A surface

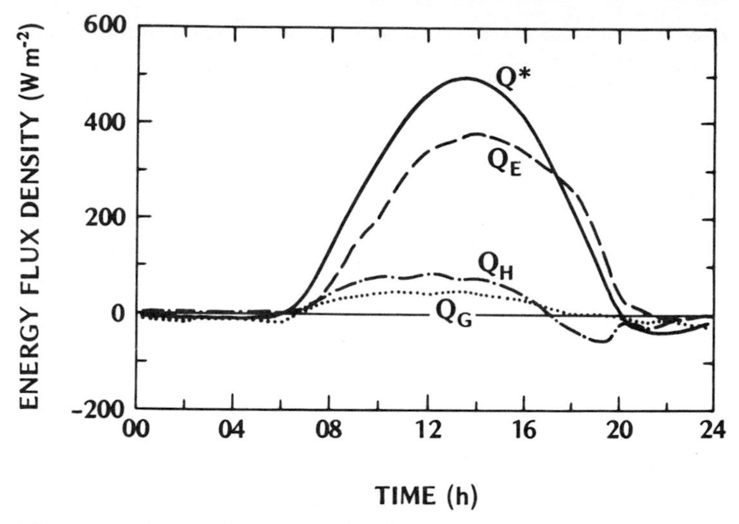

Figure 1.11 Energy balance components for 25 July 1976, with cloudless skies at Pitt Meadows, B.C. (49°N) over a 0·25 m stand of irrigated mixed orchard and rye grass (after Black and Goldstein, 1977).

with a low albedo will absorb well, but unless it possesses channels for rapid heat dissipation this will result in a high surface temperature. Thus the large K^* gain will be matched, at least in part, by a large L^* loss.

TABLE 1.3 Energy cascade for an irrigated field of mixed orchard and rye grass. Data are daily totals (MJ m^{-2} day^{-1}) for 25 July 1976, at Pitt Meadows, British Columbia (49°N); after Black and Goldstein, 1977.

		Derived terms	
Q^*	14·2	β†	0·16
Q_E	11·7	E (mm)	4·81
Q_H	1·8	Q_E/Q^*	0·86
Q_G	0·6		

† Dimensionless

Appendix A2 provides examples of the currently utilized methods (measurement and calculation) for the determination of the surface radiation balance fluxes, and the relevant surface radiative properties.

The net all-wave radiation flux is not only the end result of the radiation budget but also the basic input to the surface energy balance. Figure 1.11 shows the typical diurnal variation of the components of the surface energy balance at an 'ideal' site, and Table 1.3 summarizes the daily energy totals. At any given time it can be seen that any surface radiative imbalance is accounted for by a combination of convective exchange to or from the atmosphere, either as sensible (Q_H) or latent heat (Q_E), and conduction to or from the underlying soil (Q_G). Thus the surface energy balance is:

$$Q^* = Q_H + Q_E + Q_G \qquad (1.16)$$

The sign convention employed in Figure 1.11 (and throughout the remainder of the book) is that non-radiative fluxes directed away from a surface (or system) are positive. Thus the terms on the right-hand side of equation 1.16 are positive when they represent losses of heat for the surface (or system), and negative when they are gains. On the left-hand side Q^* is positive as a gain and negative when a loss. When both sides of equation 1.16 are positive it describes how the available radiative surplus is partitioned into sub-surface and atmospheric energy sinks; and this is usually the situation by day. When both sides are negative the equation states how the surface radiative deficit is partitioned between heat gain from available sub-surface and atmospheric sources; and this is the normal nocturnal situation. The flux of momentum is an exception to this convention.

The exact partitioning of the radiative surplus or deficit (between Q_H, Q_E and Q_G) is governed by the nature of the surface, and the relative abilities of the soil and atmosphere to transport heat. The particular apportionment arrived at by a surface is probably the most important determinant of its microclimate.

The diurnal course of Q^* in Figure 1.11 is very similar to that of Figure 1.10. *Under the given conditions* the daytime Q^* is dissipated as Q_E, Q_H and Q_G in descending order of importance. The dominant role of Q_E is a result of the free availability of soil and plant moisture at this irrigated site. If water became more restricted we might expect the role of Q_E to drop, and of Q_H to rise. No matter which dominates it is clear that convection is the principal means of daytime heat transport away from the interface.

At night on the other hand the situation is reversed. The nocturnal Q^* loss is most effectively replenished by conduction upwards from the soil, and the convective contribution is least effective from Q_E. The essential difference between the two convective situations is due to the fact that by day free convection is enhanced, but by night it is damped by the atmospheric temperature stratification (p. 48). The size of Q_G is not greatly different between day and night. In fact, although Q_G is a significant energy source, or sink, on an hourly basis, when integrated over the full day its net effect is not large (Table 1.3). In summer the daytime storage slightly exceeds the nocturnal output and the soil gradually warms. The reverse is true in winter.

Figure 1.12 provides a convenient schematic summary of the terms involved in the surface radiation and energy budgets of an 'ideal' site.

Before concluding this section a few remarks should be added concerning the effects of cloud, and non-uniform surface properties, upon

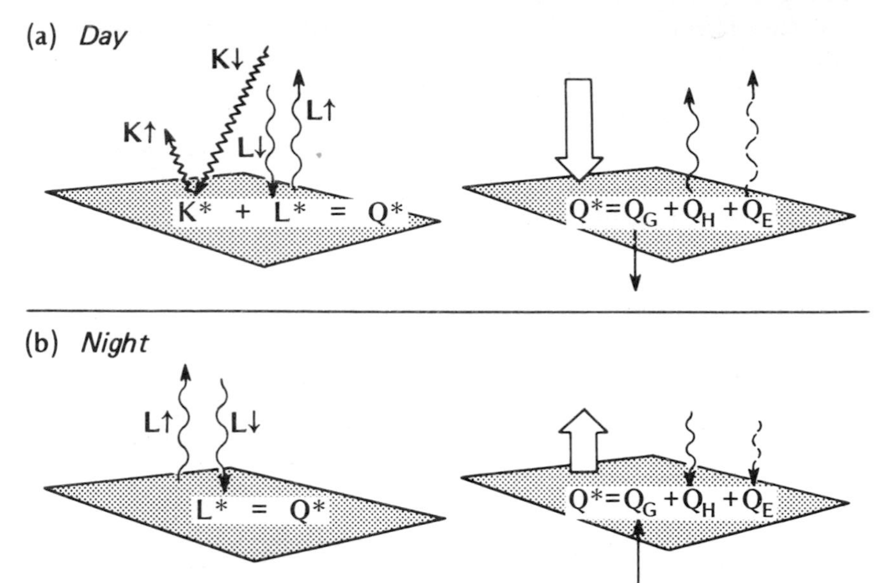

Figure 1.12 Schematic summary of the fluxes involved in the radiation budget and energy balance of an 'ideal' site, (a) by day and (b) at night.

the 'ideal' situation described above. Clouds exert a major influence on the exchanges of short- and long-wave radiation. The surface receipt of $K\downarrow$ is reduced because of cloud absorption and the reflection from cloud tops (average cloud albedo is approximately 0·55), and the proportion of D relative to S is increased because of cloud scattering and reflection. The surface long-wave budget is profoundly affected because clouds are almost black bodies (p. 17) and thus absorb and emit very efficiently. Clouds therefore absorb much of $L\uparrow$ from the surface and re-emit it back so that $L\downarrow$ is enhanced, and L^* reduced. The arrival of cloud explains the abrupt increase of $L\downarrow$ noted in Figure 1.10 after 18 h. The cloud emission depends on the cloud-base temperature, therefore the effect of Stratus (low, relatively warm) is much greater than Altus or Cirrus (high, cold) cloud. The net result of cloud is to damp the diurnal surface radiation budget variation, and serves to reduce the diurnal temperature range. This explains why cloudy weather is associated with comparatively uniform temperatures because daytime solar heating and night-time long-wave cooling are both reduced.

If the site is not sufficiently extensive it is possible that heat exchange can occur between it and upwind surfaces possessing different energy partitioning. For example, of our grass site was downwind of a hot dry soil surface it is possible that horizontal airflow could carry air with greater sensible heat content (and therefore higher temperature) across the grass.

This would alter the atmospheric conditions and give rise to an adjustment in the surface energy fluxes. In the example used it would tend to suppress the local surface value of Q_H and to augment Q_E (this is explained more fully in Chapter 5). The net horizontal convective heat transport (both sensible and latent) is called *advection* (ΔQ_A). Unless specifically noted it may be assumed that the surface climates outlined in Part II are advection-free. In reality this is rarely completely true.

(d) ATMOSPHERIC MOTION

Horizontal temperature variations in the E-A system give rise to horizontal pressure differences, which result in motion (winds). In this way thermal energy from the solar energy cascade is converted into the kinetic energy of wind systems. The energy then participates in the *kinetic energy cascade* involving the transfer of energy to increasingly small scales of motion by turbulence. This is the sequence depicted in Figure 1.1. Kinetic energy enters the cascade at a size-scale governed by the forces generating the motion. The energy is then passed down to smaller-sized eddies until it eventually reaches the molecular scale and is dissipated as heat (i.e. it returns to the thermal portion of the solar cascade). This energy does not appear in Figure 1.8 because on an annual basis there is a balance between kinetic energy production and dissipation.

In the boundary layer we are concerned both with motion generated on the micro- and local scales, and with the modification of existing airflow generated on scales larger than those of the boundary layer. In the first category we are concerned with wind systems generated by horizontal thermal differences in the boundary layer. These local scale thermal winds are especially prevalent across the boundaries between contrasting surface types. Examples include the breezes occurring at land/sea (lake), mountain/valley, forest/grassland, and urban/rural interfaces. In the second category we are concerned with the role of surface roughness in shaping the variation of wind speed with height, and with the way in which uneven terrain (e.g. hills and valleys) and isolated obstacles (e.g. a tree or a building) perturb existing flow patterns. All of these aspects are considered in Chapter 5.

4. Mass cascades

(a) PROPERTIES OF WATER

Water possesses a number of unusual properties which make it an important climatological substance. One important thermal property is its high *heat capacity* (see p. 38 and Table 2.1). This effectively means that in comparison with most other natural materials it takes much more energy input to cause a similar rise in the temperature of water. Equally, subtraction of energy does not cause water to cool as rapidly. This property makes water a good energy storer, and a conservative thermal influence.

If, as above, the addition or subtraction of energy to a body is sensed as a rise or fall in its temperature then it is referred to as *sensible heat*. On the other hand to enable a substance to change from liquid at a given temperature, to vapour *at the same temperature*, requires the addition of heat. This heat which is not sensed as a temperature change is called *latent heat*. It is locked-up within the substance and is available for release should the substance revert to its former state. Energy is taken up to move in the direction of a higher energy state (e.g. solid to liquid, or liquid to vapour) and released in moving in the opposite direction.

Water is the only substance that exists in all of its states at temperatures normally encountered in the E-A system. In changing between ice, water and water vapour, latent heat is taken up or liberated and as a result the energy and water cascades become enmeshed. The energy required to effect a change between the ice and water phases (i.e. consequent upon melting or freezing) is $3{\cdot}33 \times 10^5$ J kg^{-1} at 0°C, and is called the *latent heat of fusion* (L_f). The change between liquid water and water vapour (i.e. consequent upon evaporation or condensation) requires almost 7·5 times more energy, so that at 10°C the value of the *latent heat of vaporization* (L_v) is $2{\cdot}48 \times 10^6$, at 20°C $2{\cdot}45 \times 10^6$, and at 30°C $2{\cdot}43 \times 10^6$ J kg^{-1} (Appendix A3). In the event that the water changes directly between the ice and vapour phases (i.e. sublimates) the *latent heat of sublimation* (L_s) is the algebraic sum of L_f and L_v, and at 0°C it is $2{\cdot}83 \times 10^6$ J kg^{-1}. To gain some measure of the energy amounts involved it should be realized that the energy locked-up in evaporating 1 kg of water is roughly equivalent to that necessary to raise 6 kg of water from 0°C to 100°C.

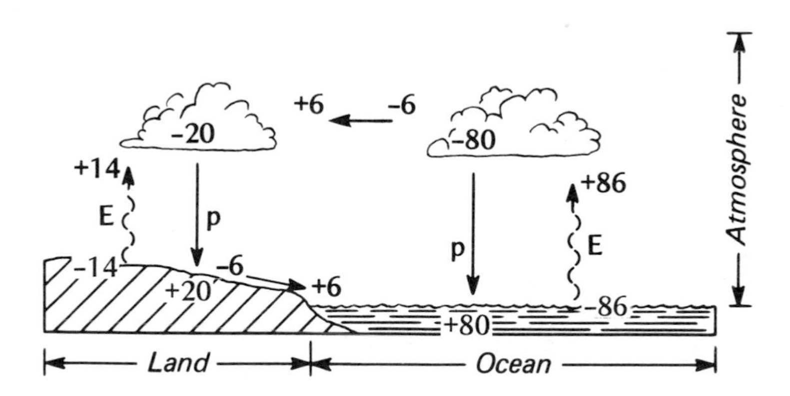

Figure 1.13 Schematic diagram of the average annual hydrologic cascade of the Earth-Atmosphere system. Values expressed as percentages of the mean annual global precipitation of 1040 mm (data from Chow, 1975).

(b) WATER CASCADE

The annual global cascade of water in the E-A system is given in Figure 1.13. All quantities of water are represented as percentages of the mean annual global precipitation, which is approximately 1040 mm. Utilizing energy provided by the energy cascade (Figure 1.8) water is evaporated from open water surfaces (oceans and lakes) and the soil, and is transpired from vegetation. The composite loss of water to the air from all sources is then called *evapotranspiration* (E). The water vapour is carried up into the Atmosphere by unstable air masses, and mechanical convection. Eventually the vapour is cooled to its dew-point (p. 54), and it condenses as a cloud droplet, or ice crystal. Under favourable conditions the cloud droplets or crystals may grow to a size where they can no longer be held in suspension and they fall to the Earth as *precipitation* (p). Near the surface water may also be deposited by direct condensation or sublimation as dew, hoar frost, and rime, or be impacted as fog-drip. Over land areas p is greater than E, and the excess is transported as streamflow to the oceans where E is less than p (Figure 1.13). So that for the Land and Ocean sub-systems their annual water balance may be written:

$$p = E + \Delta r$$

where, Δr – *net run-off*, which may have a positive or a negative sign. For the total E-A system on an annual basis the balance is even simpler:

$$p = E$$

because the system is closed to the import or export of mass and hence all horizontal transfers (such as run-off or ocean currents) are internal, and the net storage change for the system is zero.

We will accept the macro-scale conditions as given, and concentrate on small-scale surface/atmosphere interaction over relatively short time periods. Let us return to the case of our 'ideal' grassed site with a moist soil on level terrain. If we consider the water exchanges through the surface plane (Figure 1.14a) then we can formulate the surface water balance equation as:

$$p = E + f + \Delta r \qquad (1.17)$$

where, f-infiltration to deeper soil layers. Infiltration is not easily determined so for practical purposes it is better to consider a column (Figure 1.14b) which extends from the surface to a depth where significant vertical exchanges are absent (i.e. where $f \rightarrow 0$), then the water balance is given by:

$$p = E + \Delta r + \Delta S \qquad (1.18)$$

where, ΔS – the net change in *soil moisture content*. The soil moisture

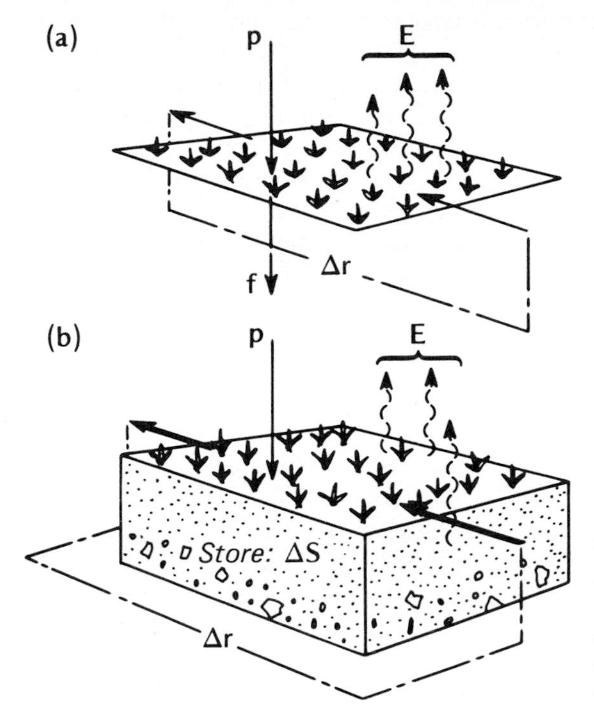

Figure 1.14 Diagrammatic representation of the components of the water balance of (a) a natural surface, and (b) a soil-plant column.

content is a measure of the mass of water stored in a soil in the same way as soil temperature is a measure of the soil heat content. Equation 1.18 shows how the water storage in the system is dependent upon the water input which is usually mainly p, and the water output via E and Δr. Input could also be supplied by irrigation which would require an additional term on the left-hand side of equation 1.18, or it could be as dewfall (i.e., convective transfer from the air to the surface, $-E$). Evapotranspiration consists of evaporation of free surface water (e.g. puddles), and soil pore water, and water transpired from vegetation.

The time scale over which equation 1.18 is valid proves awkward if we are properly to integrate it with the surface energy balance (equation 1.16) on time periods of a day or less. This arises because the input/output processes are fundamentally different in nature. Precipitation usually occurs in discrete, short-period bursts, whereas evaporation is a continuous and variable function. Thus for example, during periods with no precipitation water input is zero but the soil moisture store is being almost continually depleted by evapotranspiration. In these circumstances equation 1.18 effectively reduces to:

$$E = \Delta S \qquad (1.19)$$

because Δr is negligible on level terrain. Therefore, unlike the annual situation where net water storage is zero, on the short time-scale ΔS is non-zero and very important.

Soil moisture is significant in surface energy balance considerations because it is capable of affecting radiative, conductive and convective partitioning. For example the addition of moisture can alter the surface albedo thereby changing K^* and Q^*. Equally the thermal properties of a soil are changed by adding water so that heat transfer and storage are affected. Most important however are the potential latent heat effects of soil moisture. The common term in the water and energy balance equations is evaporation. The fluxes of mass (E) and energy (Q_E) associated with evaporation are linked by the relation:

$$Q_E = L_v E \qquad (1.20)$$

where the units of E are kg m^{-2} s^{-1} (mass transport through unit surface area in unit time). Therefore if the energy scale of Figure 1.11 were divided by L_v the curve of Q_E becomes the diurnal course of E. This has been done in Table 1.3 to yield the total daily loss of water to the atmosphere (given as an equivalent height of water.[1] It is also satisfying to note that conversion of the Q_E annual energy term in Figure 1.8 to an equivalent height of water evaporated gives a value of E very close to that in Figure 1.13. Thus loss of water to the air not only depletes the mass store (soil moisture) but also the energy store (soil and air temperature) as a result of taking up latent heat. Condensation operates in the reverse sense by adding to both the mass and energy stores. Melting and freezing are energetically less significant, but still of importance especially in soil climate.

The measures of soil moisture content, and the processes of soil moisture movement and evaporation, are outlined in Chapter 2.

(c) OTHER MASS CASCADES

The cascades of energy and water are by far the most important in explaining surface climates, but it should be noted that there are other natural mass cascades operating in the E-A system on similar space and time scales. Examples of these cascades include those of carbon, nitrogen, oxygen and sulphur. Of these the CO_2 portion of the carbon cycle is of most immediate interest to this book because it interacts with both the solar energy and water cascades in the process of photosynthesis. This is dealt with in Chapter 4 and an example of the diurnal flux of CO_2 is given in Figure 4.8.

[1] Since L_v and the density of water change rather slowly with temperature the following conversion is accurate to $\pm 2{\cdot}5\%$ in the range from 0 to 40°C; 1 mm evaporation = $2{\cdot}45 \times 10^6$ J m^{-2}, or an evaporation rate of 1 mm h^{-1} = 680 W m^{-2}.

CHAPTER 2

Physical basis of boundary layer climates

1 Sites of energy exchange

(a) THE 'ACTIVE' SURFACE

Frequent reference has already been made to the Earth's 'surface'. In technical terms a surface is a plane separating two different media. Of itself it contains no energy or mass, but it is the site of very important energy and mass exchange and conversion.

Problems arise in determining the position of the surface in many natural environments. For example the surface may be in motion (e.g. water), or be semi-transparent to radiation (e.g. ice, snow, water, plant cover), or be composed of many elemental surfaces (e.g. leaves on a tree), or may be serrated by large roughness units (e.g. trees in a forest, buildings in a city). For climatic purposes therefore we define the *'active' surface* as the principal plane of climatic activity in a system. This is the level where the majority of the radiant energy is absorbed, reflected and emitted; where the main transformations of energy (e.g. radiant to thermal, sensible to latent) and mass (change of state of water) occur; where precipitation is intercepted; and where the major portion of drag on airflow is exerted. Where appropriate the position of the active surface will be identified for each of the environments treated in this book.

Viewed in detail no surface is flat, but is composed of some form of roughness element, therefore some degree of simplification is necessary as a working basis. For climatic purposes it seems best to retain a pragmatic

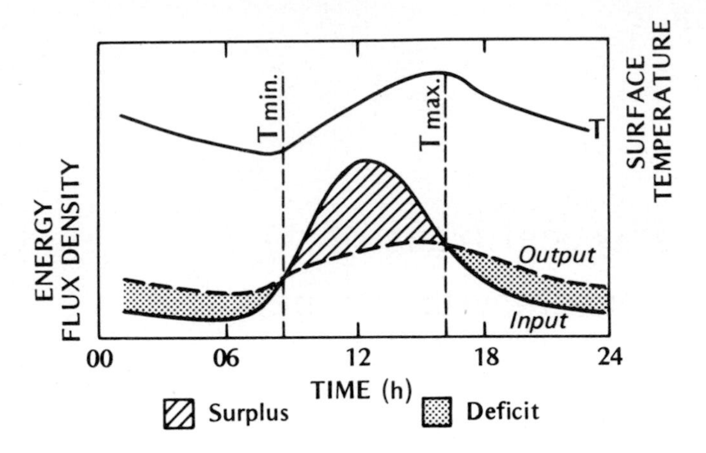

Figure 2.1 The relationship between surface energy exchange and the diurnal surface temperature regime.

concept of the active surface which is adjustable to allow for the scale of inquiry and the entity under study.

The importance of the active surface is demonstrated by considering the relationship between the surface energy balance and the surface temperature. Figure 2.1 illustrates the case of an 'ideal' bare soil site during a cloudless summer day. During most of the daytime energy is arriving faster than it can be dissipated (i.e. input exceeds output). This results in the accumulation of an energy surplus which causes the surface temperature to increase. Note however that the maximum temperature does not coincide with the time of maximum energy input. The temperature continues to rise after the time of maximum input because for a few hours it still exceeds the loss. Thus the energy budget is still accumulating a surplus. The maximum temperature occurs at the time when input and output are equal. Thereafter more heat is being extracted than is being added (output exceeds input) and the temperature starts dropping. It continues to drop as long as the rate of loss is greater than the rate of gain. The minimum temperature also occurs at the time when input and output balance. This explains why minimum temperatures are recorded just after sunrise, and maxima in mid-afternoon.

The surface being the site of the net radiant absorption by day has the greatest energy surplus and hence the highest temperature in the soil/atmosphere system. Conversely, by night it is the site of net radiant emission and experiences the greatest energy deficit and the lowest temperature. The effects diminish with distance away from the interface resulting in the following important general features of the vertical distribution (*profile*) of temperature near the surface:

(i) By day temperatures decrease with height in the lower atmosphere, and also with depth in the soil. This condition is referred to as a *lapse* profile, the gradient has a negative sign and this rate of decrease with height is known as the lapse rate ($\Delta T/\Delta z$, with units of °C m^{-1}).

(ii) By night temperatures increase with height in the atmosphere, and also with depth in the soil. This is termed an *inversion* profile, and the temperature gradient has a positive sign.

(iii) The surface experiences the greatest diurnal range of temperature, being the hottest by day and the coldest by night.

(iv) The temperature gradient is greatest near the surface and decreases with height and depth (Figure 2.2).

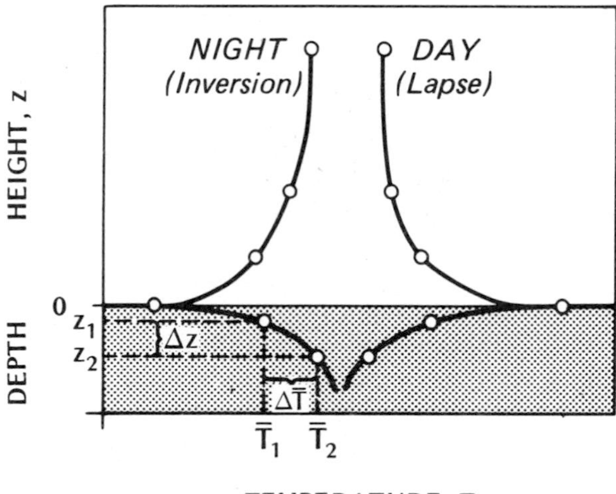

Figure 2.2 Idealized mean profiles of air and soil temperature near the soil/atmosphere interface in fine weather.

All interfaces with the atmosphere have a thin layer of air adhering to them called the laminar boundary layer (Figure 2.3, see also Figures 4.6 and 6.13). The thickness of this layer depends mainly on the roughness of the surface and the external wind speed. With high wind speeds the layer becomes very thin or can even be temporarily absent. Within the layer any motion is *laminar* (i.e. in streamlines parallel to the surface with no cross-stream component), and thus convection is not present. This means that all non-radiative transfer across this layer is by means of molecular diffusion, and since the molecular diffusivity of air is very small (Appendix A3) this layer provides an important insulating barrier between the surface and the bulk of the atmosphere.

At the top of the laminar boundary layer the flow becomes unstable and

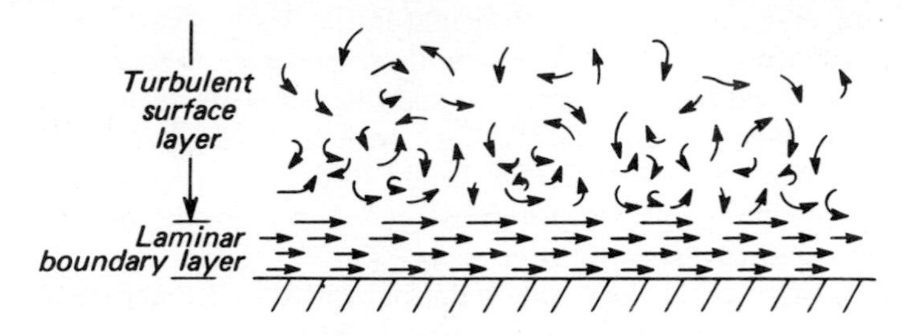

Figure 2.3 Laminar and turbulent air motion near a surface.

breaks down into an apparently haphazard jumble of swirling eddies. This is the turbulent surface layer (Figures 1.2 and 2.3) which may extend up to about 50 m above the surface. In this layer transfer is by turbulence (free and forced convection) which is many orders of magnitude more efficient than molecular diffusion.

(b) ENERGY EXCHANGE IN A VOLUME

The formulation of the surface energy balance (equation 1.16):

$$Q^* = Q_H + Q_E + Q_G$$

is consistent with equation 1.1 (Energy Input = Energy Output) because the surface is considered to be a massless plane which itself has no heat content. In many of the systems we are about to analyse it is more appropriate to view the energy balance as relating to a volume (or layer). In these situations it is necessary to include changes of energy storage (ΔQ_S) as described by equation 1.2 (Energy Input − Energy Output − Energy Storage Change = 0), so that we should write:

$$Q^* = Q_H + Q_E + Q_G + \Delta Q_S \tag{2.1}$$

The term ΔQ_S arises because of energy absorption or release by the volume. This means that the input and output of *at least* one of the individual fluxes (Q^*, Q_H, Q_E and Q_G) do not balance. This concept is illustrated in Figure 2.4 where the arrows represent fluxes of energy into (Q_{in}), and out of (Q_{out}), the system volume. The length of the arrows represents the magnitude of the flux. There are three possibilities:

(i) Q_{in} exceeds Q_{out} – from simple systems considerations if the input of energy exceeds the output then there must be a net energy storage gain ($+\Delta Q_S$) in the volume. This situation is one of *flux convergence*, and will result in a warming of the volume (Figure 2.4a). The amount of the temperature change per unit of energy storage change will depend upon the thermal properties of the materials of the volume.

(ii) Q_{in} is less than Q_{out} – if input is less than output the volume must be

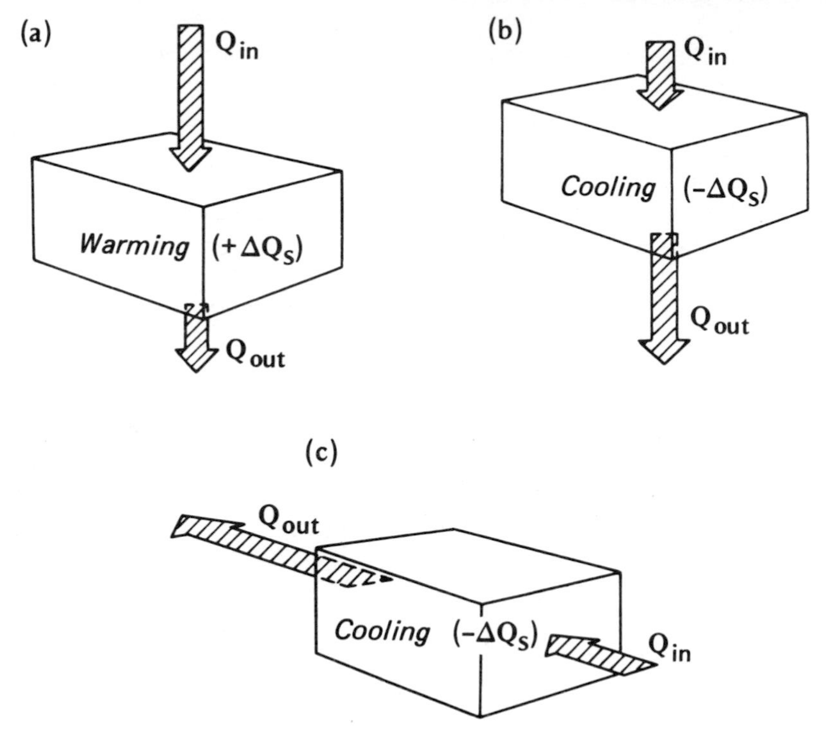

Figure 2.4 Schematic depiction of (a) vertical flux convergence, (b) vertical flux divergence and (c) horizontal flux divergence, in a volume.

losing energy, thereby depleting its energy store ($-\Delta Q_S$) and hence cooling. This is the case of *flux divergence* (Figure 2.4b).

(iii) $Q_{in} = Q_{out}$ – if input equals output there is no net change in the energy status of the volume (i.e. $\Delta Q_S = 0$), or its temperature.

The concept of convergence and divergence is more general than depicted in Figure 2.4. For example the direction of the flux is unimportant, it is the net change in the flux that influences the energy balance of the volume. Convergence and divergence of horizontal fluxes can also contribute to warming and cooling (Figure 2.4c). This is advection (ΔQ_A) and its effects are dealt with in Chapter 5. Thus it is often necessary to consider a full three-dimensional balance. Moreover the flux arrows in Figure 2.4 could represent a conductive, radiative or convective transport, or any combination of the three. Finally, the concept is easily extended to include the mass balance of the volume (e.g. water or pollution) by considering the appropriate mass fluxes. The case of the water balance of a volume has already been outlined (Figure 1.14b) and equation 1.18 is clearly analogous to equation 2.1. The general case of pollution was mentioned previously (p. 8).

A few examples of convergence and divergence involving different transport mechanisms and different media may help to illustrate this rather abstract concept. First, imagine that the arrow in Figure 2.4a represents Q_G, and the volume is composed of soil. The downward flux is typical of conditions during the day. If Q_G entering the soil (top of the box) is greater than that arriving at some depth (base of the box), then energy has been retained in the layer, and the soil warms up due to conductive flux convergence (i.e. $+\Delta Q_S$ represents $+\Delta Q_G$).

The situation is the same if we imagine the flux to be that of the incident solar radiation ($K\downarrow$), and the volume to be a layer of air. If the air were perfectly clear the flux would be constant with height. Pollutants, however, may scatter, reflect or absorb some of the radiation. If the sum of the short-wave radiation leaving all sides of the box (i.e. after scattering, reflection or transmission) does not amount to that which entered then the layer is warmed by short-wave radiative flux convergence (i.e. $+\Delta Q_S$ represents $+\Delta K\downarrow$).

As a final but more complicated example we will let the box represent a volume of rural air near the ground, and consider how this air warms by day and cools by night. The temperature (energy status) of the air can be changed as a result of the convergence or divergence of one or more of four main energy exchanges (Q^* including K^* and/or L^*, Q_H, Q_E and Q_A). If we restrict consideration to the 'ideal' site and weather conditions outlined in Chapter 1 then we may neglect horizontal fluxes (i.e. $\Delta Q_A \rightarrow 0$), and if the air inside the volume remains unsaturated we may ignore latent heat effects (i.e. $\Delta Q_E \rightarrow 0$). This effectively restricts consideration to the convergence and divergence of radiation and sensible heat as the agents of air temperature change.

By day when Q_H is directed upwards from the heated surface, and the atmosphere is fully turbulent, it is normally assumed that the warming of the lower atmosphere is entirely due to the vertical convergence of $Q_H(+\Delta Q_H)$. Turbulence so effectively mixes radiatively-important agents such as water vapour, CO_2 and dust that vertical variations of Q^* are unlikely to be significant.

At night a similar situation normally holds except that Q_H is directed downwards and the air volume is cooled by vertical divergence of $Q_H(-\Delta Q_H)$. But if the air is almost calm and turbulence becomes very weak then concentrations of radiatively-important constituents may build up so that internal atmosphere radiative exchanges are capable of dominating the change of air temperature. On these occasions the net long-wave loss increases with height so that the air is cooled by the vertical divergence of L^*. In fact in very favourable conditions radiative cooling has been observed to be equivalent to cooling rates as great as $10°C\ h^{-1}$. Since actual cooling rates are rarely greater than $3°C\ h^{-1}$ it is probable that some of the radiative cooling ($-\Delta L^*$) is offset by turbulent *warming* ($+\Delta Q_H$).

Thus we see that the lower atmosphere warms and cools as a result of the variation of fluxes with height. This is fairly evident on an instantaneous basis but if the fluxes are averaged over time (say 30 minutes) the vertical differences are very small indeed. For example, if the time-averaged flux density of Q_H decreases by just 1 W m^{-2} for each metre of height in our volume it will produce a warming rate of approximately 3°C h^{-1} *if no other processes are involved.* With most modern instrumentation it is impossible to measure vertical differences of this magnitude. Thus it is usual to assume that measurements of fluxes made anywhere in the lowest 50 m over an ideal site are equal to their surface values. Accordingly the turbulent surface layer (Figure 2.3) is also referred to as the *'constant-flux' layer.*

2 Sub-surface climates

(a) SOIL HEAT FLUX (Q_G) AND SOIL TEMPERATURE (T_s)

Consideration of the situation in Figure 2.2 reveals that by day soil temperatures decrease with depth. Hence we expect heat to flow downwards into the soil by thermal conduction (i.e. from high to low temperature). At night when the surface is cooling the sign of the temperature gradient is reversed, and heat may be expected to flow upwards. The rate of these heat flows depends on the strength of the mean temperature gradient ($\Delta\overline{T}/\Delta z$), and the ability of the particular soil to transmit heat:

$$Q_G = -k_s \frac{(\overline{T}_2 - \overline{T}_1)}{(z_2 - z_1)} = -k_s \frac{\Delta\overline{T}_s}{\Delta z} \qquad (2.2)$$

where, k_s – *thermal conductivity* (W m^{-1} K^{-1}), the overbars represent time-averaged quantities, and the subscripts refer to the soil depths (Figure 2.2). The negative sign indicates the direction of flow in accordance with the convention set out on p. 24. Therefore by day when $\Delta\overline{T}_s/\Delta z$ is negative equation 2.2 gives a positive value of Q_G, indicating a flux away from the surface, or a surface energy loss.

Before considering soil heat in more detail it is worthwhile to pause and draw attention to the form of equation 2.2. The equation expresses the relationship between the flux of heat and the gradient of temperature. This general form also holds for the transport of other entities such as water vapour and carbon dioxide with the appropriate gradient substituted. So that in general:

$$\text{Flux of an entity} = \text{Ability of the medium to transport the entity} \times \text{Gradient of a relevant property}$$

We will return to this *flux-gradient* form on a number of occasions.

The thermal conductivity of a substance (here soil) is a measure of its

TABLE 2.1 Thermal properties of natural materials

Material	Remarks	ρ Density ($kg\ m^{-3} \times 10^3$)	c Specific heat ($J\ kg^{-1}\ K^{-1} \times 10^3$)	C Heat capacity ($J\ m^{-3}\ K^{-1} \times 10^6$)	k Thermal conductivity ($W\ m^{-1}\ K^{-1}$)	κ Thermal diffusivity ($m^2\ s^{-1} \times 10^{-6}$)
Sandy soil (40% pore space)	Dry	1·60	0·80	1·28	0·30	0·24
	Saturated	2·00	1·48	2·96	2·20	0·74
Clay soil (40% pore space)	Dry	1·60	0·89	1·42	0·25	0·18
	Saturated	2·00	1·55	3·10	1·58	0·51
Peat soil (80% pore space)	Dry	0·30	1·92	0·58	0·06	0·10
	Saturated	1·10	3·65	4·02	0·50	0·12
Snow	Fresh	0·10	2·09	0·21	0·08	0·10
	Old	0·48	2·09	0·84	0·42	0·40
Ice	0°C, pure	0·92	2·10	1·93	2·24	1·16
Water*	4°C, still	1·00	4·18	4·18	0·57	0·14
Air*	10°C, still	0·0012	1·01	0·0012	0·025	20·50
	Turbulent	0·0012	1·01	0·0012	~125	$\sim 10 \times 10^6$

* Properties depend on temperature, see Appendix A3.
Sources: van Wijk and de Vries (1963), List (1966).

ability to transmit heat. It is formally defined as the quantity of heat (J) flowing through unit cross-sectional area (m^2) of the substance in unit time (s), if perpendicular to it there exists a temperature gradient of 1 degree m^{-1}. Typical values for a range of natural materials are listed in Table 2.1; of these motionless air is the most notable because it is such a very poor conductor of heat (i.e. it is a good insulator).

Unfortunately k_s is not a simple constant for a given soil. It varies both with depth and with time. However if we restrict ourselves to bulk averages k_s depends upon the conductivity of the soil particles, the soil porosity, and the soil moisture content. Of these the soil moisture content is the only short-term variable for a given soil. The addition of moisture to an initially dry soil increases its conductivity (Figure 2.5a). This happens for two reasons. First, coating the soil particles increases the thermal contact between grains. Second, since the soil pore space is finite the addition of pore water must expel a similar amount of pore air. From Table 2.1 we can see that this means replacing soil air with a substance whose conductivity is more than an order of magnitude greater.

The soil heat flux given by equation 2.2 refers to the value at a single depth (or the surface). On p. 36 we noted that vertical changes of Q_G lead to soil warming (convergence) or cooling (divergence). The actual temperature change resulting from these processes depends not only on the amount of energy added or subtracted, but also upon the *volumetric heat capacity*

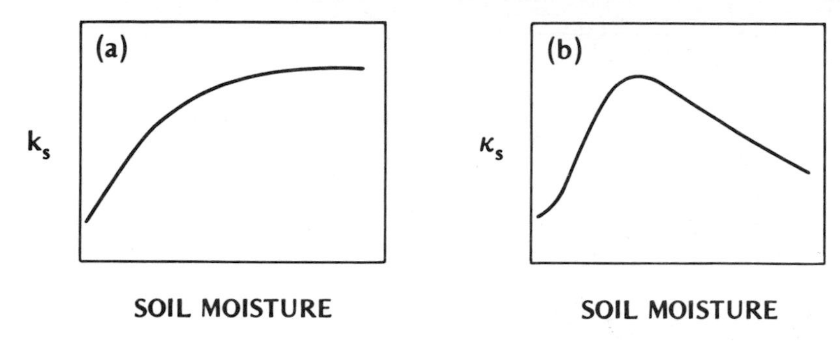

Figure 2.5 Relationship between soil moisture content and (a) soil thermal conductivity, k_s and (b) soil thermal diffusivity, κ_s for most soils.

(C) of the soil. The heat capacity is defined as the amount of heat (J) necessary to raise the temperature of unit volume (m^3) of a body by 1 degree (K). Typical values for some natural materials are listed in Table 2.1, from which it is clear that compared with most other materials water requires a very large heat input to effect a given change in temperature (p. 26), whereas air requires very little. Therefore the result of adding water to a soil, and thereby excluding a proportionate volume of soil air, is to increase the soil's heat capacity and to reduce its thermal sensitivity.

The total thermal response of a soil is directly proportional to its ability to transmit heat (k_s), but inversely proportional to the amount of heat required to effect temperature change (C_s). These relationships are stated as:

$$\kappa_s = k_s/C_s$$

where, κ_s – soil *thermal diffusivity* ($m^2\ s^{-1}$), and typical values are given in Table 2.1. This property may be viewed as a measure of the time required for temperature changes to travel within the soil. For example, in the morning the surface heat input penetrates rapidly in a soil where k_s is large, but can be slowed if it takes large amounts of heat to warm intermediate layers because C_s is also large.

The value of κ_s is obviously affected by the same soil properties that influence k_s and C_s, especially soil moisture (Figure 2.5b). Note that adding moisture to a dry soil initially produces a sharp increase in κ_s by increasing thermal contact and expelling soil air (i.e. increasing k_s as in Figure 2.5a). However, in most soils beyond about 20% soil moisture content by volume κ_s begins to decline. This happens because whereas k_s levels off (Figure 2.5a) the value of C_s continues to increase at higher moisture contents.

Soils with high diffusivities allow rapid penetration of surface temperature changes and permit these effects to involve a thick layer. Thus for the same heat input their temperature regimes are less extreme than for soils

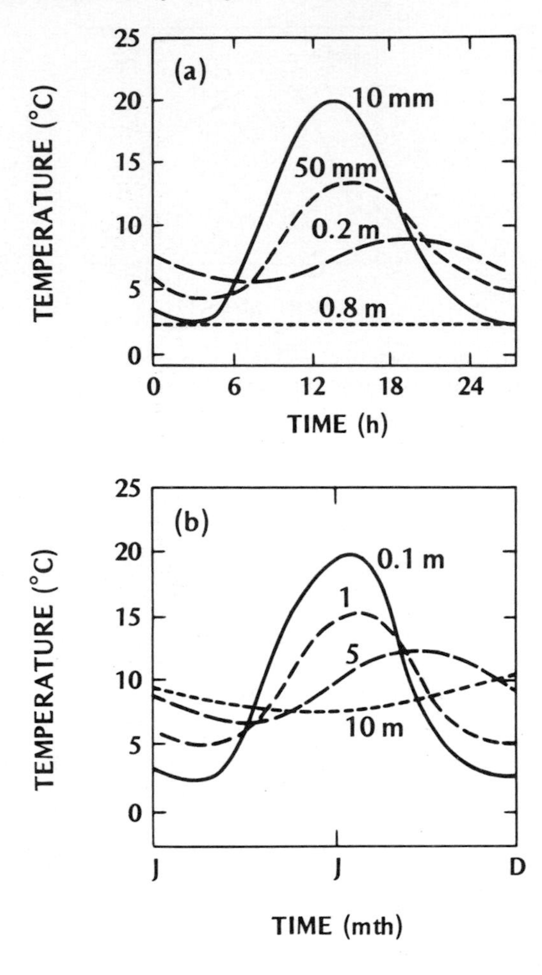

Figure 2.6 Generalized cycles of soil temperature at different depths for (a) daily and (b) annual periods.

with low diffusivities. By day the surface heating is used to warm a thick layer of soil, and at night the surface cooling can be partially offset by drawing upon heat from a similarly thick stratum. Soils with poor diffusivities concentrate their thermal exchanges in the uppermost layer only, and consequently experience relatively extreme diurnal temperature fluctuations. Therefore in general a wet clay is conservative, whereas an almost dry peat is extreme in its thermal climate.

The course of soil temperature with time is very regular in comparison with any other atmospheric element. Typical soil temperature variations at a number of depths on a cloudless day are given in Figure 2.6a. The near-surface temperature variation is wave-like and agrees closely with that of the surface (Figure 2.1). The wave penetrates downward to lower depths, but in doing so its amplitude decreases, and the times of maximum and minimum temperature are lagged (shift to the right in time). Both features depend on κ_s. The wave amplitude at any depth $(\Delta \overline{T}_s)_z$ is given by:

$$(\Delta \overline{T}_s)_z = (\Delta \overline{T})_0 e^{-z(\pi/\kappa_s P)^{1/2}} \qquad (2.3)$$

where, $(\Delta \overline{T})_0$ – surface temperature wave amplitude, e – base of Naperian logarithms, P – wave period (s). This shows that the diurnal temperature range decreases exponentially with depth. In most soils the daily surface temperature wave is only discernible to a depth of about 0·75 m. In soils with low κ_s values it is even less, indicating that flux convergence has extinguished Q_G in a thin near-surface layer. The time lag for the wave crest (maximum) and trough (minimum) to reach lower depths is given by:

$$(t_2 - t_1) = \frac{(z_2 - z_1)}{2}(P/\pi\kappa_s)^{1/2} \qquad (2.4)$$

where t_1 and t_2 are the times at which the wave crest or trough reaches depths z_1 and z_2. Note that because of this time lag, at any given time the soil may be cooling in its upper layers but warming at only a short distance beneath, and vice versa when the upper layers are warming (Figure 2.6a).

The annual soil temperature regime (Figure 2.6b) follows a wavelike pattern entirely analogous with the diurnal one. The wave period is of course dependent upon the annual rather than the daily solar cycle. With that adjustment, equations 2.3 and 2.4 still apply to the wave amplitude and lag with depth. With a longer period the wave amplitude decreases less rapidly with depth, and the depth of the affected layer is much greater than for the diurnal case. In typical soils the annual wave may penetrate to about 14 m, which would then be termed the depth of zero annual range. The temperature at this depth is sometimes used as a surrogate for the average annual air temperature of the site. This is based on the premise that long-term thermal equilibrium exists between the soil and the atmosphere. During the warm season soil temperatures decrease with depth and the associated downward heat flux builds up the soil's heat store. In the cold season the gradient is reversed and the store is gradually depleted. The spring and autumn are transitional periods when the soil temperature gradient reverses sign. These reversals (or 'turnovers') are important biological triggers to soil animals and insects. In the spring they may come out of hibernation, and/or move upwards towards the warmer surface layers. In the autumn they retreat to depths where soil warmth is more equable.

The effects of cloud on the diurnal soil temperature pattern are fairly obvious. With overcast skies absolute temperatures are lower by day but warmer at night, and the wave amplitude is smaller; variable cloudiness induces an irregular pattern upon the diurnal wave. Rainfall is capable of either increasing or decreasing soil temperatures depending upon the temperature of the rain in comparison with the soil. It is also capable of transporting heat as it percolates down through the soil. The effects of a snow or vegetation cover over the soil are dealt with in Chapters 3 and 4, respectively.

Appendix A2 provides examples of the instruments for measuring soil heat flux and soil temperature in the field.

(b) SOIL WATER FLOW (M) AND SOIL MOISTURE (S)

Soil moisture is usually expressed in one of two ways. *Soil moisture content* (S) is a measure of the actual water content, and is defined as the percentage volume of a moist soil occupied by water. This is particularly pertinent in water balance studies (p. 28) where changes in mass are important. *Soil moisture tension* (m) on the other hand is an indirect measure of water content, and may be visualized as the energy necessary to extract water from the soil matrix. The units of m are those of pressure (Pa = mb $\times 10^{-2}$), which can also be expressed as a head of water displaced (1 m head of water = $1{\cdot}0 \times 10^4$ Pa). This concept is of value in estimating the availability of water for plant use, and in calculating moisture movement. Methods for determining these two terms are given in Appendix A2.

The tension forces which bind soil water are related to the soil porosity and the soil water content (Figure 2.7). The forces are weakest in the case of

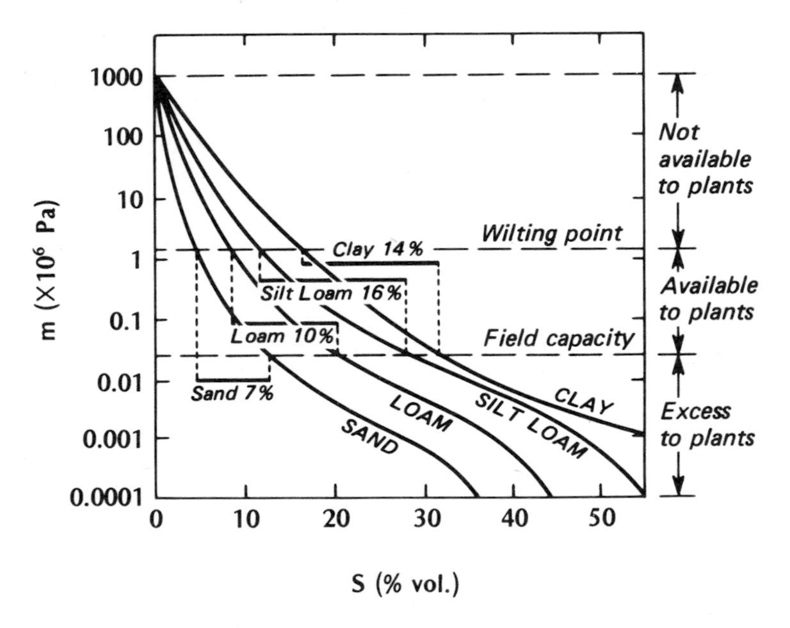

Figure 2.7 Relationship between soil moisture tension (m) and soil moisture content (S) in soils with different textures. Horizontal bars show the volumetric water available to plants (modified after Buckman and Brady, 1960).

open textured, wet soils, and greatest for compact dry soils. Thus at a given value of S the tension is greatest for a clay, least for a coarse sand, and intermediate for a loam. Similarly in a given soil the tension increases as S decreases, but not in a linear fashion. Thus it is relatively easy to extract moisture from a wet soil but as it dries out it becomes increasingly difficult to remove additional units. Figure 2.7 shows that in the range of tensions which permit plants to extract soil water, sand has the least available water (7% vol.) and a silt loam the most (16% vol.).

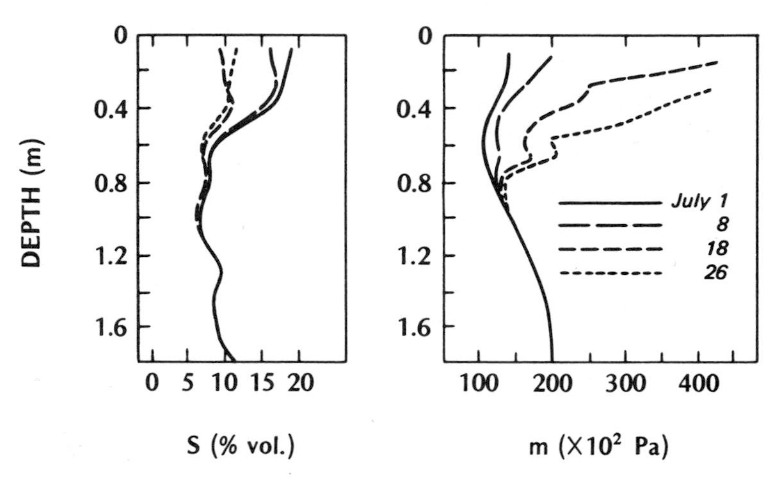

Figure 2.8 Profiles of soil moisture content (S) and soil moisture tension (m) in a sandy loam during a drying phase (after Rouse and Wilson, 1972).

Figure 2.8 conveniently illustrates these two concepts of soil moisture. It shows profiles of both S and m at the same site during a period of one month when the soil was almost continually drying-out. As S decreased throughout the period, the value of m increased. It is possible to utilize such changes in soil moisture content to estimate evapotranspiration losses to the atmosphere by integrating the area between successive profiles in time. In this case it appears as though the soil down to a depth of 0·8 m participated in the soil-air exchange.

The vertical flux of soil water (M) in the absence of percolating rain, is composed of both liquid (M_1) and vapour (M_v) flow. The movement of liquid moisture in a saturated or unsaturated soil may be considered analogous with the flux-gradient relationship for heat (equation 2.2). In this case the flux of water is related to the vertical tension gradient by Darcy's Law:

$$M_1 = \psi \frac{\Delta \overline{m}}{\Delta z} \qquad (2.5)$$

where, ψ – hydraulic conductivity. The effect of evapotranspiration is to create a tension gradient which becomes greater than the opposing gravitational gradient and encourages the upward movement of water from low to high tension potential. Unfortunately in unsaturated soils ψ is not a constant but depends on both S and m, as well as other soil factors.

The flow of water vapour similarly obeys a flux-gradient relationship:

$$M_{v} = -\rho\kappa_{w}\frac{\Delta\bar{q}}{\Delta z} \tag{2.6}$$

where, ρ – air density (kg m^{-3}), κ_{w} – molecular diffusivity for water vapour ($\simeq 0{\cdot}24 \times 10^{-4}$ m^{2} s^{-1}, Appendix A3), and q – specific humidity (p. 55) or water vapour concentration. The air in pore spaces of moist soils is in close contact with soil water and is therefore commonly close to being saturated. The saturation vapour concentration is directly related to temperature (Figure 2.15), being higher at higher temperatures. This creates a water vapour concentration gradient and a corresponding vapour flux from high to low soil temperatures. Thus, noting the temperature distributions in Figures 2.2 and 2.6a, there is a tendency for vapour to flow down into the soil by day and up towards the surface at night. The nocturnal vapour flux commonly results in condensation upon the cold soil surface known as *distillation*. This is not the same process as dewfall (p. 57) which involves the turbulent transfer of atmospheric vapour down onto the surface.

3 Lower atmosphere climates

(a) MOMENTUM FLUX (τ) AND WIND (u)

The wind field in the boundary layer is largely controlled by the frictional drag imposed on the flow by the underlying rigid surface. The drag retards motion close to the ground and gives rise to a sharp decrease of mean horizontal wind speed ($\bar{u}$) as the surface is approached (Figure 2.9). In the absence of strong thermal effects the depth of this frictional influence depends on the roughness of the surface (Figure 2.9a). The profiles in this figure are based on measurements in strong winds, and the height z_{g} is the top of the boundary layer above which $\bar{u}$ is approximately constant with height (i.e. surface drag is negligible). The depth of this layer increases with increasing roughness. Therefore the vertical gradient of mean wind speed ($\Delta\bar{u}/\Delta z$) is greatest over smooth terrain, and least over rough surfaces. In light winds the depth z_{g} also depends upon the amount of thermal convection generated at the surface. With strong surface heating z_{g} is greater than in Figure 2.9a, and with surface cooling it is less (see also p. 53).

The force exerted on the surface by the air being dragged over it is called the *surface shearing stress* (τ) and is expressed as a pressure (Pa, force per unit surface area). This force is equally opposed by that exerted by the

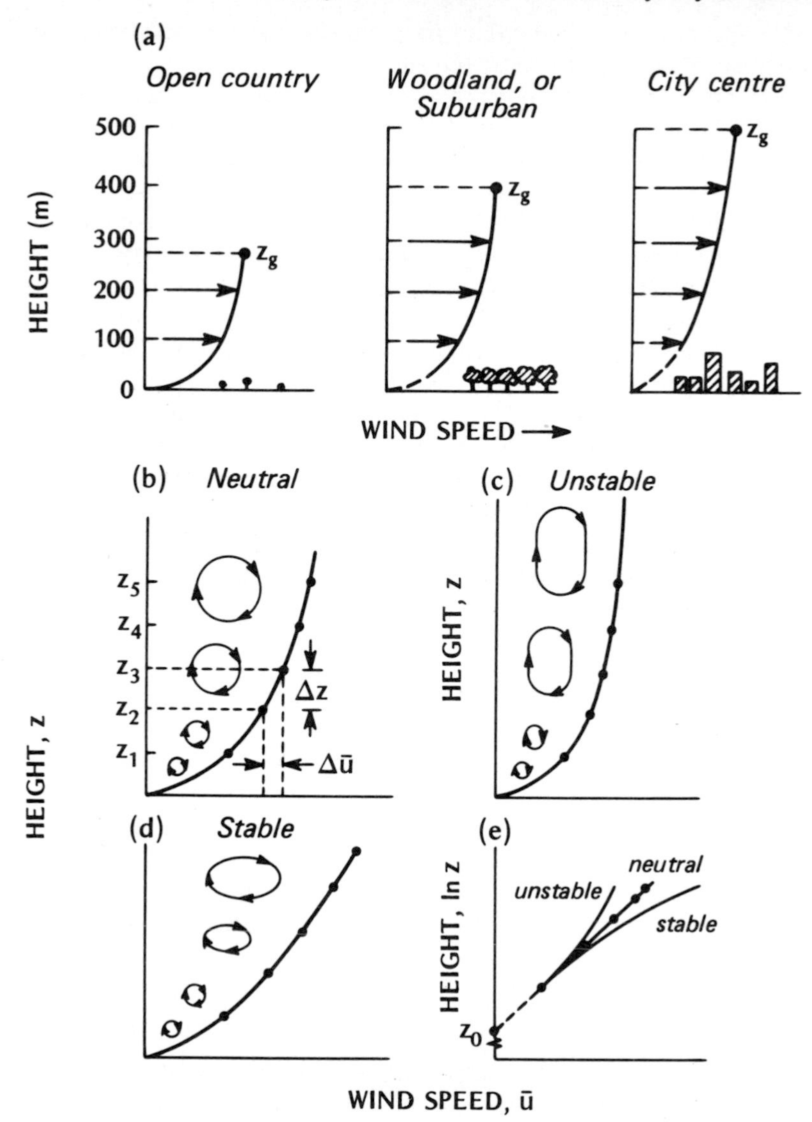

Figure 2.9 The wind speed profile near the ground including: (a) the effect of terrain roughness (after Davenport, 1965), and (b) to (e) the effect of stability on the profile shape and eddy structure (after Thom, 1975). In (e) the profiles of (b) to (d) are plotted with a natural logarithm height scale.

surface on the atmosphere. However since air is a fluid it only acts on the lower boundary and not throughout the total bulk of the Atmosphere. The surface layer of frictional influence generates this shearing force and transmits it downwards as a *flux of momentum*. In the 'constant-flux' layer

(p. 37) τ does not vary by more than 10% and hence atmospheric values are approximately equal to those at the surface.

The flux of momentum is not as easy to visualize as that of heat or water vapour, but the following simple conceptualization may help. The momentum possessed by a body is given by the product of its mass and velocity. In the case of air the mean horizontal momentum of unit volume is therefore given by its density (ρ) multiplied by its mean horizontal wind speed (i.e., momentum $= \rho\bar{u}$). Since for practical purposes we may consider air density to be a constant in the surface layer (see Appendix A3) the mean horizontal momentum possessed by different levels is proportional to the profile of wind speed (i.e., it increases with height). Now consider the situation at level z_3 in Figure 2.9b. Due to the effects of forced convection generated by the surface roughness, and the mutual shearing between air layers moving at different speeds, turbulent eddies are continually moving up and down through z_3. An eddy arriving at z_3 having originated at z_4 above will, upon mixing, impart a net increase in velocity (and hence momentum). A fast-response wind speed sensor at z_3 would therefore see this downdraft as an increase in wind speed, or a 'gust'. Conversely an updraft from z_2 would be sensed as a 'lull' in horizontal wind speed. Notice that due to the increase of wind with height the net effect of *both* updrafts and downdrafts is always to sustain a net flux of momentum downwards.

In the case of the molecular conduction of heat and water through solids (e.g., soil) and still fluids (e.g. pore air, laminar boundary-layer of air) the flux-gradient relationship works well (equations 2.2, 2.6). In the atmosphere however transfer is dominated by convective activity and true molecular transfer is negligible. Nevertheless meteorologists have attempted to extend an analogy between the role played by molecules in conduction to that of eddies in convection. Thus just as the transfer of horizontal momentum in the laminar boundary layer is given by:

$$\tau = \rho\kappa_M \frac{\Delta\bar{u}}{\Delta z} \tag{2.7}$$

where, κ_M – kinematic viscosity ($\simeq 0{\cdot}15 \times 10^{-4}$ m^2 s^{-1}, see Appendix A3), so in the turbulent surface layer above it is given as:

$$\tau = \rho K_M \frac{\Delta\bar{u}}{\Delta z} \tag{2.8}$$

where, K_M – eddy viscosity (m^2 s^{-1}). Equation 2.8 therefore relates the flux (τ) to the gradient of horizontal momentum ($\rho \cdot \Delta\bar{u}/\Delta z$) and the ability of the medium to transfer momentum (K_M). In absolute magnitude K_M is 4 to 6 orders of magnitude greater than κ_M, but is variable in both time and space.

The molecular analogy is a long-standing feature of micro-meteorology. It was introduced in order to aid measurement of the fluxes (Appendix A2) but it is only an approximation of the process and leads to problems if implemented in detail. The analogy is used here, and later only because of its simple basis, and unifying pedagogic appeal.

In terms of the general characteristics of the interaction between different surfaces and airflow we may expect that for a surface of given roughness τ will depend upon $\bar{u}$. As $\bar{u}$ at a reference level increases so will τ, and so will the depth of the forced convection layer. Similarly if we considered the value of $\bar{u}$ to be constant over surfaces with different roughnesses, the magnitude and depth of forced convective activity would be greatest over the roughest surface (Figure 2.9a).

The actual form of the wind variation with height under neutral stability (Appendix A1) has been found to be accurately described by a logarithmic decay curve. Thus using the natural logarithm of height (ln z) as the vertical co-ordinate the data from Figure 2.9b fall upon a straight line in Figure 2.9e. This provides the basis for the logarithmic wind profile equation:

$$\bar{u}_z = \frac{u_*}{k} \ln \frac{z}{z_0} \qquad (2.9)$$

where, $\bar{u}_z$ – mean wind speed (m s^{-1}) at the height z, u_* – *friction velocity* (m s^{-1}), k – von Karman's constant ($\simeq$0·40), z_0 – *roughness length* (m). It has been found that the shearing stress is proportional to the square of the wind velocity at some arbitrary reference height. Thus we introduce u_* for which this square law holds exactly so that:

$$u_*^2 = \tau/\rho \qquad (2.10)$$

This is helpful because u_* can be evaluated from wind profile measurements (the slope of the line in Figure 2.9e is k/u_*) and therefore we can obtain τ, which can be used in evaluating other fluxes (Appendix A2).

The length z_0 is a measure of the aerodynamic roughness of the surface. It is related, but not equal to, the height of the roughness elements. It is also a function of the shape and density distribution of the elements. Typical values of z_0 are listed in Table 2.2. This term is defined as the height at which the neutral wind profile extrapolates to a zero wind speed (Figure 2.9e). An alternative means of evaluation is given in Table 8.5.

The foregoing discussion relates to neutral conditions where buoyancy is unimportant. Such conditions are found with cloudy skies and strong winds, and in the lowest 1 to 2 m of the atmosphere. Cloud reduces radiative heating and cooling of the surface; strong winds promote mixing and do not permit strong temperature stratification to develop; and in the lowest layers forced convection due to frictionally-generated eddies is dominant. In the simplest interpretation these eddies may be conceived as being circular and to increase in diameter with height (Figure 2.9b). In reality they are three-dimensional and comprise a wide variety of sizes.

In unstable conditions (Appendix A1) the vertical movement of eddies

TABLE 2.2 Aerodynamic properties of natural surfaces

Surface	Remarks	z_0 Roughness length (m)	d Zero plane displacement* (m)
Water†	Still – open sea	0.1–10.0×10^{-5}	–
Ice	Smooth	0.1×10^{-4}	–
Snow		0.5–10.0×10^{-4}	–
Sand, desert		0·0003	–
Soils		0·001–0·01	–
Grass†	0·02–0·1 m	0·003–0·01	⩽0·07
	0·25–1·0 m	0·04–0·10	⩽0·66
Agricultural crops†		0·04–0·20	⩽3·0
Orchards†		0·5–1·0	⩽4·0
Forests†	Deciduous	1·0–6·0	⩽20·0
	Coniferous	1·0–6·0	⩽30.0

* Calculated as $d \simeq \frac{2}{3}h$ (see p. 98)
† z_0 depends on wind speed (see p. 119)
Sources: Sutton (1953), Szeicz *et al.* (1969), Kraus (1972).

(and therefore the momentum flux) is enhanced. Near the surface mechanical effects continue to dominate but at greater heights thermal effects become increasingly more important. This results in a progressive vertical stretching of the eddies and a reduction of the wind gradient (Figure 2.9c). Conversely strong stability dampens vertical movement, progressively compresses the eddies and steepens the wind gradient (Figure 2.9d).

Stability effects on turbulence are further illustrated in Figure 2.10. This is a graph of wind inclination (roughly corresponding to vertical winds, because the scale refers to the tilt angle of a horizontal vane) over a period of 3 minutes. The upper trace is from lapse (unstable), and the lower trace from inversion (stable) conditions, over the same grass site with approximately equivalent horizontal wind speeds. Thus differences between the two traces are due to stability differences. In the unstable case two types of fluctuation are evident. First, there are long-period 'waves' lasting about 1 to 1·5 minutes. These are relatively large buoyancy-generated eddies bursting up through the measurement level (positive values) and being replaced by sinking air parcels (negative values). Superimposed on this pattern are a second set of much shorter-period fluctuations. These are the small roughness-generated and internal shearing eddies. Therefore Figure 2.10 visually presents the two elements of turbulence – free convection (large) and forced convection (small). The combination of these two elements (upper trace) provides a very efficient means of both vertical transport and mixing. This is the 'ideal' daytime mixed convection situation. The stable case (lower trace) by contrast only exhibits the short-period eddies due to forced convection because buoyancy is absent. This is the 'ideal' nocturnal situation, and is not conducive to vertical exchange.

In summary we may say that below approximately 2 m the effects of forced convection dominate even in non-neutral conditions as long as there is a reasonable airflow. Above this height the relative role of free convection grows and the possibility of stability effects on momentum transfer increases. These effects are manifested as curvature in the wind profile (Figure 2.9b–e). Strong instability weakens the wind gradient by promoting vertical exchange over a deep layer, and thereby mixing the greater momentum of faster-moving upper air with that nearer the surface. Strong stability on the other hand strengthens the wind gradient. It therefore follows that since there is a characteristic diurnal cycle of stability there is an associated diurnal variation of wind speed in the surface layer (see pp. 53 and 67).

The study of momentum exchange and the form of the wind profile is important because of what it tells us about the state of turbulence. This is central to questions concerning the transport of Q_H and Q_E, and to the dispersal of air pollutants.

(b) SENSIBLE HEAT FLUX (Q_H) AND AIR TEMPERATURE (T_a)

A portion of the daytime radiative heat surplus is carried into the atmosphere as sensible heat (Q_H). This heat must pass through the laminar boundary layer by molecular conduction:

$$Q_H = -\rho c_p \kappa_H \frac{\Delta \overline{T}}{\Delta z} \tag{2.11}$$

where, c_p – specific heat of air at constant pressure (J kg^{-1} K^{-1}), and κ_H – molecular diffusion coefficient for heat, or the thermometric conductivity ($\simeq$0·21 m^2 s^{-1}, see Appendix A3). Since Q_H may be as high as 400 W m^{-2} at midday over a bare soil, from equation 2.11 we can see that this requires $\Delta \overline{T}/\Delta z$ to be equivalent to $\simeq 16 \times 10^3$°C m^{-1} in the laminar boundary layer! In reality of course the lapse rate only applies to a very thin layer and so the actual temperature difference is more reasonable. In the example cited above if the layer were 5 mm deep then the difference would be $\simeq$8°C (see also Figure 6.12).

In the overlying turbulent surface layer the flux of sensible heat is given by:

$$Q_H = -\rho c_p K_H \left(\frac{\Delta \overline{T}}{\Delta z} + \Gamma\right) \tag{2.12}$$

where, K_H – eddy conductivity (m^2 s^{-1}), and Γ – dry adiabatic lapse rate = $9{\cdot}8 \times 10^{-3}$ °C m^{-1} (Appendix A1). The form of equation 2.12 is directly analogous to that of equation 2.8 for momentum transfer, and the value of K_H is similar to that of K_M. The direction of the heat transfer (sign of Q_H) is determined by the sign of the temperature gradient. By day the gradient is negative (lapse) and Q_H is positive (i.e. directed from the surface into the lower atmosphere). By night the gradient is positive (inversion) and Q_H is negative. The term Γ is included to correct the observed temperature gradient for the effects of vertical atmospheric pressure changes. It is only of importance if Δz is large (e.g. greater than 2 m).

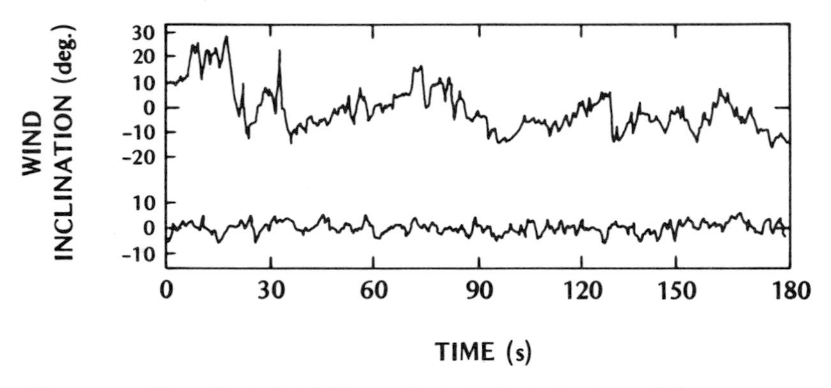

Figure 2.10 The effect of stability on the turbulent structure of the atmosphere. Wind inclination fluctuations at a height of 29 m during unstable (upper trace) and stable (lower) conditions over a grass site with winds of 3 to 4 m s^{-1} (after Priestley, 1959).

The vertical transfer of sensible heat by eddies can be visualized with the aid of Figure 2.11. This shows the variation of air temperature (T), vertical velocity (w) and the associated instantaneous flux of heat over a period of 120 s from fast response instruments placed at a height of 23 m over a grass surface at Edithvale, Australia. The data are from a daytime unstable period, and the vertical wind velocity pattern clearly resembles the upper trace of Figure 2.10. The simultaneous record of air temperature exhibits the same pattern, and moreover its fluctuations are closely in phase with those of the vertical wind. Thus in unstable conditions an updraft (positive w) is associated with an increase of T, and downdraft (negative w) with a decrease of T, relative to its mean value. This occurs because unstable conditions are associated with a lapse T profile, and an updraft through the measurement level has originated closer to the ground where it is warmer. Conversely a downdraft comes from higher levels where it is cooler. For both situations (up- and downdraft) the net sensible heat transfer is therefore upwards. The instantaneous heat flux (lowest trace in Figure 2.11) also shows that most of the transfer tends to occur in 'bursts' coinciding with the upward movement of a buoyant thermal. Closer to the surface this pattern is less evident because of the greater influence of the

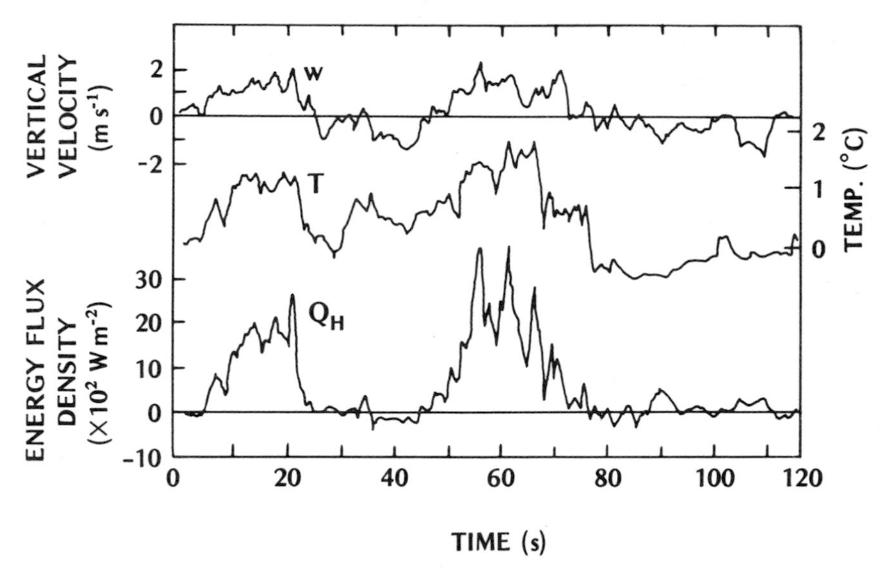

Figure 2.11 The relationships between vertical velocity (w) and air temperature (T) fluctuations, and the instantaneous sensible heat flux (Q_H). Results from fast-response instruments at a height of 23 m over grass in unstable conditions (after Priestley, 1959).

frictionally-generated small eddies. The heat flux (Q_H) given by equation 2.12 should correspond to the time-average of the instantaneous flux.

A similar set of observations under stable conditions would show that both w and T have traces similar in form to that of the lower trace in Figure 2.10 (i.e. small fluctuations with no longer-period buoyant waves). The w and T fluctuations would tend to be in antiphase with each other due to the inverted T profile. Using similar reasoning to that for the unstable case it can be seen that both up- and downdrafts will tend to result in a net downward heat flux through the measurement level.

In neutral conditions the w trace would again only be composed of small forced convection fluctuations, but the T trace would show virtually no variation with time. This is because although eddies are moving through the measurement level they thermodynamically adjust their temperature during ascent or descent so that they are always at the same value as the mean environmental temperature (Appendix A1). The net heat flux is therefore zero.

The diurnal surface temperature wave (Figure 2.1) penetrates up into the atmosphere (Figure 2.12) mainly via vertical turbulent transfer (Q_H). The upward migration of this wave is analogous to that in the soil (Figure 2.6) in that there is a time lag for the wave to reach greater distances from the surface, and the wave amplitude decreases. There is however a considerable

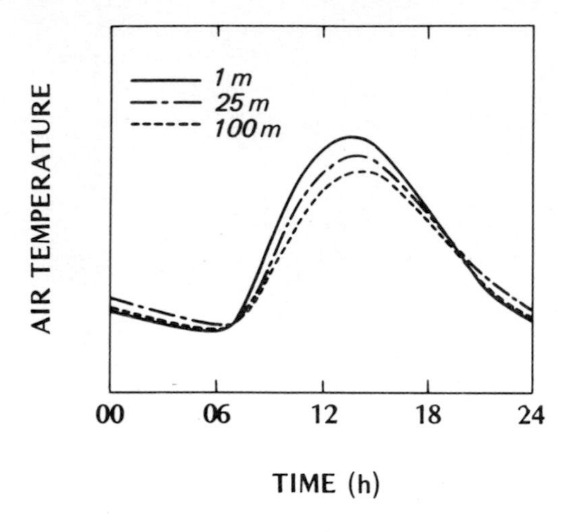

Figure 2.12 Generalized daily cycle of air temperature at three different heights in the atmosphere on a cloudless day.

difference between the rates and distances travelled in the two media. In the soil these are controlled by the value of κ_s (equations 2.3 and 2.4) but in the atmosphere by K_H. The latter is very much more efficient (see Table 2.1). This explains why the air temperature wave in Figure 2.12 penetrates to a height of 100 m with only a slight lag and little reduction in amplitude. On an unstable afternoon surface heated air parcels may reach as high as 2 km.

Figure 2.13 illustrates a typical sequence of air temperature profiles through the period of a fine day. Before sunrise (left-hand profile), near the time of the surface temperature minimum, the lowest layers of the

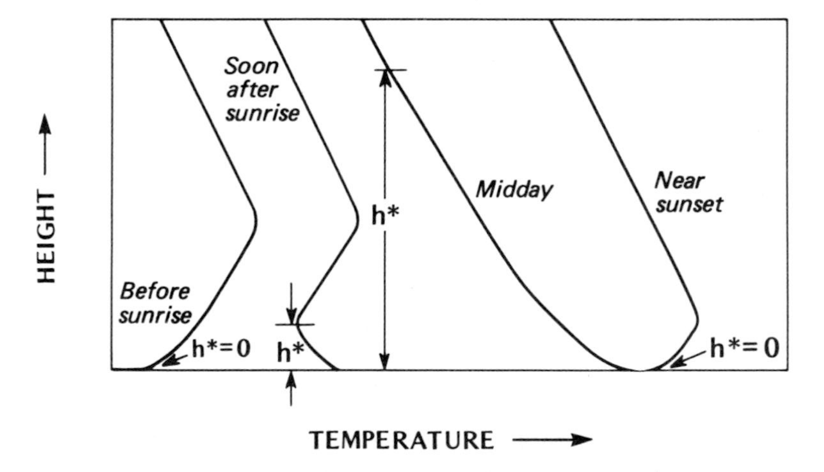

Figure 2.13 Generalized form of the air temperature profile in the boundary layer at several times on a day with fine weather at a rural location, including the depth of the surface mixed layer, h^*(modified after Shaw and Munn, 1971).

atmosphere (less than 100 m) are characterized by an inversion. This is a *radiation inversion* produced as a result of cooling of the ground by long-wave radiation emission through the atmospheric 'window'. The low surface temperature induces a downward flux of sensible heat (Q_H) from the lowest layers and therefore they also cool. This sensible heat divergence may be joined or even superseded by radiative flux divergence in the air if winds become calm (p. 36). Soon after sunrise (next profile) solar heating generates a surface radiative surplus, and convergence of the upward sensible heat flux warms the lowest layers. This process progressively erodes the nocturnal inversion from below, but its remnants remain aloft for a few hours. The convective warming produces an unstable layer next to the ground which grows in depth with time. By midday (next profile) the warming allows a lapse profile to extend throughout the boundary layer. Near sunset (right-hand profile) the surface cooling re-establishes a surface-based radiation inversion but above this stable layer the air remains slightly unstable although it is now severed from its surface heat source. The inversion intensifies, and the stable layer grows through the night until sunrise whereupon the cycle starts again.

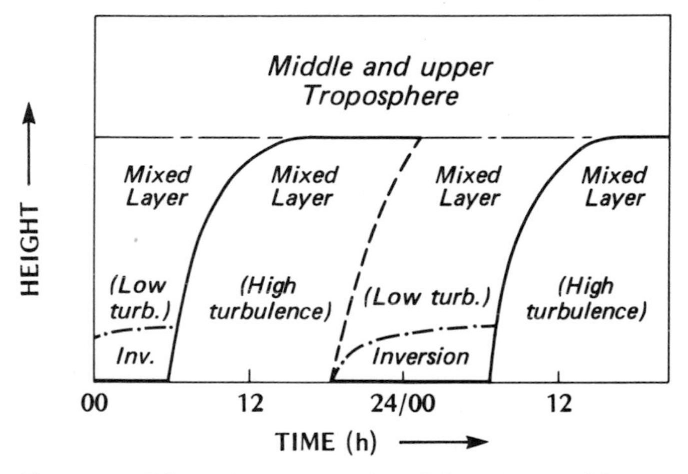

Figure 2.14 Schematic representation of the structure of the lower troposphere during a period of fine weather. The solid line shows the depth of the mixed layer, h^* (modified after Munn and Bolin, 1971).

The temporal dynamics of this diurnal sequence are shown in Figure 2.14. By day the surface-generated instability allows vertical exchanges to penetrate well up into the atmosphere. This permits greater momentum transport from faster moving air aloft and increases surface winds. The combined effect is to provide very effective overturning throughout a deep layer which is consequently known as the *mixed layer*. The depth of this layer (h^*) increases rapidly during the morning period so that by mid-afternoon it is anywhere from 0·5 to 2 km in height. By night the surface inversion stratification dampens vertical exchange and hinders momentum transport to the surface. With light winds and weak turbulence the layer of surface influence contracts to a depth of 50 m or less, and there is no mixed layer in contact with the ground (i.e. $h^* = 0$).

(c) WATER VAPOUR AND LATENT HEAT FLUXES (E, Q_E) AND ATMOSPHERIC HUMIDITY (q)

Prior to a discussion of the transfer of water vapour between the surface and the lower atmosphere it is necessary to introduce ways of representing the vapour content of air.

The *total atmospheric pressure* (P) commonly exhibits values in the approximate range of 97 to 103 kPa. This value represents the sum of the partial pressures exerted by its constituent gases. The *vapour pressure* (e) is that portion due to the water vapour content and is relatively small (less than 4 kPa). The maximum amount of water vapour that air can 'hold' in the vapour phase is called the *saturation vapour pressure* ($e^*_{(T)}$) and this value is related to the temperature (T) as shown in Figure 2.15 (see also Appendix A3). This relationship is valid for air with respect to a plane surface of pure liquid water. The value of $e^*_{(T)}$ is slightly lower if the surface is a solution, or a pure ice surface (see inset Figure 2.15). Consider a sample A of air at 15°C with a vapour pressure $e = 1000$ Pa as plotted on Figure 2.15. Clearly the sample is unsaturated because it lies beneath the $e^*_{(T)}$ curve, but it could be brought to saturation, whereupon condensation would occur, in two ways. First, if more water vapour is added to the sample its vapour pressure would increase until it equalled $e^*_{(T)}$. If no change in temperature occurred during this addition the sample would become saturated at about 1700 Pa (i.e. the point of intersection between a vertical line drawn through A and the $e^*_{(T)}$ curve). The second way would be to cool the sample and hence diminish its ability to 'hold' water vapour. If there is no change in vapour content the sample would become saturated at about 7°C (i.e. the point of intersection between a horizontal line drawn through A and the $e^*_{(T)}$ curve). This temperature is referred to as the *dew-point temperature* of the sample (or *frost-point* if it is equal to or less than 0°C). Thus sample A has a dew-point of 7°C and a saturation vapour pressure of 1700 Pa. Ways of measuring e in the field are outlined in Appendix A2.

In this book we will use two other measures of atmospheric humidity – *vapour pressure deficit* (vpd) and *specific humidity* (q). The vapour pressure deficit is defined as the difference between the actual vapour pressure (e) and the maximum vapour pressure possible at the temperature of the sample:

$$\text{vpd} = (e^*_{(T)} - e) \qquad (2.13)$$

This is a measure of the vapour necessary to cause a sample to become saturated. In the case of sample A in Figure 2.15 its vpd is (1700 − 1000) = 700 Pa. The vpd of air above a wet surface provides some indication of the 'drying power' of the air. It is however not a good measure to use if moisture content comparisons between two or more samples are required because vpd varies with T. In fact on sunny summer days vpd follows the course of T showing maximum values in mid-afternoons (e.g. Figure 4.21c).

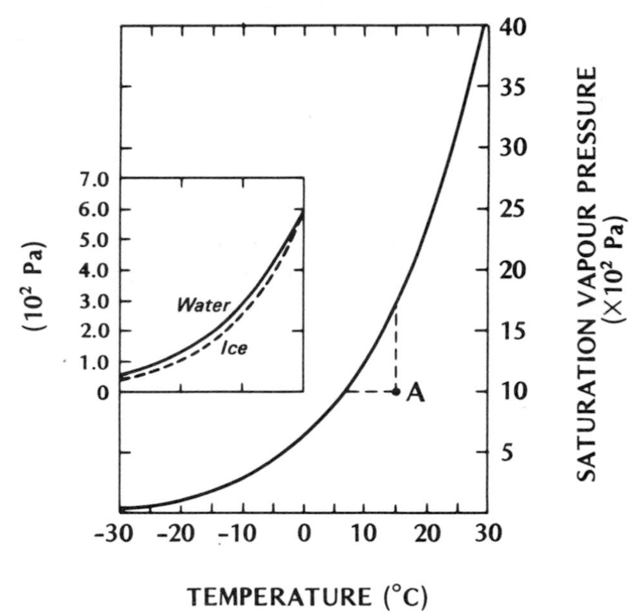

Figure 2.15 Relationship between saturation vapour pressure ($e^*_{(T)}$) over a plane surface of pure water, and temperature. Inset: Saturation vapour pressure over water and ice at temperatures below 0°C (after Byers, 1965). (Note: 1 Pa = 10^{-2} mb.)

The specific humidity is defined as the ratio of the mass of water vapour to the mass of moist air in a sample. It is related to e by:

$$q = \frac{0{\cdot}622e}{P - 0{\cdot}378e} \simeq \frac{0{\cdot}622e}{P} \qquad (2.14)$$

and has units of kg kg^{-1} (more conveniently quoted as g kg^{-1} to avoid very small numbers). At an atmospheric pressure of 100 kPa sample A in Figure 2.15 would have a specific humidity of approximately 6·2 g kg^{-1}. This measure is particularly useful in dealing with the mass transfer of water to and from the atmosphere.

Evaporation from the surface passes through the laminar boundary layer according to equation 2.6, and in the turbulent surface layer this mass flux is given by:

$$E = -\rho K_W \frac{\Delta \bar{q}}{\Delta z} \tag{2.15}$$

where, K_W – eddy diffusivity for water vapour ($m^2\ s^{-1}$). The energy required to vaporize the water is considerable (p. 27) and the accompanying flux of latent heat is given by substituting equation 1.20 in 2.15:

$$Q_E = -\rho L_v K_W \frac{\Delta \bar{q}}{\Delta z} \tag{2.16}$$

The evaporation process depends not only upon the availability of water but also upon the availability of energy to enable change of state; the existence of a vapour concentration gradient; and a turbulent atmosphere to carry the vapour away.

The exchange of moisture between the surface and the atmosphere determines the humidity, just as the sensible heat flux largely governs the temperature in the lowest layers. However, whereas heat is pumped into the air by day and returned to the surface by night, the flux of water is overwhelmingly upward. The evaporative loss is strongest by day, but often continues at a reduced rate throughout the night. Under certain conditions this loss may be halted and water is returned to the surface as *dew*, but in comparison with the daytime mass flow it is almost negligible. The water put into the atmosphere is of course returned by the process of precipitation rather than by turbulence.

By day the profile of vapour concentration lapses with height away from the surface moisture source (Figure 2.16a and b) in the same manner as the temperature profile (Figure 2.2). Vapour is transported upwards by eddy diffusion in a process analogous to that for sensible heat. If the temperature (T) trace in Figure 2.11 were replaced with that for specific humidity (q) then the associated flux would be that of water vapour (E), or with appropriate modification, of latent heat (Q_E).

In the morning hours the evapotranspiration of surface water (dew, soil water, and plant water) into a moderately unstable atmosphere adds moisture (by flux convergence) to the lower layers and the humidity increases quite sharply (Figure 2.16b). By the early afternoon, although E is at a peak (e.g. Q_E in Figure 1.11), the humidity concentration drops slightly. This is the result of convective activity having penetrated to such heights in the boundary layer that the vapour concentration becomes diluted by mixture with descending masses of drier air from above. This feature is best seen at continental or desert stations where air masses are dry, and surface heating is strong. (The rural data in Figure 8.17 are an example from a continental station.) In the late afternoon surface cooling is strong and the

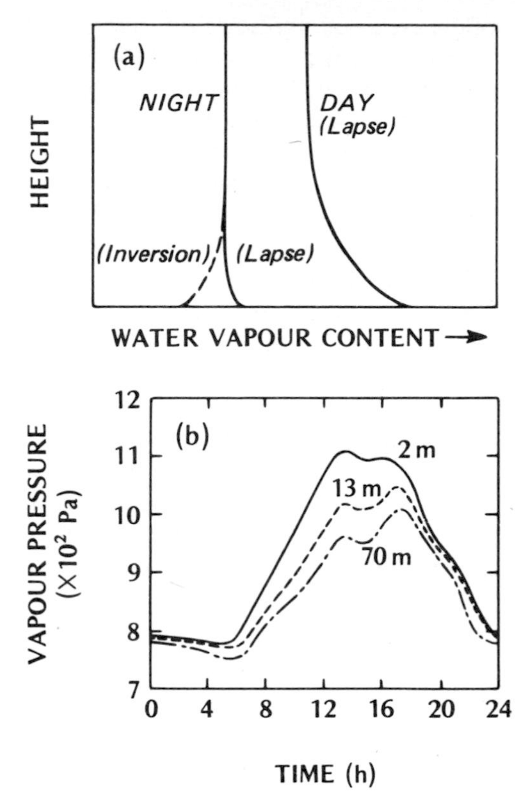

Figure 2.16 (a) Idealized mean profiles of water vapour concentration near the ground's surface, and (b) the diurnal variation of vapour pressure at 3 heights at Quickborn, Germany on cloudless days in May (after Deacon, 1969, using data of Frankenberger).

lowest layers become stable. Thus the ability to transport vapour to higher layers is less than the rate at which it continues to be added from the surface. Moisture converges into the lowest layers and a second humidity maximum is observed (Figure 2.16b). Thereafter evapotranspiration declines into the night period. Under certain conditions (see below) the vapour profile may become inverted near the surface (Figure 2.16a) so that vapour is transferred downwards as dewfall. This depletes the moisture in the lowest layers and humidities decrease (Figure 2.16b), until after sunrise when the cycle re-commences.

Radiative cooling at night may cause the surface temperature to fall below that of the contacting moist air. The ensuing condensation on the surface gives rise to an inverted lapse rate so that turbulence leads to a downward flux and further deposition. The process of *dewfall* is therefore a quasi-turbulent phenomenon requiring wind speeds to lie within a critical range. If the air is calm the loss of moisture to the ground cannot be replenished from more humid layers above and dewfall ceases. On the other hand if winds are too vigorous the surface radiative cooling (L^*) is offset by turbulent warming (Q_H), and the vapour inversion is destroyed. The critical wind speed depends upon the roughness of the surface. For a short grass surface a minimum wind speed of 0·5 m s^{-1} at 2 m is required (Monteith, 1957). If the dew freezes, or if the vapour originally sublimates rather than condenses upon the surface, the deposit is called *hoar frost*.

Radiation or ground fog is another humidity-related phenomenon observed on cloudless nights with light winds. It commonly forms over moist or marshy ground on such nights, and as with dewfall it is the result of a fine balance between radiative cooling and turbulent warming of an air layer (volume) near the surface. The process is particularly aided if the air is humid and close to saturation in the evening, and if the air aloft is relatively dry. Under these conditions the moist surface air layer has a strongly negative long-wave radiation budget because it radiates more energy than it receives from the colder surface beneath. Similarly it emits more than it receives in its exchange with the air above because its vapour content gives it a greater emissivity. The layer therefore cools by long-wave radiative flux divergence ($-\Delta L^*$). On cooling to its dewpoint it becomes saturated and fog droplets develop. The fog formation is aided by light winds which enhance the loss of sensible heat from the layer to the surface (Q_H), but beyond a certain limit increased winds thwart fog formation by increasing turbulent mixing which weakens the inversion strength and dilutes the moisture concentration.

Once a fog bank has formed the active radiating surface becomes the fog top and not the surface of the ground because the water droplets are almost full radiators for long-wave radiation (p. 17). They therefore absorb and emit very efficiently at these wavelengths and hence $L\uparrow$ from the fog top continues to cool it and helps the fog to become progressively thicker.

Radiation fogs usually linger for a few hours after sunrise, and sometimes can last all day aided by the high albedo of the fog top. Fog dissipation does not usually result from solar heating of the droplets, but by convection generated at the surface or by increased wind speeds. In both cases the mixing of the fog with drier air is the cause of its disappearance.

(d) CONCLUDING REMARKS ON CONVECTIVE EXCHANGE

Convection is the principal means of transporting the daytime energy surplus of the surface away from the interface. The relative importance of sensible versus latent heat is mainly governed by the availability of water for evaporation, although the relative strengths of the atmospheric heat and water vapour sinks are also important. For example if an abnormally cold and moist air mass settles over a region it would strengthen the daytime surface-air temperature gradient, and diminish the vapour gradient. Inspection of equations 2.12 and 2.16 shows that this would favour Q_H rather than Q_E.

The energy partitioning between Q_H and Q_E has direct relevance to boundary layer climates. The ratio of these two fluxes is called Bowen's ratio (β) so that:

$$\beta = Q_H/Q_E \tag{2.17}$$

Thus if β is greater than unity, Q_H is larger than Q_E as a channel for dissipating heat. This may be found over surfaces where water is to some extent limited. Since a majority of the heat being convected into the atmosphere is in the sensible form the climate is likely to be relatively warm. On the other hand if β is less than unity, Q_E is larger than Q_H, and the heat input to the atmosphere is mainly in the latent form. This will not directly contribute to warming of the lower atmosphere, but may increase its humidity. Therefore the climate is likely to be relatively cool and moist. Negative β values merely indicate that the two fluxes have different signs. This is common at night when the sensible heat flux is downwards (negative), but evaporation continues so that Q_E is away from the surface (positive). Typical average values of β are 0·1 for tropical oceans; 0·1 to 0·3 for tropical wet jungles; 0·4 to 0·8 for temperate forests and grassland; 2·0 to 6·0 for semi-arid areas; and greater than 10·0 for deserts.

Although we have concentrated on the turbulent transport of heat, water vapour and momentum, other substances are also convected to and from the atmosphere. For example the flux of carbon dioxide (C) may be represented by the flux-gradient equation:

$$C = -K_C \frac{\Delta \bar{c}}{\Delta z} \tag{2.18}$$

where, C – flux density of CO_2 (kg m^{-2} s^{-1}), K_C – eddy diffusivity for CO_2 (m^2 s^{-1}), and c – CO_2 concentration (kg m^{-3}). Similar relationships could be constructed for carbon monoxide, ozone, pollen, spores, dust, etc. The major requirements are that the substances should be inert (so that they do not decay quickly), and lightweight (so that gravitational settling does not deplete the concentration).

The measurement of convective fluxes is exceedingly difficult. An introduction to some of the modern approaches to this problem is given in Appendix A2, but a few general comments may be useful here. For example the set of flux-gradient equations used to describe these transfers (equations 2.8, 2.12, 2.15, 2.16 and 2.18) potentially provide a means of evaluation. If both the appropriate mean gradient ($\Delta \bar{i}/\Delta z$), and the appropriate eddy diffusion coefficient (K_i), can be measured then the flux can be determined. However, although the gradient is fairly easily assessed, the diffusion coefficient is not. The problem can be simplified considerably if the so-called *principle of similarity* is invoked. This involves the assumption that the diffusion coefficients (K_M, K_H, K_W, K_C etc.) are equal for a given state of the atmosphere. This assumes that an eddy is non-discriminatory with regard to the property being transported, so that an eddy will carry heat, water vapour, momentum, carbon dioxide etc. with equal facility. Thus the determination of one coefficient determines them all for the same site, time period and height. There remains considerable argument as to whether this is valid, but it provides the basis for two of the approaches (aerodynamic and Bowen ratio) outlined in Appendix A2.

PART 2

Natural atmospheric environments

In this part of the book we will consider the boundary layer climates associated with a wide range of natural surfaces and systems. The text is organized in a progression from relatively simple surfaces to more complex systems. Thus we start with environments where the surface is relatively flat, uniform in character, and extensive (e.g. bare soil and sandy desert). Then we consider systems where this straightforward situation is complicated by the fact that the surface is semi-transparent to radiation (e.g. snow and ice), and the system is able to transmit heat by internal convection (e.g. water). Next we introduce a layer of vegetation between the soil and the atmosphere, and then the complicating effects of sloping and hilly terrain, and the advective interaction between the climates of adjacent contrasting surfaces. Finally in this part we consider the climates of animals. These represent some of the most complex climatic systems because they are able to move from one environment to another, and they carry with them their own internal energy supply (metabolic heat).

CHAPTER 3

Climates of simple non-vegetated surfaces

This chapter investigates the atmospheric systems associated with simple, flat, non-vegetated surfaces (bare soil, desert, snow, ice and water). In each case we will initially evaluate any special properties of the surface (or system volume) not previously encountered. Then we will examine the cycling of energy and water typically encountered (i.e. depending upon the particular mix of radiative, thermal, moisture and aerodynamic properties of the system), and finally we will discuss the characteristic climate which results.

The simple case of a bare soil surface was discussed in some detail in Chapter 2 and little further will be said here; however it is important to realize that significantly different climates exist in relation to different soils. Probably the most important variables governing these differences are the soil albedo (controlling short-wave radiation absorption); the soil texture (determining the porosity and therefore the potential soil, air and water contents, that in turn control the thermal properties of the soil); and soil moisture availability (governing the partitioning of sensible and latent heat, and the thermal response of the soil).

To illustrate some of these relationships consider the case of a dry peaty soil. Peat has a high porosity and hence when dry contains a lot of air. As a result it has a very low diffusivity, as low as that of fresh snow (Table 2.1). The albedo of peat is also somewhat extreme, being rather low in comparison with other soils. Thus on a sunny day a dry peaty soil is a good absorber of solar energy, but it is not well suited to transmit this heat to deeper soil layers. As a result of soil heat flux convergence (Figure 2.4a) a thin surface layer becomes very hot. But since the soil is dry latent heat losses are negligible and the main modes of heat dissipation are via long-wave emission (L^*) and sensible heat transfer by convection (Q_H). At night the surface continues to lose heat by L^* but finds it difficult to draw upon the soil heat reservoir because of the low diffusivity. Thus as a result of soil heat flux divergence (Figure 2.4b with the arrow directed upwards) the surface becomes cold and a strong inversion develops. This soil therefore promotes an extreme thermal climate. It might also be mentioned that adding water does not help very greatly in getting the heat gain (or loss) to be spread over a deeper soil layer. Note that saturating a peaty soil only marginally increases its diffusivity (Table 2.1) because the water greatly increases its heat capacity. The addition of water would however help to alleviate the high daytime surface temperature by permitting energy loss through evaporation (Q_E).

Moist sands and wet clay soils on the other hand have relatively conservative thermal climates. Their porosity and moisture characteristics give them good diffusivities (Table 2.1) so that soil heat transfer is well developed. This permits a large volume of soil to be involved in diurnal energy exchanges and surface temperature fluctuations are dampened. The daytime radiant surplus can also be dissipated by Q_E, and this serves to cool the surface and thereby reduce the daytime maximum temperature.

1 Sandy desert

(a) ENERGY AND WATER BALANCES

The sandy desert is the classic example of a thermally extreme climatic environment. To a large extent this can be explained in terms of the lack of water, both in the ground and in the atmosphere.

The radiation budget (equation 1.14) of a desert is characterized by large radiant input and output. Except for dust the desert atmosphere is usually very clear because water vapour content is low, and cloud is generally absent. As a result it is quite common for 80% of the extra-terrestrial short-wave radiation to reach the desert surface. Combined with the fact that the Sun is often close to the zenith in the sub-tropics, this produces very strong solar input ($K\downarrow$). However, the impact of $K\downarrow$ is somewhat ameliorated by the fact that most sandy deserts have relatively high albedos (Table 1.1) so that the short-wave loss ($K\uparrow$) is also considerable. Mainly as a result of 'soil' factors, however, the desert surface does become very hot (p. 66), and hence the output of long-wave radiation ($L\uparrow$) is also great. As a result of the large reflection and emission the net radiation (Q^*) absorbed by a desert is really not as large as might at first be anticipated. For example in the relatively low latitude desert example given in Figure 3.1 the maximum midday value of Q^* is approximately 600 W m^{-2} in mid-summer. This is only slightly greater than the values for mid-latitude grass and crops (Figures 1.10, 1.11, 4.13), less than mid-latitude coniferous forests (Figures 4.20, 4.21) in the same season, and less than the value for a mid-latitude water surface (Figure 3.11) in late summer. Despite their smaller solar radiation input these latter surfaces are able to retain a greater proportion because of their lower albedos and cooler surface temperatures. The nocturnal net radiation (L^*) in a desert is clearly negative because the clear skies keep the atmospheric 'window' open.

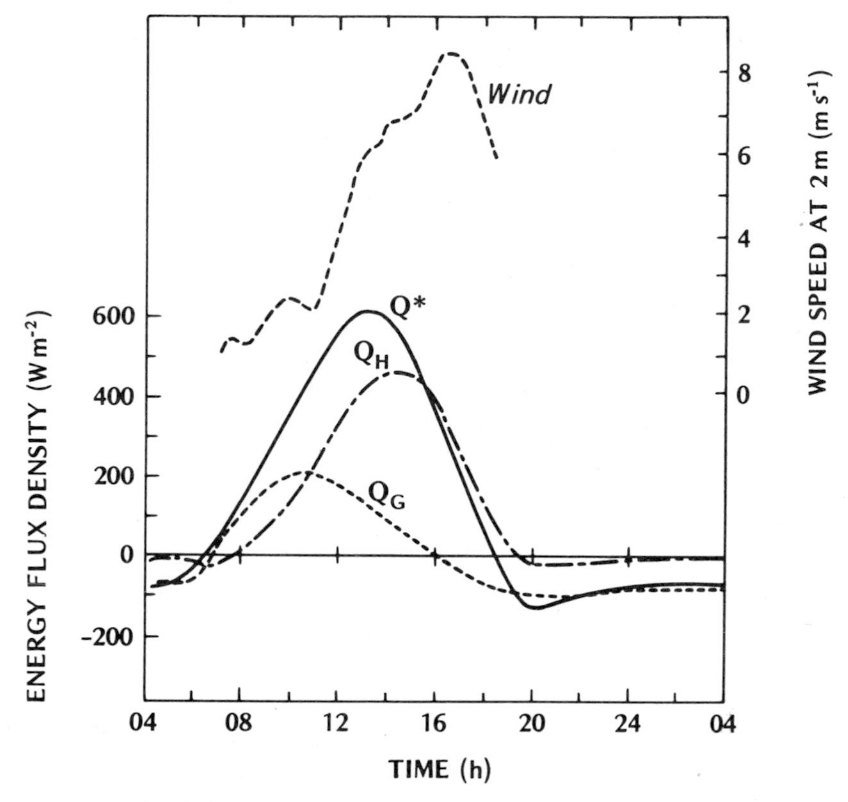

Figure 3.1 Energy balance components and wind speed at a dry lake (desert) surface on 10–11 June 1950, at El Mirage, California (35°N) (after Vehrencamp, 1953).

All of the available radiant energy in a desert must be dissipated as sensible heat (i.e. to warm the air or the soil) because evaporation is almost negligible. The majority of the daytime radiative surplus is carried into the atmosphere by turbulence (Q_H). In fact during the full 24 h period in Figure 3.1, Q_H consumed about 90% of Q^* leaving only 10% to heat the soil (Q_G) because evaporation was considered negligible.

On an hourly basis Q_G is very much more significant in the balance. In the early morning, and throughout the night, Q_G is the most important means of off-setting the radiative imbalance at the surface. These are times when winds are light and turbulent transfer is relatively restricted. In the late morning and afternoon thermodynamic instability and the associated increase in wind speed combine to pump the excess surface heat up into the atmosphere. It is these changing roles of conduction versus convection that explains the asymmetric shape of the Q_G and Q_H curves relative to that of Q^* (Figure 3.1).

In a desert it is to be expected that Q_G will diminish rapidly with depth because dry soil and sand have low diffusivities (Table 2.1). By day this leads to strong heat flux convergence (Figure 2.4a) in a thin near-surface layer and therefore very strong surface heating. Conversely at night this layer is the site of heat flux divergence (Figure 2.4b) and strong surface cooling. Deeper layers are not involved in significant energy exchange with the surface because of this poorly transmitting 'buffer' layer, and hence show relatively little temperature change.

(b) CLIMATE

As a result of the lack of moisture, and the concentration of heat in the uppermost sand layer, the daytime surface temperature of a desert is high. In the case of the pumice desert in Figure 3.2a the surface temperature at midday is 49°C, but in more extreme cases desert surface temperatures can reach at least 70°C. Even on a dry sandy beach in the mid-latitudes surface temperatures can be too hot for walking with bare feet. Notice also that in the intertidal zone of a beach, where water is available, evaporative cooling greatly reduces the surface temperature. Both immediately above and beneath the surface, temperatures drop sharply, producing very steep lapse rates. At 2 m above the surface of the two desert sites where the data in Figures 3.1 and 3.2 were observed, the air temperature was 28 to 29°C below that at the surface at midday. In southern Arabia Griffiths (1966) reports having measured a difference of 27°C in the lowest 50 mm over a sandy desert. This is equivalent to an environmental lapse rate that is 55,000 times greater than the dry adiabatic rate. It is hardly surprising therefore that the lower atmosphere over a desert is convectively very unstable (Chapter 2 and Appendix A1).

Such conditions lead to the development of special instability phenomena such as miniature whirlwinds known as 'dust devils', and a whole range of unusual optical effects. These include the 'shimmering' of objects viewed through the lower atmosphere, and the well-known mirage, both of which are due to the refraction of light as it passes through media of different density. Shimmering is caused by multiple refraction of light as it passes from the object through a field of vertically-arranged filaments of air of differing density. A mirage is caused by the refraction of light from the sky as it passes through the horizontal temperature (density) stratification of the lower atmosphere, and the amount of bending depends on the lapse rate.

The nocturnal temperature profile over and beneath the surface of a desert is similar to that above bare soil (Figure 2.2). As the surface cools an inversion develops and the lower layers become stable. The relatively unrestricted radiative cooling causes temperatures to drop markedly, and in terms of human comfort the desert may be a distinctly cold environment at night. Thus another important feature of deserts is their large diurnal range of temperature. At weather screen height (1·5 m) the diurnal range is commonly 40°C, and has been found to be as great as 56°C at Tucson, Arizona. Consideration of Figure 2.2 reveals that this range will be even greater at the surface itself. Plants and animals able to survive such extreme thermal shifts usually exhibit physiologic or behavioural adaptations (Chapter 6). Humans feel overheated by day and chilly at night.

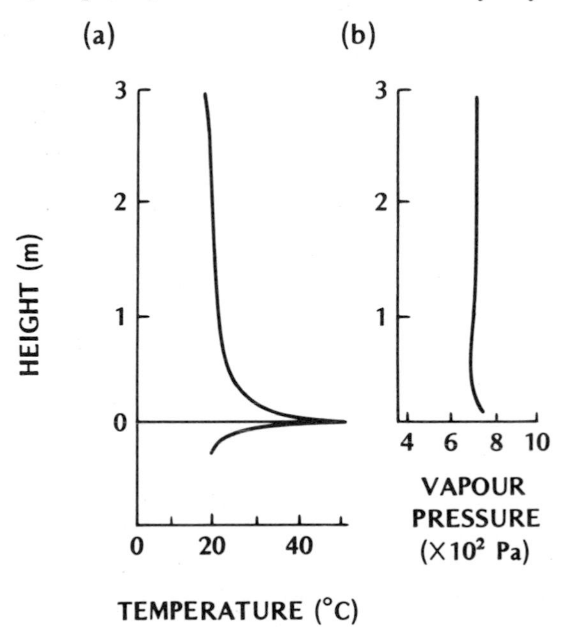

Figure 3.2 Profiles of temperature and vapour pressure at midday at a pumice desert in Oregon (after Gay, 1970).

There is little to be said about the humidity profile (Figure 3.2b) except that it is characterized by low absolute moisture content, is far from saturation, and has a very weak lapse rate on most occasions.

The strong diurnal shift in stability gives rise to a pronounced and regular diurnal pattern of wind speeds which can be explained as follows (Figure 3.1). It was pointed out on p. 49 that instability promotes vertical eddy exchange and allows horizontal momentum to be readily transferred towards the surface. Thus the strong daytime convection of deserts allows the higher momentum possessed by faster moving upper air layers to be brought down and mixed into the surface layer. This downward transport of greater horizontal momentum results in higher surface layer wind speeds. Conversely, at night stability weakens this transport of momentum, and the surface layer becomes partially decoupled from upper layers, and winds near the ground subside. By mid-afternoon desert winds are often sufficiently strong to lift sand grains and the resulting abrasive action can be geomorphologically effective in eroding upstanding objects.

2 Snow and ice

(a) RADIATION BUDGET

Radiatively, snow and ice surfaces are very much more complex than the surfaces we have considered previously. One of the most important differences is that snow and ice both allow some transmission of short-wave radiation. This means that the short-wave radiation incident at any depth can be transmitted, reflected or absorbed according to equation 1.6:

$$t + \alpha + a = 1$$

and that radiation absorption occurs within a volume rather than at a plane. This is another example of flux convergence (Figure 2.4a) because the short-wave radiation incident at the surface ($K\downarrow_0$) is greater than that found at any depth below. The decay of the flux with distance into the snow or ice follows an exponential curve (Figure 3.3) so that the amount of short-wave radiation reaching any depth z is given by:

$$K\downarrow_z = K\downarrow_0 e^{-az} \tag{3.1}$$

where $K\downarrow_z$ – short-wave radiation incident at depth z, e – base of natural logarithms, a – extinction coefficient (m^{-1}). Equation 3.1 is known as *Beer's Law* and strictly is applicable only to the transmission of individual wavelengths in a homogeneous medium, but it has been used with success for fairly wide wave-bands (especially the short-wave) in meteorological applications. The extinction coefficient depends on the nature of the transmitting medium, and the wavelength of the radiation. It is greater for snow than for ice and hence the depth of penetration is greater in ice (Figure 3.3). The depth of short-wave penetration can be as great as 1 m in snow, and 10 m in ice. The exponential form of the depletion means that absorption is greatest near the surface and tails off at lower depths.

The internal transmission of radiation through snow and ice gives problems in formulating the surface balance and in observation. For example measurements of reflected short-wave from an instrument mounted above the surface include both surface *and* sub-surface reflection. The albedo calculated from such measurements is therefore a volume not a surface value. Consider also the practical problem of an instrument buried

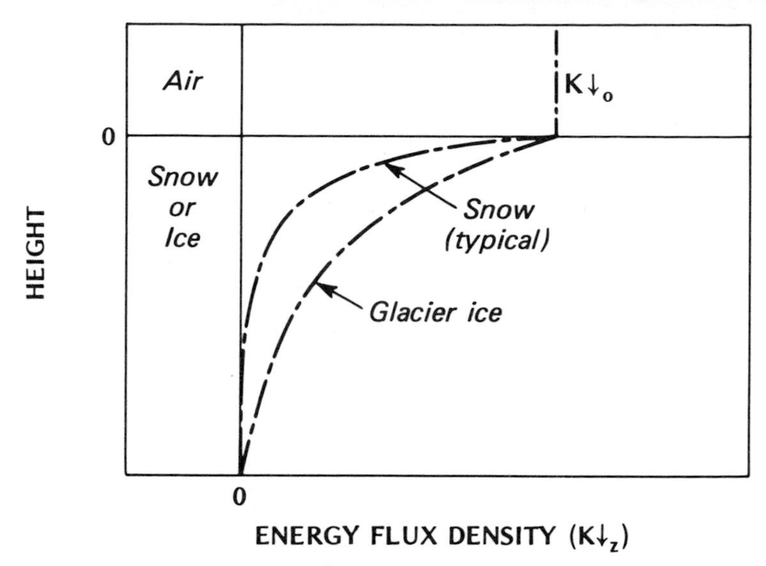

Figure 3.3 Typical profiles of solar radiation within snow and ice illustrating the exponential attenuation with depth (after Geiger, 1965).

within snow or ice to measure the temperature, or heat flow at depth. Such a body is likely to absorb transmitted radiation causing it to warm up and become an anomalous thermal feature. It therefore records its own response and not that of the surrounding environment.

One of the most important characteristics of snow and ice is their high albedo (α, see Table 1.1). Their rejection of such a large proportion of $K\downarrow$ is of primary importance in their overall low energy status. The introduction of even a thin snow cover over the landscape has dramatic effects. In a matter of a few hours a natural landscape can experience a change in albedo from approximately 0·25 to perhaps 0·80. Thereafter α declines as the snow pack ages (becomes compacted, and soiled), but with a fresh snowfall it rapidly increases again. The albedo of most surfaces exhibit a diurnal variation with higher values in the early morning and evening, and a minimum near midday. This is usually attributed to the changing angle of solar incidence (for a fuller discussion see p. 110). In the case of ice and snow this variation may also be due to physical changes in the state of the surface especially if conditions are conducive to surface melting. In the early morning and evening the surface is frozen and this, combined with the high zenith angle of the Sun, gives a relatively high albedo. In the afternoon even a thin film of meltwater on the surface serves to reduce α to a value closer to that of water (Table 1.1). It should also be noted that the albedo of snow varies with wavelength, being highest for the shortest wavelengths and decreasing to quite low values in the near infra-red. This is almost the reverse of the case for soil and vegetation surfaces.

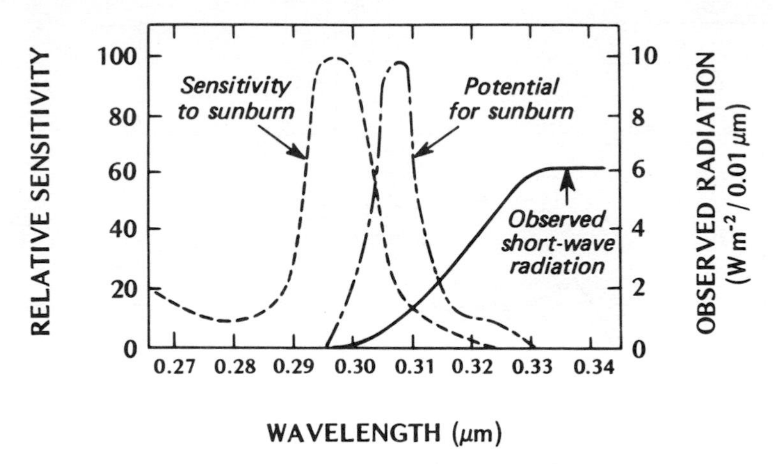

Figure 3.4 Potential for human sunburn at Davos, Switzerland (47°N). The potential is the product of the sensitivity spectrum for human skin (left), and the observed spectrum of short-wave (especially ultra-violet) radiation (right), at Davos (after Urbach, 1969).

The wavelength dependence of the albedo of snow helps to explain the ease with which skin becomes sunburnt especially on sunny days on snow-covered mountains. The human skin is very sensitive to ultra-violet light, with a peak sensitivity at about 0·295 μm (Figure 3.4). Comparison of the sensitivity spectrum with the typical short-wave radiation distribution at a mountain station (Figure 3.4) shows that the potential for sunburn stretches from 0·295 to 0·3300 μm, with a peak at about 0·3075 μm. The lower limit is governed by the almost total absorption of these wavelengths by ozone in the high atmosphere (Figure 1.9). With a fresh snow cover the receipt of potentially burning ultra-violet radiation by exposed skin is almost doubled, because in addition to that received from the incoming beam there is a very significant proportion received after surface reflection (due to the high value of α at these short wavelengths). The upward flux is responsible for sunburn on earlobes, throat and within nostrils, areas which are sensitive and normally in shade. Suntan oil is useful in protecting the skin by increasing its albedo, and by providing a layer of pigment able to absorb ultra-violet radiation.

In the long-wave portion of the spectrum, ice and snow (and especially fresh snow) are almost full radiators. However, although their emissivity is high the absolute magnitude of $L\uparrow$ is usually relatively small because T_0 is low. One helpful simplification occurs if the surface is melting. Then T_0 is set at 0°C (273·2 K) and if it is assumed that $\varepsilon = 1{\cdot}0$ the value of $L\uparrow$ is constant at 316 W m^{-2} (i.e. substitution of T_0 and ε in equation 1.4). Since clouds are also close to being full radiators the net long-wave exchange (L^*) between a fresh snowpack and a complete overcast is simply a function of their respective temperatures (i.e. $L^* = L\downarrow - L\uparrow \simeq \sigma(T_c^4 - T_0^4)$ where T_c – cloud-base temperature). Should the cloud-base be warmer than the snow surface there will actually be a positive L^* budget at the snow surface. With clear skies L^* is always negative as with other surfaces.

In general it may be said that the daytime net radiation surplus (Q^*) of snow and ice is small by comparison with most other natural surfaces. This is directly attributable to the high surface albedo. The nocturnal deficit is similar to that of most other environments. It should perhaps be pointed out that the radiation budget data given in Figure 3.6 and Table 3.2 are not representative of all snow and ice surfaces. These data are from a melting glacier whose surface was pocked with 0·5 m high ice hummocks, with the relatively low albedo of 0·25. As a result the radiation budget components agree surprisingly well with those of a cropped surface with a similar radiant input and albedo (e.g. Table 4.4).

(b) ENERGY AND WATER BALANCE

The energy balance of snow is complicated not only by the penetration of short-wave radiation into the pack but also by internal water movement, and phase changes. Water movement inside a snowpack may be due to the percolation of rainfall, or of meltwater. If the water temperature is significantly different to that of deeper layers it will involve heat as well as mass transport. In the case of rain percolation, this represents an additional heat source for the pack; in the case of meltwater it merely involves an internal re-distribution of heat. Phase changes of water within the pack (e.g. freezing, melting, sublimation, evaporation or condensation) involve energy uptake or release at that location. For example if rainwater percolating through the pack freezes it will release latent heat of fusion which is available to warm the surrounding snow.

All of these features make it difficult to formulate an accurate surface energy or water balance for snow or ice. A better approach is to consider a volume balance and to treat all fluxes as equivalent flows through the sides of the volume. If we ignore horizontal energy transfers and define the volume as extending from the surface to a depth where there is no significant vertical heat flux (Figure 3.5a, b) then equation 2.1 may be simplified to read:

$$Q^* = Q_H + Q_E + \Delta Q_S \qquad (3.2)$$

The net heat storage term (ΔQ_S) then represents the convergence or divergence of vertical heat fluxes within the volume. Noting the problems expressed earlier, this term includes internal energy gains or losses due to variations of radiation, heat conduction, and phase changes of water.

The energy balance of a snow volume is quite different depending on whether it is a 'cold' (less than 0°C) or a 'wet' (0°C, often isothermal) pack.

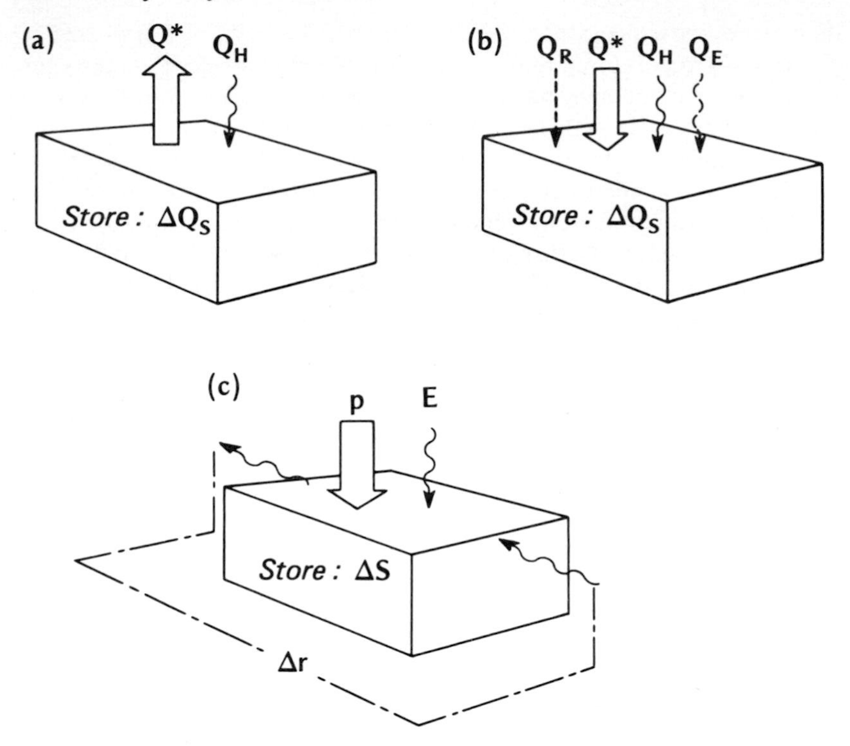

Figure 3.5 Schematic depiction of the fluxes involved in the (a, b) energy and (c) water balances of a snowpack volume. The energy balances are for (a) a 'cold' or frozen pack and, (b) a 'wet' or melting pack.

Let us consider the case of a 'cold' snowpack typical of high latitudes in winter with little or no solar input. Under these conditions Q_E is likely to be negligible because there is no liquid water for evaporation, little atmospheric vapour for condensation or sublimation, and both the precipitation and the contents of the snowpack all remain in the solid phase. Similarly heat conduction within the snow will be very small because of the very low conductivity of snow (Table 2.1) and the lack of any solar heating. As a result the energy balance (Figure 3.5a) is basically between a net radiative sink (Q^*) and a convective sensible heat source (Q_H). The radiation budget is negative because it is dominated by long-wave exchanges, and the outgoing flux ($L\uparrow$) is readily able to escape through the atmospheric 'window' (if there is no cloud) because of the lack of atmospheric water vapour and other absorbing agents (e.g. pollutants). A radiative inversion exists because of the surface cooling, and hence any mixing acts to transfer heat from the atmosphere to the snow surface.

It is also possible to find 'cold' snowpacks where Q^* and Q_E are significant energy sources. For example on a summer day the snow cover on a high latitude ice cap or glacier may be in receipt of considerable amounts of radiation. This will result in radiation absorption in the upper layers and generate a heating wave which will be transmitted downwards by conduction, so that ΔQ_S becomes significant over short periods. Taken overall, however, these gains are not sufficient to raise temperatures above 0°C. In cloudy, moist areas it is possible that a 'cold' snowpack can receive energy via Q_E. With the surface temperature below 0°C vapour may sublimate directly onto the surface as hoar frost or *rime*. In these circumstances we may modify equation 1.20 to read:

$$Q_E = L_s E \tag{3.3}$$

where $L_s E$ – the latent heat released due to the sublimation of vapour onto the surface at the rate E. On mountain tops where the water distribution within the cloud is essentially uniform with height the amount of rime accretion is related to the rate of vapour supply. Since this is a simple function of wind speed, with a steady wind the ice loading on an object increases with height (cf. Figure 2.9), and is greatest in the windward direction.

With a 'wet' snowpack during the melt period the surface temperature will be held very close to 0°C but the air temperature may be above freezing. Precipitation may then be as rain and the energy balance (Figure 3.5b) becomes:

$$Q^* + Q_R = Q_H + Q_E + \Delta Q_S \tag{3.4}$$

where Q_R – heat supplied by rain with a temperature greater than the snow. In some mid-latitude locations Q_R can be a significant energy source for melt, especially where the area is open to storms originating over warm oceans. Some melting snowpacks are isothermal throughout a deep layer, others have a temperature stratification similar to that in a soil. In the former by definition heat transfer is zero, but in the latter percolating rain and meltwater, and its subsequent re-freezing, are the primary means of heat transfer (i.e. conduction is small). During active melting both radiation (Q^*) and convection ($Q_H + Q_E$) act as energy sources (Figure 3.5b) to support the change of phase (ice to water). The temperature of the snow changes very little in this process therefore the large change of energy storage (ΔQ_S) is due to latent rather than sensible heat uptake. The role of Q_E for 'wet' snow is interesting. The surface vapour pressure of a melting ice or snow surface is of course the saturation value ($e^*_{(T)}$) at 0°C which equals 611 Pa (Figure 2.15 and Appendix A3). In absolute terms this is a low value (e.g. it is lower than the values given for a dry pumice desert, Figure 3.2b) and it is very common to find that the warmer air above the surface has a greater vapour pressure. Therefore an air-to-surface vapour pressure gradient would exist so that turbulence results in a downward flux of moisture, and condensation on the surface. Since the latent heat of vaporization (L_v) released upon condensation is 7·5 times larger than the latent heat of fusion (L_f) required for melting water (p. 27), for every 1 g of water condensed sufficient energy is supplied to melt a further 7·5 g. Under these conditions Q_E is an important energy source. Of course should the air be drier than 611 Pa the vapour gradient would be reversed, evaporation would occur and Q_E would be an energy sink in the balance.

TABLE 3.1 Measured and estimated energy balance components (W m^{-2}) over melting snow on Lake Mendota, Wisconsin (43°N), 26 March 1971 (based on data from Hicks and Martin, 1972).

Variable (units)	1130–1230 h	1340–1440 h	1450–1550 h	1600–1700 h	Overall average	Remarks
$K\downarrow$	875	715	442	205	560	Estimated
K^*	438	358	221	103	280	Assumed $\alpha = 0{\cdot}50$
$L\downarrow$	216	217	218	218	217	$L\downarrow = L\uparrow - L^*$
$L\uparrow$	316	316	316	316	316	σT_0^4 with $T_0 = 0$°C
L^*	−100	−101	−102	−102	−101	Equation 3 in Table A2.2
Q^*	338	257	119	1	179	$Q^* = K^* + L^*$
Q_H	−12	−7	−4	−12	−9	Measured
Q_E	22	30	22	10	21	Measured ($=L_vE$)
ΔQ_S	328	234	101	3	167	Residual in equation 3.2
$\bar{u}$ (m s^{-1})	2·65	3·08	2·66	2·48	2·72	Measured at 3·2 m
$\Delta\bar{T}$ (°C)	2·2	2·8	3·5	3·5	3·0	Measured, surface to 2 m
$\Delta\bar{q}$ (g kg^{-1})	−0·87	−0·91	−1·01	−1·14	−0·98†	Measured, surface to 2 m

† Equivalent to a mean vapour pressure gradient of 190 Pa.

There are little data available with which to illustrate these energy balance relationships, but Table 3.1 shows results from a study over melting snow on an ice-covered lake. Although the radiative terms were estimated, the turbulent fluxes of heat and water vapour were measured directly. The day was sunny with light winds, and during the experiment the surface was melting (i.e. at 0°C). On this basis the value of $L\uparrow$ was assumed to be constant at 316 W m^{-2}. Note that because of the high albedo and the constant long-wave drain the net radiant absorption (Q^*) never exceeds an efficiency of 31% in capturing the radiant input ($K\downarrow + L\downarrow$). In fact in the period between 16 and 17 h the total input was 423 W m^{-2} but after accounting for reflection and emission the net absorption was essentially zero. Throughout the period the lowest layers were characterized by a temperature inversion (the air at 2 m was approximately 3°C warmer than the surface) but the air was dry enough to give a lapse vapour profile. Thus on average the atmosphere was a source of sensible heat for the surface (Q_H

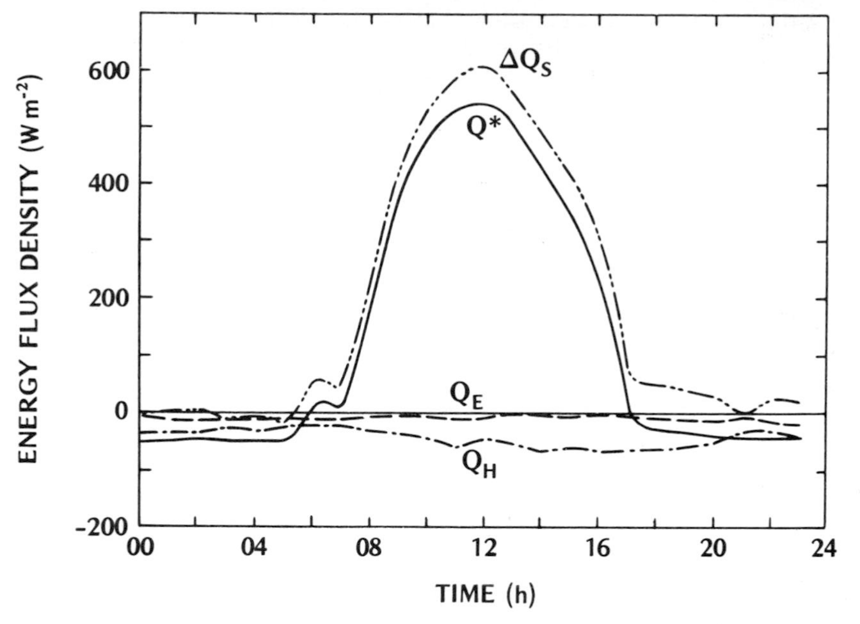

Figure 3.6 Energy balance components for the surface of a melting glacier at Peyto Glacier, Alberta on 29 August 1971 (data after Munro, 1975).

$= -9\ \mathrm{W\ m^{-2}}$), but a sink for vapour and therefore for latent heat (Q_E $= 21\ \mathrm{W\ m^{-2}}$). Since these terms approximately offset each other virtually all of the available energy was consumed in sensible and latent heat storage (i.e. $Q^* \simeq \Delta Q_S$) to melt the snow and to warm the snow, lake ice and lake water.

Figure 3.6 and Table 3.2 give the results of an energy balance study over a mid-latitude alpine glacier during an almost cloudless day in the summer melt period. As has been noted (p. 71) the albedo of this glacier allows a good deal of the solar radiation to be absorbed. Throughout the period both the air temperature and vapour pressure gradients remained inverted. Thus the surface was continually in receipt of sensible and latent heat from the atmosphere. At night the combined convective transfer ($Q_H + Q_E$) was sufficient to allow the net radiant emission to be offset or surpassed and therefore was able to support a small amount of ice melt. By day the convective gain supplemented that from radiation and permitted an augmented rate of melting. Over the complete day the convective supply provided 29% of the total energy used in the melt (23% Q_H, 6% Q_E), and the remaining 71% came from net radiation absorption. This ranking of heat sources during the melt season agrees with the review by Paterson (1969) of 32 glacier energy balance studies, and is also likely to apply to snow melt over other open surfaces (e.g. tundra and prairie sites), but in

TABLE 3.2 Radiation and energy balance components of an alpine glacier in the summer melt period. Data are daily totals (MJ m^{-2} day^{-1}) for Peyto Glacier, Alberta (51°N) on 29 August 1971 (constructed from data in Munro, 1975).

Radiation budget		Energy balance		Derived terms	
$K\downarrow$	21·4	Q_H	−3·7	α‡	0·25
$K\uparrow$	5·3	Q_E	−0·9	β‡	4·11
K^*	16·2	ΔQ_S†	16·0		
$L\downarrow$	22·4				
$L\uparrow$	27·3				
L^*	−4·8				
Q^*	11·4				

† Assumed to be used in melting ice. Obtained by residual in equation 3.2.
‡ Dimensionless

forested areas the trees are efficient absorbers of short-wave radiation and act as sources of sensible heat and long-wave radiation for the surrounding snow (p. 126).

The water balance of a snow or ice volume with its upper side at the snow or ice/air interface, and with its lower side at the depth of negligible water percolation (Figure 3.5c) is given by:

$$\Delta S = p - E + \Delta r \qquad (3.5)$$

Hence the net change of mass storage (ΔS) is due to the precipitation input (snow or rainfall); the net turbulent exchange with the atmosphere (input as condensation and sublimation, or output as evaporation and sublimation); and the net surface and sub-surface horizontal exchange (surface snow drifting and meltwater flow or sub-surface meltwater throughflow). In the simple case, ΔS can be related to snow depth so that it increases after a storm, and decreases as a result of melting. But *ablation* (lowering of snow depth) could also be due to an increase in snow density (i.e. a decrease in volume with no change in mass). Density changes are a feature of snow ageing and could be due to simple compaction, re-freezing, and metamorphism of ice crystals. Therefore rather than snow depth it is more pertinent to measure ΔS in terms of an equivalent depth of water (i.e. the depth of water obtained by melting unit volume of the ice or snow, or weighing a snow core of known cross-sectional area). As mentioned this will vary with density but as an approximate rule-of-thumb 100 mm of snow is equivalent to 10 mm of water. It is then an easy task to convert the water equivalent to the energy required to evaporate or melt this depth of water using the appropriate latent heat (p. 27).

For the case of Lake Mendota (Table 3.1) the latent heat flux (Q_E) is equivalent to an evaporation (E) of 0·29 mm of water over a 10 h day. With

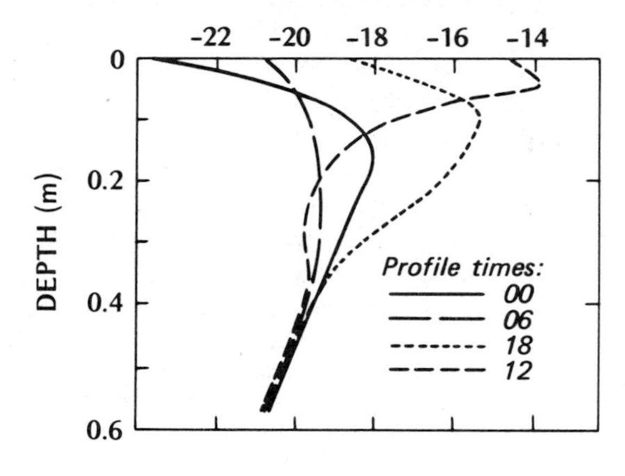

Figure 3.7 Diurnal sequence of snow temperature profiles from Devon Island ice cap (after Holmgren, 1971).

typical snow densities this should have produced a snow lowering of 0·83 mm. More significantly, if we assume that all of ΔQ_S was used to melt snow it would have produced 17 mm of meltwater, or an ablation of 50 mm. For the Peyto Glacier (Table 3.2, Figure 3.6) the corresponding melt due to ΔQ_S is 48 mm and this would have been slightly offset by a 2·7 mm *addition* of mass due to E.

(c) CLIMATE

The sub-surface temperature profiles typical of a deep snow pack are unlike those of soil (Figure 3.7), because of the occurrence of a temperature maximum just *beneath* the surface. This feature is a result of the fact that by day radiative heat transfer dominates over heat conduction in the upper 0·5 m of snow, and the upper 5 m in ice (Schwerdtfeger and Weller, 1967), and also because short-wave radiation is transmitted very much more readily than long-wave radiation in these media. If we ignore conduction the pattern of energy gain/loss by the upper layers of the snow pack is given by the vertical profile of Q^* as illustrated in Figure 3.8. The radiative input (both short- and long-wave) to the pack from above is absorbed in general accord with Beer's Law (equation 3.1, Figure 3.3). The long-wave portion is relatively quickly absorbed, but short-wave penetrates to much greater depths. The radiative loss consists of short-wave reflection, and that long-wave emission able to escape to the atmosphere. The strong absorptivity of snow in the infra-red only allows this loss to occur from a thin surface layer. Therefore the net radiation at any depth (Q_z^*, the difference between these gains and losses) shows a maximum absorption just below the surface during the day. This level, and not the snow surface, is the site of maximum heating, and therefore has the highest temperature (Figure 3.7). If Q^* dominates the melt at a site it is therefore most effective below the surface and this accounts for the 'loose' or 'hollow' character of the surface of a melting snowpack. At night with only long-wave radiative exchange the active surface is at, or very near, the actual surface. The lowest nocturnal temperatures occur at the snow surface, and the daytime sub-surface temperature maximum migrates downward by conduction.

As was mentioned previously buried instruments may become anomolously warm (compared with the snow or ice at the same depth) because their opacity and low albedo dictate that they preferentially absorb short-wave radiation reaching that depth. All buried objects will act in this way, and some of the most common include stones, twigs, leaves, dirt layers, and the bases of trees or fences. The radiant heating of these objects may be sufficient to melt the overlying or surrounding snow. This gives rise to micro-relief features such as melt-holes corresponding to the shape of the object. If the snow cover is less than 0·15 m deep, absorption by the underlying surface (e.g. soil) may become significant in helping to melt the layer from *below*.

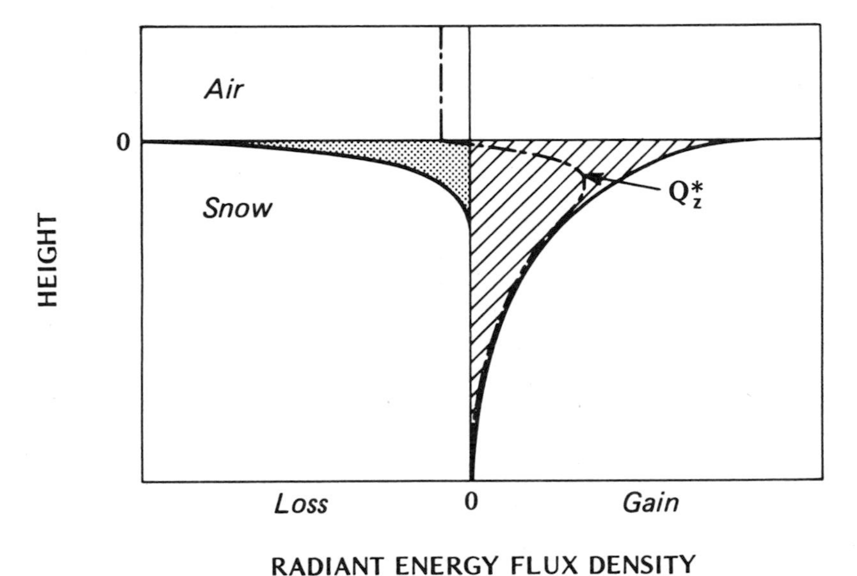

Figure 3.8 Vertical variation of radiative loss and gain, and the resulting profile of net all-wave radiation (Q_z^*), in the upper layer of a snowpack.

The very low conductivity and diffusivity of snow (especially when fresh) makes it an effective insulating cover for the ground beneath. This is especially true at night when radiative exchange is concentrated in the surface layer of the snow. Then as little as 0·1 m of fresh snow will insulate the ground from snow surface temperature changes, and thereby help to conserve soil heat. Figure 3.9 shows an example where even though snow

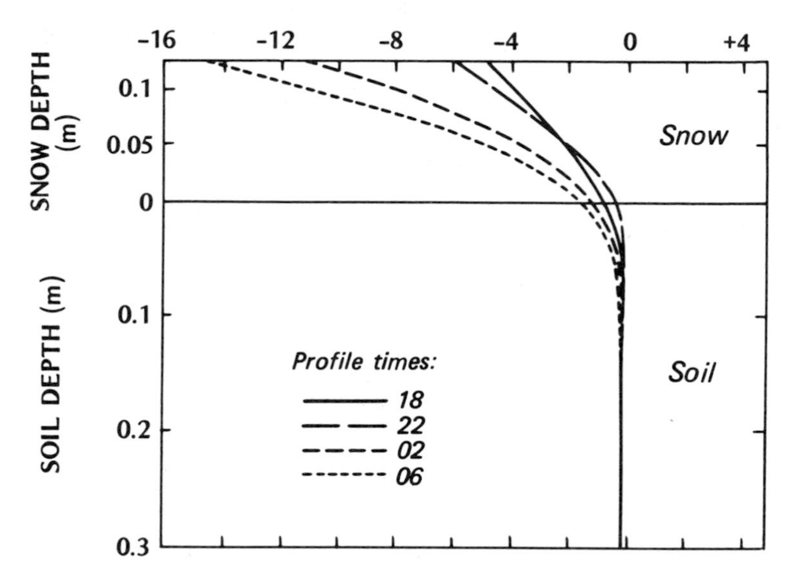

Figure 3.9 Sequence of nocturnal temperatures in a fresh snow cover and the underlying soil at Hamilton, Ontario (data from Oke and Hannell, 1966).

surface temperatures dropped by 10°C during the night the soil surface temperature only changed by about 1°C. However, not only did the snow protect the soil by moving the 'active' surface upwards, but it also conserved the latent heat released in the soil. Notice that through the upper 0·3 m of the soil the temperature was very close to 0°C. All the way down to the lower limit of this almost isothermal layer (the 'frost line') pore water was very close to becoming frozen. However, when freezing occurred heat (L_f) was released, thereby warming the surrounding area, and tending to slow down the freezing process. Thus until all the water changed to ice there was a self-equilibrating process which kept temperatures hovering near 0°C. Thus the presence of a snow cover serves to trap the latent heat released in a soil and to prevent, or at least delay, soil freezing compared with a snow-free site (p. 205).

In many locations farmers are keen to retain a deep snow cover over their land during the winter and early spring for three reasons. First, the snow minimizes frost penetration (Figure 7.4) and thus hastens spring warm-up ready for seed germination. Second, snow meltwater may be a significant source of soil moisture. Third, the snow cover may provide thermal protection for early seedlings. Plants within snow are in a conservative if cool environment, but if portions protrude above the surface they are open both to a wider range of temperature fluctuations, and to abrasion by blowing snow and ice pellets. A particularly harmful situation occurs when the root zone is frozen but the exposed shoots are warmed by the Sun. If transpiration occurs the plant is unable to replace plant water losses via its root system and it dies from dessication.

At night the poor diffusivity of snow results in fast surface cooling and the development of intense inversions based at the surface. Over snow-covered surfaces at high latitudes winter radiative cooling is almost continuous which leads to semi-permanent inversion structures. In these circumstances it is not uncommon to encounter a temperature increase of 20°C in the first 20 m above the surface. These inversions are not usually destroyed by convective heating generated at the surface. Their breakdown is due to mechanical mixing induced by an increase in wind speed. Any increase in turbulence results in an enhanced transfer of sensible heat towards the surface from the relatively warmer higher air layers. This accounts for areas of relative warmth observed to occur downwind of isolated obstacles, or zones of increased roughness.

In absolute terms the amount of water vapour in air over extensive snow-covered surfaces is very small. This is due to the lack of local moisture sources (if the surface temperature is below freezing), and to the low saturation vapour pressure of cool or cold air (Figure 2.15). However, even if the surface is melting the surface cannot exceed a saturation vapour pressure of only 611 Pa, and therefore a strong evaporative gradient cannot develop unless the air is exceptionally dry.

3 Water

The thermal and dynamic properties of water bodies (oceans, seas, lakes, etc.) makes them very important stores and transporters of energy and mass. The exchanges occurring at the air/water interface are, however, complicated by the fact that water is a fluid. This means that heat transfer within water is possible not only by conduction and radiation, but also by convection and advection. As in the atmosphere these modes of transfer greatly facilitate heat transport and mixing, and thereby allow heat gains or losses to be spread throughout a large volume. Although water is not compressible like air, it can be deformed, giving surface waves.

(a) RADIATION BUDGET

Short-wave radiation can be transmitted within water, and its variation with depth is well approximated by Beer's law (equation 3.1), with the extinction coefficient dependent upon both the nature of the water and the wavelength of the radiation. It depends upon the chemical make-up, plankton growth, and *turbidity* (amount of suspended material) of the water, and increases with wavelength towards the infra-red. This spectral dependence was noted in the case of snow and ice and is in accord with the absorptivity spectrum of water vapour and cloud noted in Chapter 1 (see Figure 1.9 and discussion).

Obviously the more absorbing substances there are in the water the greater is the extinction coefficient, and the less the penetration. In most water bodies short-wave radiation is restricted to the uppermost 10 m, but in some very clear tropical waters it has been observed to reach 700 to 1000 m. The different colours of lakes (especially blue and green combinations) are a result of different values of the extinction coefficient in the visible portion of the electromagnetic spectrum, which in turn are a result of differences in lake water composition.

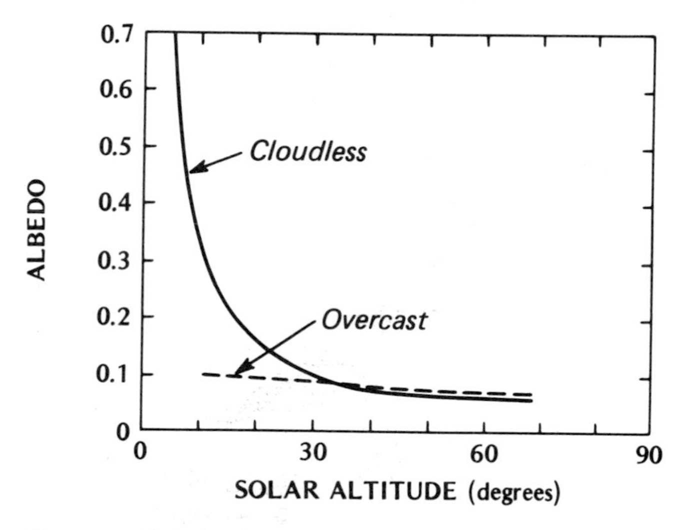

Figure 3.10 Relation between solar altitude and the albedo of lake water for clear and cloudy days over Lake Ontario (after Nunez *et al.*, 1972).

The albedo of a water surface, like that of snow, is not constant. In particular it depends upon the angle at which the direct beam (S) strikes the surface (Table 1.1 and Figure 3.10). With cloudless skies and the Sun at least 30° above the horizon, water is one of the most effective absorbing surfaces ($\alpha = 0{\cdot}03$ to $0{\cdot}10$), but at lower solar altitudes its reflectivity increases sharply. When the Sun is close to the horizon near sunrise and sunset the reflection is mirror-like, and this accounts for the dazzling effect at these times. Under cloudy skies the diffuse-beam (D) forms a larger proportion of $K\downarrow$, and the effect of solar altitude is considerably dampened (Figure 3.10). The altitude dependence is also modified by the roughness of the water surface. With a roughened surface (waves) and high solar altitudes there is a greater probability that the incident beam will hit a sloping rather than a horizontal surface, thereby tending to increase α;

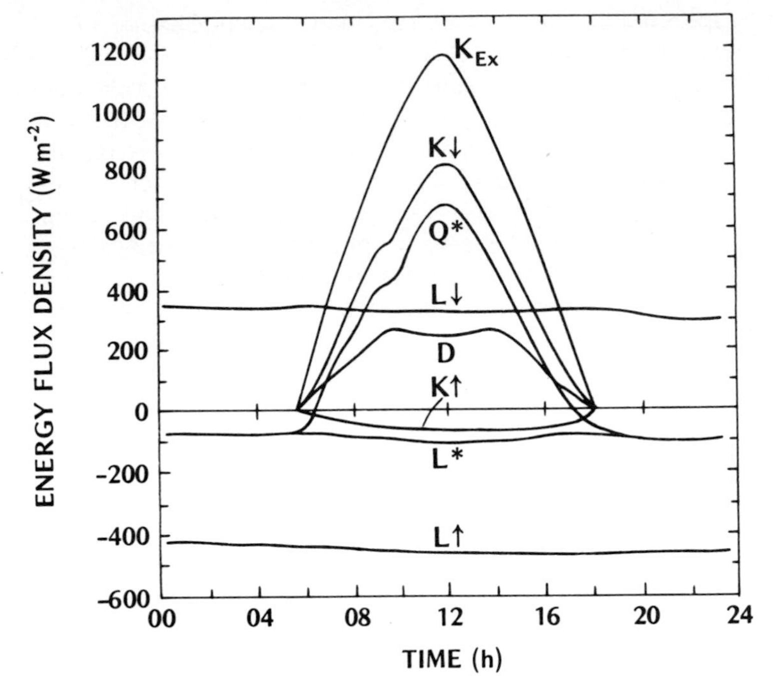

Figure 3.11 Variation of the radiation budget components for Lake Ontario on 28 August 1969, with cloudless skies (after Davies *et al.*, 1970).

whereas at low altitudes instead of grazing the surface the beam is likely to encounter a wave slope at a local angle which is more conducive to absorption, thereby decreasing α in comparison with smooth water. In all cases the albedo includes reflection from within the water as well as from the surface.

Long-wave radiation from the atmosphere ($L\downarrow$) is almost completely absorbed at a water surface with no significant reflection or transmission. The outgoing long-wave flux ($L\uparrow$) from a large water body is distinguished from that of most other natural surfaces by being virtually constant through the day. This is due to the very small diurnal range of surface water temperature (see Figure 3.14 and discussion).

Figure 3.11 and Table 3.3 present results from one of the few studies over a water body in which almost all of the radiation budget components were measured. The observations were taken on an almost cloudless day, from a tower in Lake Ontario. The following general features emerge:

(i) The extra-terrestrial solar input, K_{Ex} (computed not observed) describes a smooth symmetrical curve with a peak input at solar noon of slightly less than 1200 W m^{-2}. This is less than the value of the solar

TABLE 3.3 Radiation budget components for Lake Ontario near Grimsby, Ontario (43°N) on 29 August 1969. Data are daily totals (MJ m^{-2} day^{-1}) (after Davies *et al.*, 1970).

Short-wave		Long-wave		All-wave		Derived term	
K_{Ex}	34·2	$L\downarrow$	30·7	Q^*	15·2	α†	0·07
$K\downarrow$	22·0	$L\uparrow$	35·9				
S	14·4	L^*	−5·2				
D	7·6						
$K\uparrow$	1·6						
K^*	20·4						

† Dimensionless

constant (1353 W m^{-2}) because of the date and the latitude of the site. These factors determine that at solar noon the Sun is still 33·5° from the zenith (i.e. solar altitude = 56·5°).

(ii) The surface receipt of short-wave ($K\downarrow$) follows the same pattern as K_{Ex} but atmospheric attenuation (absorption, scattering and reflection) reduces the flux by about one-third. Of the receipt, 25–30% is as diffuse-beam (D) and the remainder as direct-beam (S) (not plotted) at midday. This proportion increases to as high as 75% at lower solar altitudes due to the increased path length through the atmosphere.

(iii) The reflected short-wave radiation ($K\uparrow$) is relatively small due to the very low albedo of water (daily average α = 0·07, for diurnal variation at the same site see Figure 3.10). The diurnal course of the net short-wave radiation (K^*) is not plotted on Figure 3.11 but would obviously describe a curve equivalent to about 93% of $K\downarrow$.

(iv) Both of the long-wave fluxes ($L\uparrow$ and $L\downarrow$) are relatively constant with time due to the small diurnal temperature variation of lake surface and bulk air temperatures, respectively. Consequently the net long-wave balance (L^*) shows an almost constant energy loss throughout the period.

(v) The net all-wave budget (Q^*) is dominated by K^* by day, and of course is equal to L^* at night. The daytime budget is notable for its high energy absorption values. At midday Q^* is almost 700 W m^{-2} due both to the low surface albedo (high K^*) and the relatively low surface temperature (low L^*).

In summary we may note that water surfaces are excellent absorbers of radiation, and that short-wave absorption occurs within a considerable volume. The convergence of the net radiative flux in the upper water layers leads to warming, but observed temperature variations are slight for the reasons outlined on p. 88.

(b) ENERGY BALANCE

The energy balance of the surface layer of a water body (ocean, lake, pond or puddle) extending to a depth where there is no vertical heat transfer is given by:

$$Q^* = Q_H + Q_E + \Delta Q_S + \Delta Q_A \qquad (3.6)$$

where, ΔQ_S – change of heat storage in the layer, ΔQ_A – net horizontal heat transfer due to water currents. The schematic heat balance (Figure 3.12) shows that ΔQ_A is a form of horizontal heat flux convergence or divergence. If the water depth is small it is possible that net heat transfer by rainfall (Q_R) could be significant and should also be added to equation 3.6.

On an annual basis for large water bodies ΔQ_S can be assumed negligible (i.e. zero net heat storage). The energy balances of the major oceans are then as given in Table 3.4 which strikingly illustrates the dominant role played by evaporation (Q_E) as an energy sink for a water body. On an annual basis approximately 90% of Q^* is used to evaporate water, and this leads to characteristically low Bowen ratio (β) values of approximately 0·10.

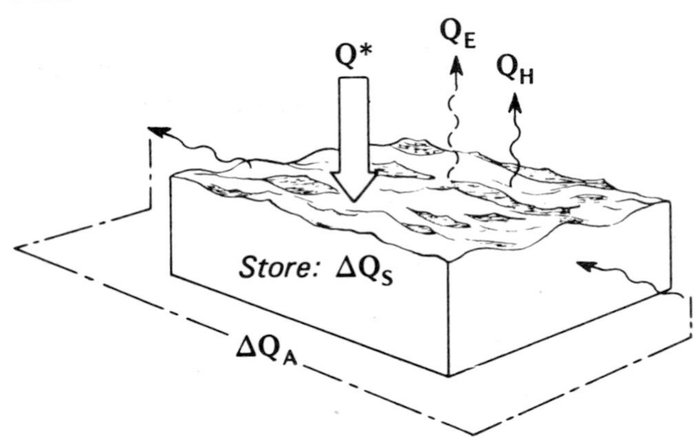

Figure 3.12 Schematic depiction of the fluxes involved in the energy balance of a water volume.

The diurnal pattern of energy partitioning by water bodies is given in Figure 3.13. The data in Figure 3.13a are from a shallow layer of water typical of a flooded rice paddy field. In this case ΔQ_A has been ignored, but heat conduction into or out of the underlying soil (Q_G) has been included. Radiation absorption by the water is strong and it should be noted that for large portions of the daytime more of this energy is being used to heat the water (ΔQ_S), or is being conducted to the underlying soil (Q_G), than is being carried into the air by convection ($Q_H + Q_E$). It is probable that Q_G was principally due to absorption of short-wave radiation transmitted through

TABLE 3.4 Component fluxes of the annual energy balance of the oceans. Data are daily totals (MJ m^{-2} day^{-1}) (after Budyko, 1963).

Ocean	Q^*	Q_E	Q_H	ΔQ_A	β $(=Q_H/Q_E)$
Atlantic	9·4	8·2	1·0	0·3	0·12
Indian	9·7	8·8	0·8	0·1	0·09
Pacific	9·8	8·9	1·0	0	0·11
All oceans	9·4	8·5	1·0	0	0·11

the shallow water layer. If the water were deeper all of this energy would have been absorbed by the water volume (i.e. entered water heat storage). In the late afternoon and at night the water and soil became the most important heat sources for the system. In fact the release of the stored daytime heat is sufficient not only to offset the net long-wave radiative loss at the surface, but also to support continued evaporation throughout the night. Integrated over the complete day the sinks and sources would approximately balance. The turbulent losses to the atmosphere are predominantly as latent rather than sensible heat (by day $\beta \simeq 0{\cdot}20$ to $0{\cdot}25$). The diurnal pattern of turbulent heat loss shows a late afternoon peak. This probably coincides with the time of maximum surface water temperature (and therefore of surface saturation vapour pressure) and maximum vpd in the air (p. 55). The resulting water-to-air vapour pressure difference would create a strong evaporative demand by the air.

There are few studies of the diurnal energy balance over an ocean. The results in Figure 3.13b are from a ship stationed in the tropical Atlantic Ocean, and are average data for a ten-day period. The degree of detail is less than that of the paddy field because the frequency of observation was less. The net radiation data are based on measured short-wave values and an estimated long-wave radiation term. The water heat storage change data are calculated from measured temperature profile changes over time (see Figure 3.14) in a manner similar to that outlined on p. 43 for calculating evapotranspiration from profiles of soil moisture content. Initially it is assumed that ΔQ_A is negligible, and that all the energy input to the water is contained in the uppermost 27 m layer. The latter assumption is based on the almost zero diurnal range of water temperature at this depth (Figure 3.14). The combined turbulent transport of sensible and latent heat to the atmosphere $(Q_H + Q_E)$ is then obtained as a residual in equation 3.6.

The diurnal pattern of energy exchange for the ocean is remarkably similar to that of the paddy field. By day the primary energy sink is ΔQ_S, as the radiant input is largely absorbed by the water layer leaving little for transport into the atmosphere until late afternoon. At night this energy store becomes the source of energy (approximately 300 W m^{-2}) which

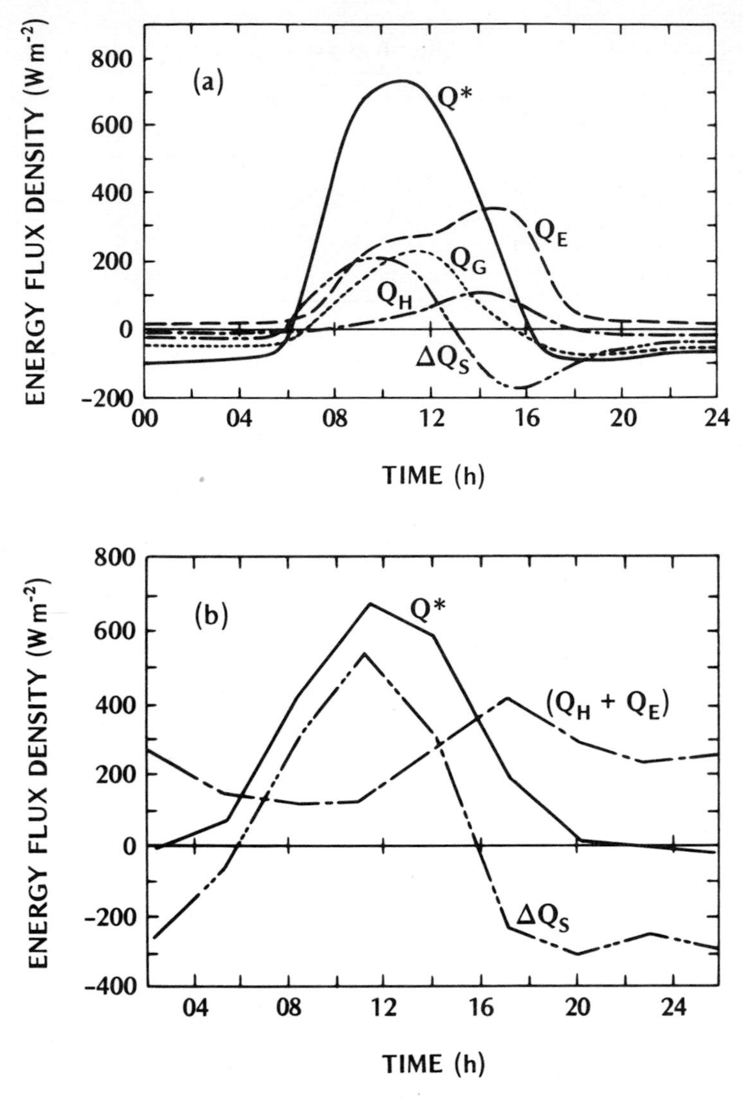

Figure 3.13 Diurnal variation of the energy balance components in and above (a) a shallow water layer on a clear September day in Japan (after Yabuki, 1957), and (b) the tropical Atlantic Ocean based on measurements from the ship *Discoverer* in the period 20 June to 2 July 1969 (after Holland, 1971).

sustains an upward flow of heat to the atmosphere throughout the period. The ocean therefore acts as a major heat sink by day, and a major heat source at night.

On a daily basis the complete energy balance (as per equation 3.6) can be

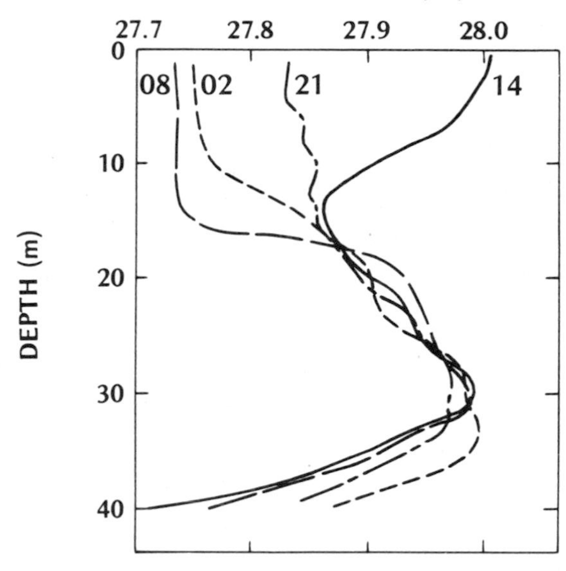

Figure 3.14 Diurnal sequence of ocean temperature profiles for the tropical Atlantic Ocean from measurements made in the period 20 June to 2 July 1969 (after Holland, 1971).

obtained by including an estimated advective component (ΔQ_A) and a seasonal storage change (ΔQ_S) as given in Table 3.5. This again highlights the dominance of Q_E as an energy sink, especially relative to the role of Q_H ($\beta = 0{\cdot}07$). This leads to the situation where it is suggested that the buoyancy of air in the lowest layers over tropical oceans is due more to their moisture, rather than their sensible heat content. This arises because the density of saturated air is less than that of dry air at the same temperature (e.g. compare $\rho_{a^*_{(T)}}$ and ρ_a at the same temperature in Appendix A3).

Differences from the above pattern of energy partitioning could arise as a result of localized advection by currents, or changes in air mass characteristics. A dry air mass enhances the evaporation rate (because the water-to-air humidity gradient is increased), and humid air suppresses it. Similarly the introduction of a relatively cold air mass enhances Q_H (because it causes the water-to-air temperature profile to become more lapse, and accordingly increases convective instability), whereas a warm one has a dampening effect. Phillips (1972) reports an example of enhanced Q_H over Lake Ontario in January. At this time the lake water is considerably warmer than the cold continental air traversing it. Climatological calculations show that Q_H may be as large as 20 MJ m^{-2} day^{-1} and since Q^* is very small at this time the energy output to the atmosphere must be derived from lake heat storage (ΔQ_S).

(c) CLIMATE

The thermal climate of a water body is remarkably conservative. This fact

TABLE 3.5 Component fluxes of the energy balance of the tropical Atlantic Ocean. Data are daily totals ($MJ\ m^{-2}\ day^{-1}$) based on data from the ship *Discoverer* gathered in the period from 20 June to 2 July 1969 (after Holland, 1971).

		Derived terms	
Q^*	22·2		
Q_H	1·2	β†	0·07
Q_E	16·2	E (mm)	6·6
ΔQ_S	1·2	Q_E/Q^*†	0·73
ΔQ_A	3·6		

† Dimensionless

is clearly demonstrated by the water temperature profiles given in Figure 3.14. These observations are from the same experiments used to describe the diurnal energy balance of a tropical ocean. A discernible diurnal heating/cooling cycle is evident but the maximum diurnal temperature range is only 0·275°C at the surface. On an annual basis the maximum range of sea surface temperature is 8°C at latitude 40°, and at the Equator it is only 2°C.

This presents a paradox: on the one hand, of all natural surfaces water bodies are noted to be about the best absorbers of radiation, but on the other they exhibit very little thermal response. The lack of response can be attributed to four characteristics:

(i) penetration – since water allows short-wave radiation transmission to considerable depths (p. 81) energy absorption is diffused through a large volume;
(ii) mixing – the existence of convection and mass transport by fluid motions also permits the heat gains/losses to be spread throughout a large volume;
(iii) evaporation – unlimited water availability provides an efficient latent heat sink, and evaporative cooling tends to destabilize the surface layer and further enhance mixing (see below);
(iv) thermal capacity – the thermal capacity of water is exceptionally large (Table 2.1) such that it requires about three times as much heat to raise a unit volume of water through the same temperature interval as most soils.

These properties contrast with those of land surfaces. As noted previously it is not uncommon to find surface temperature ranges of at least 20°C for soils, and this is almost two orders of magnitude greater than for water. Shorelines therefore demarcate sharp discontinuities in surface thermal climate, which leads to horizontal cross-shoreline interaction (Chapter 5). Figure 5.4 vividly illustrates the difference in thermal climate between land and water surfaces. It shows a surface temperature transect across a prairie landscape including a lake and some ponds. Clearly by day the water surfaces are very much cooler than the surrounding land. In addition lake temperatures show much less spatial variability.

Figure 3.14 shows that the upper 30 m of the ocean is most active in diurnal heat exchange. Below this depth temperatures decrease rapidly. This zone is known as the *thermocline* and it divides the upper active mixed layer from the more stable layer beneath.

In lakes the upper layer is called the *epilimnion* and the lower one the *hypolimnion*. This division is important biologically because it tends to stratify habitats for thermally sensitive aquatic organisms. During the summer the epilimnion is warmer than the hypolimnion and species preferring cool water stay at depth. In cold winter climates the changeover occurs very rapidly due to the density characteristics of water. Pure water reaches its maximum density at about 4°C (Table 2.1 and Appendix A3). In the spring, if the surface water temperature is below this value any warming serves to *increase* its density. Therefore warming of the surface leads to instability, the surface water sinks, and convective mixing raises the heat content of the upper layer relatively rapidly. After the surface has warmed beyond 4°C further warming only increases stability and restricts mixing (except by vigorous wave action) to the epilimnion. In the autumn, surface cooling again increases density, and therefore instability, and the epilimnion cools rapidly. It can also be seen that in the summer evaporative cooling of the surface destabilizes the upper layer so that overturning brings warmer water to the surface and helps to maintain the almost constant temperature situation described above. This is an example of a negative feedback process.

Figure 3.15 shows the climatic characteristics of the atmosphere above the tropical ocean at a location close to that of the experiments in Figures 3.13b and 3.14. The data are from three different instrument systems and different time periods, and this accounts for the profile discontinuities, but their slopes are believed to be representative. The air temperature is given as *potential temperature* (θ), and as outlined in Appendix A1 this means that the data have been corrected to allow for the dry adiabatic lapse rate (Γ). In interpretation this means that any portion of the profile sloping to the left of vertical is unstable, and to the right is stable.

In the lowest 10 m the decrease of $\bar{\theta}$ and specific humidity ($\bar{q}$), and the increase of wind speed ($\bar{u}$), all follow a straight line relationship with height. Since the height scale is logarithmic this means that the profiles are logarithmic also, and thus their form is in agreement with those over land surfaces (e.g. Figures 2.2, 2.9 and 2.16). The vertical differences are however small, especially in the 2–600 m layer. This well-mixed layer constitutes the ocean boundary layer above which large-scale subsidence gives rise to an elevated inversion (p. 276).

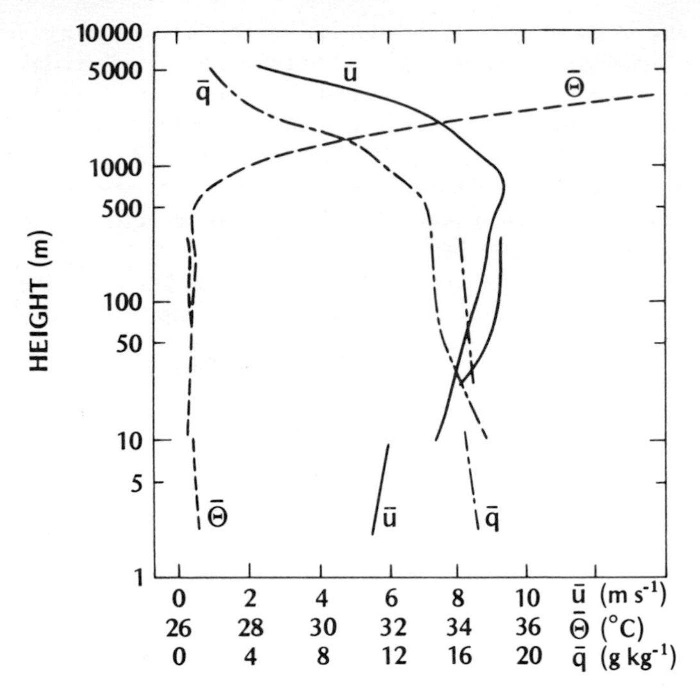

Figure 3.15 Profiles of mean wind speed ($\bar{u}$), potential temperature ($\bar{\theta}$), and specific humidity ($\bar{q}$) over the tropical Atlantic Ocean. Profile discontinuities are due to different observing systems and time periods (after Holland, 1972). Note the use of a logarithmic height scale.

In the lowest layers of the atmosphere over tropical oceans the profiles of $\bar{\theta}$ and $\bar{q}$ remain lapse throughout both day and night and the diurnal range of both is very small. At higher latitudes with cold water the sign of these profiles may well be reversed. The wind profile of course always exhibits an increase with height near the surface, and due to the small roughness of most water surfaces (Table 2.2) forced convection is relatively weak. This causes the wind gradient to be steep in the lowest 2 m because the momentum exchange is confined to a shallow layer (e.g. Figure 2.9a).

In concluding it should be pointed out that much of the preceding discussion has related to relatively large bodies of water (large lakes and oceans). In smaller systems (small lakes, ponds and puddles) the thermal inertia is reduced because of the smaller volume involved. Also border-effects become increasingly important. For example, in a shallow water body the incoming short-wave radiation can penetrate to the floor. This warms the lower border of the system and the water is warmed from below as well as by the normal processes from above. If the water contains vegetation (e.g. reeds) this warming is further enhanced due to absorption by the submerged plants. Border-effects may also occur at the sides of the water body, and the smaller the width the greater these influences become. They arise because of heat conduction between the water body and the surrounding ground, and because of the likelihood of advection across the water margins.

CHAPTER 4

Climates of vegetated surfaces

1 Special features

The introduction of a vegetation cover above an otherwise simple soil surface presents the following complicating features compared with the environments in Chapter 3.

(a) ENERGY AND WATER STORAGE IN VEGETATION SYSTEMS

As with snow, ice and water systems it is necessary to consider volume exchanges when dealing with the energy and mass balances of soil-plant (or tree)-atmosphere systems. The need to consider volume balances also means that energy and mass storage rates become potentially important (p. 34).

If we define a soil-plant-air system volume such as that in Figure 4.1a (extending from the top of the plants to a depth in the soil at which there is no significant vertical heat flux) then we may write its energy balance as:

$$Q^* = Q_H + Q_E + \Delta Q_S + \Delta Q_P \tag{4.1}$$

where, ΔQ_S – net rate of *physical heat storage* by substances in the system, and ΔQ_P – net rate of *biochemical heat storage* due to plant photosynthesis. Physical storage changes result from the absorption or release of heat by the air, soil and plant biomass (leaves, branches, stems, etc.). Changes of sensible heat content result in temperature changes, but latent heat changes do not. Biochemical storage changes are linked to the rate of CO_2 assimilation by the plant community (see next section). Depending on the

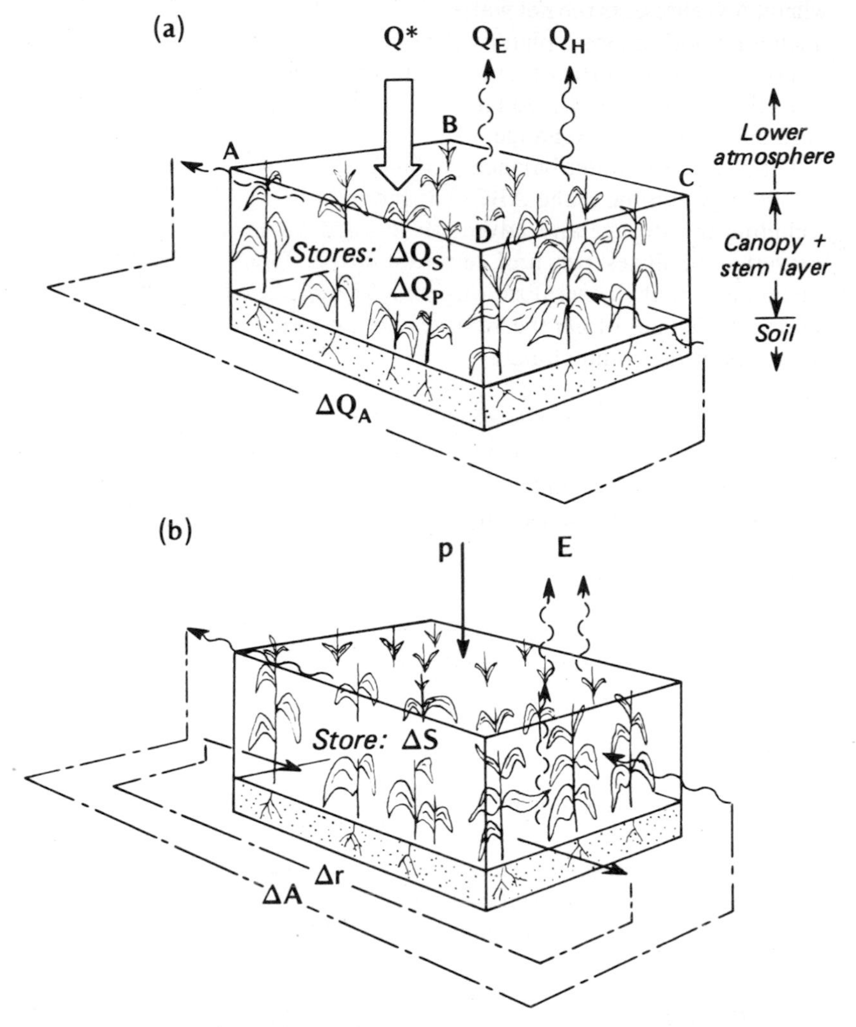

Figure 4.1 Schematic depiction of fluxes involved in (a) the energy and (b) the water balances of a soil-plant-air volume.

nature of the surrounding environment it may also be necessary to supplement equation 4.1 with an advection term (ΔQ_A) to account for net energy gain or loss due to horizontal sensible and latent heat transport (see Chapter 5).

Similarly, in accord with equation 1.18, we may write the water balance of the system (down to a depth where vertical moisture movement is absent) as:

$$p = E + \Delta r + \Delta S$$

where, ΔS represents the net water storage in the air and soil, and by plants (including both internal plant water content and the water resting on the exterior of plant surfaces due to the interception of precipitation). If advection is present the equation should include ΔA to allow for net horizontal moisture exchange.

The energy and water balances outlined above refer to the vertical fluxes passing through the plane ABCD at the top of the volume (Figure 4.1a), including the storage and advection terms if their volumetric values are converted to fluxes per unit horizontal area. Such a simplification is adequate to make generalizations about a vegetation community, but it ignores exchanges *within* the volume and gives no insight into the internal workings of the stand climate.

(b) PHOTOSYNTHESIS AND CARBON DIOXIDE EXCHANGE

Plant growth is intimately tied to the supply of solar radiation (especially the visible portion) and carbon dioxide through the processes of photosynthesis and respiration:

Gross photosynthesis

$$\begin{array}{c} \text{Carbon} \\ \text{dioxide} \end{array} + \begin{array}{c} \text{Water} \\ \text{vapour} \end{array} + \begin{array}{c} \text{Light} \\ \text{energy} \end{array} \longrightarrow \begin{array}{c} \text{Carbohydrates} \\ \text{(glucose, starch)} \end{array} + \text{Oxygen}$$

$$CO_2 + H_2O + \text{Light} \longrightarrow (CH_2O) + O_2$$

Gross respiration

$$\text{Carbohydrate} + \text{Oxygen} \longrightarrow \begin{array}{c} \text{Carbon} \\ \text{dioxide} \end{array} + \begin{array}{c} \text{Water} \\ \text{vapour} \end{array} + \begin{array}{c} \text{Combustion} \\ \text{energy} \end{array}$$

Growth depends upon the excess of dry matter gain by the assimilation of CO_2 in gross photosynthesis (P), less the CO_2 loss via plant respiration (R). Thus the net rate of CO_2 assimilation, or the *net rate of photosynthesis* (ΔP) is:

$$\Delta P = P - R \tag{4.2}$$

each having the units of weight of CO_2 per unit area per unit time (kg m^{-2} s^{-1}). The rate at which heat is stored by net photosynthesis, ΔQ_P, is therefore:

$$\Delta Q_P = \phi(\Delta P) \tag{4.3}$$

where ϕ – heat of assimilation of carbon, which is approximately $1{\cdot}15 \times 10^7$ J kg^{-1}, or about 3·2 W m^{-2} per g m^{-2} h^{-1} of CO_2 assimilation.

By day P is greater than R and therefore ΔP is positive (i.e., the crop is a net CO_2 sink). Maximum values of ΔP for crops are dependent upon species but lie in the range from 2 to 5 g m^{-2} h^{-1} (Monteith, 1973), so that the largest values of ΔQ_P are typically 6 to 16 W m^{-2}. At night R is unopposed by P (which requires solar radiation) so that ΔP is negative (i.e., the crop is a net CO_2 source), and since maximum crop respiration values are about 1 g m^{-2} h^{-1} it follows that the maximum nocturnal value of ΔQ_P is about -3 W m^{-2}. In comparison with most other terms in equation 4.1 (except ΔQ_S) the value of ΔQ_P is so small that it is often neglected in energy balance considerations. Correspondingly we will not include it further in our discussions, but we will deal with the associated mass flux of CO_2.

The daytime flux of CO_2 is supplied to the vegetation by the atmosphere and the soil. The passageway between the atmosphere and the interior of the plant or tree is the leaf stomate (Figure 4.2). These pores on the leaf surface are open during the day to capture and expel CO_2, and in this position they also expose the moist interior of the stomate to the air. Evaporation of moisture (known as *transpiration*) is therefore an inevitable by-product of photosynthesis. Transpiration is an important process in its own right however, since the water loss induces moisture and nutrient movement through the plant or tree, and the associated uptake of latent heat is a major means of dissipating the energy load on leaves.

Depending on species stomata are typically 10 to 30 μm in length, and vary in width from zero when closed to 10 μm when fully open. Their density ranges from 50 to 500 per mm^2 of leaf surface, and when open their combined area represents 0·3–1% of the total leaf area (Rutter, 1975). In some plants they occur on both leaf surfaces, in others only on the underside. The climatic significance of stomata is their ability to open and shut so that they act as regulatory 'valves' in the transfer of water vapour and CO_2 between plants and the atmosphere. At night the stomata are essentially closed, but with sunlight the *guard cells* (Figure 4.2) controlling the stomatal aperture operate to open the pore. The degree of opening depends on many factors including the light intensity, the ambient temperature and humidity and the CO_2 concentration. Stomatal closure is tied to insufficient light intensity and/or loss of water content (reduction in turgor) by the guard cells. Hence anything producing plant water stress (e.g. excessive transpiration losses, depletion of soil moisture) will close the 'valves' and hinder the flow of gases. Stomatal activity therefore provides differing degrees of resistance to the exchange of water vapour and CO_2 between the plant and the atmosphere, and makes the plant an active agent in the determination of its climate.

(c) EFFECTS OF STAND ARCHITECTURE

Considering the diversity of vegetative species it is not surprising that there is a wide range of stand structural arrangements. It is important to recognize different stand 'architectures' because they exert a considerable influence on the position of the active surface with regard to the exchanges of heat, mass and momentum.

In most cases the main sites of exchange are the leaves, and hence the

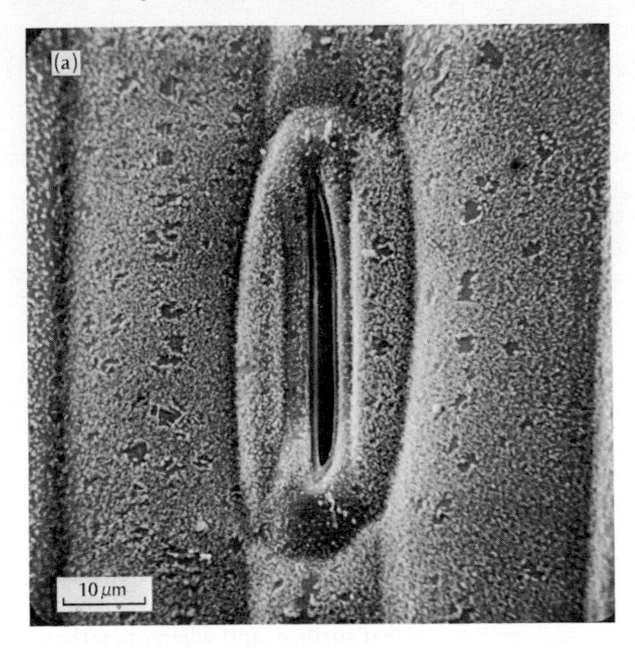

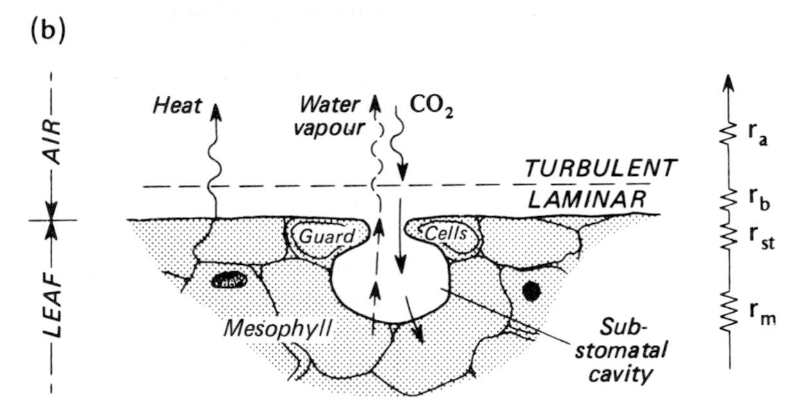

Figure 4.2 (a) View of a partially open stomate on a wheat leaf (photo by Fillery in Rutter, 1975). (b) Schematic cross-section through a portion of a leaf illustrating the exchanges of water vapour and CO_2 through a stomate, and of heat from the leaf exterior.

vertical variation of foliage density within the vegetation stand is of central interest. In a dense stand it is usually acceptable to ignore horizontal foliage differences, and hence to consider it to consist of one or more layers. In crops such as grasses and cereals where the foliage density does not vary greatly with height, the whole stand is usually considered to be the canopy. For crops like maize, sunflower and potato, and many trees, the foliage density is concentrated near the top of the stand so that it is appropriate to view the architecture as consisting of a canopy layer lying above a stem or trunk layer. Other vegetative systems have their foliage concentrated near the base of the stand, and yet others have a multi-layered structure (e.g. tropical forests). Further differences occur as a result of the orientation of the principal exchange surfaces (e.g. grass blades are close to being vertically-oriented whereas many tree leaves are horizontal). Given this complexity it is not possible to include examples of the climate of all stand architectural types here. As a compromise we will deal mainly with vertically-oriented canopies in the plant and crop section, and horizontally-layered (canopy-plus-trunk) stands in the forest section.

Although the actual height of the principal active surface depends upon species-specific factors it is a reasonable generalization to state that its position lies closer to the top than the bottom of the stand. The exact position may also depend upon the entity under consideration; that is, the levels of the effective sources and sinks of heat, water vapour, momentum and CO_2 may not necessarily coincide, nor need they agree with the heights of maximum radiation and precipitation interception, but they are all *likely* to occur in a zone near the top of the stand.

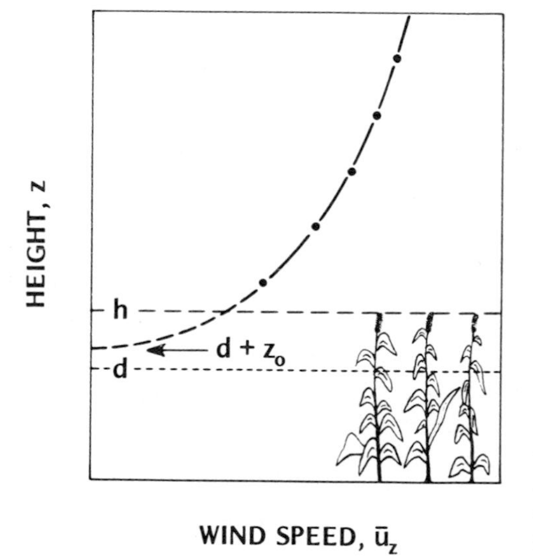

Figure 4.3 Typical wind profile measured above a vegetation stand of height *h*, illustrating the concept of a zero plane displacement at the height *d*.

The momentum exchange between the lower atmosphere and a tall vegetation stand provides a good example of the elevated position of the active surface. A plot of wind speeds measured at a number of levels above tall vegetation results in a profile such as that in Figure 4.3. Above the vegetation the wind profile is logarithmic as with other surfaces (Figure 2.9), but the extrapolation of this curve downwards shows that the flow is behaving as though the 'surface' is located at some height near the top of the stand, and not at the ground. This height is called the level of *zero plane displacement* (d), which may be visualized as representing the apparent level of the bulk drag exerted by the vegetation on the air (or the level of the apparent momentum sink). In practice for a wide range of crops and trees the value of d is approximately given by:

$$d = \tfrac{2}{3}h \tag{4.4}$$

Typical values of d will be found in Table 2.2. Equation 4.4 only applies to closely-spaced stands because d also depends on the density of the drag elements. The value of d also depends upon the wind speed because most vegetation is flexible and assumes a more streamlined form at high wind speeds. These effects on the wind profile over vegetation can be accommodated by including d in the logarithmic wind profile equation (equation 2.9) so that it reads:

$$\bar{u}_z = \frac{u_*}{k} \ln \frac{z - d}{z_0} \tag{4.5}$$

In fact of course the wind speed does not become zero near the height d but at the ground as shown in Figure 4.15a. Later it will become evident that the air temperature, humidity and CO_2 profiles similarly respond to an elevated active surface located at a level analogous to that of d for wind speed.

The three-dimensional geometry of a leaf or a canopy layer introduces a configuration we have not considered before. These shapes are particularly interesting because they have both upper and lower active surfaces. This greatly increases their effective surface area for radiative and convective exchange, and complicates the one-dimensional framework we have used previously. At the scale of the vegetation community this can be ignored by only considering the net exchanges through a plane at the top of the system volume (i.e. the plane ABCD in Figure 4.1a). However if the internal stand climate is being studied, multiple exchanges must be envisaged similar to those schematized in Figures 4.5 and 4.17.

Finally, we should note that a vegetation system is composed of many such active surfaces represented by the myriad of leaves making up the foliage, and to a lesser extent the other portions of the plant or tree structure. Faced with such a system it might perhaps seem appropriate to analyse the climate of a typical leaf and then to integrate this over the number of leaves to give the climate of the plant or tree, and then to integrate those climates to arrive at the climate of a crop or forest. Unfortunately it is not possible to make such a linear extrapolation of elemental units and thereby to combine many microclimates into a local climate. On a plant or tree the leaf is not in isolation, it is intimately linked to its total environmental setting, and the same is true of a plant or tree in a crop or forest. The effects of mutual shading, multiple reflection, long-wave radiation interaction etc. provide important feedbacks not found in the isolated case.

2 Microclimate of leaves

The radiative properties of leaves show an interesting wavelength dependence (Figure 4.4 and Table 4.1). Leaves are not opaque to short-wave radiation so that the disposition of incident radiation is given by

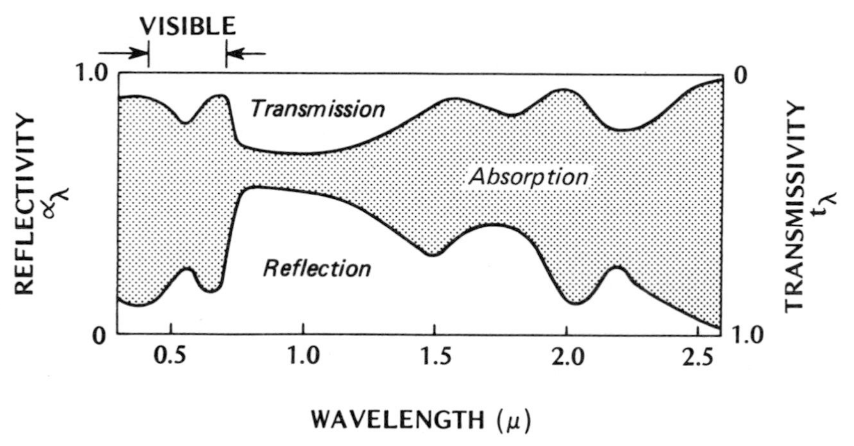

Figure 4.4 Idealized relation between wavelength and the reflectivity (α), transmissivity (t) and absorptivity (a) of a green leaf (after Monteith, 1965a).

equation 1.6, and transmission is non-zero. The relative roles of reflection (α), transmission (t) and absorption (a) are governed by the structure of the leaf interior and the radiative properties of the main plant pigments (especially chlorophyll and carotenoids). The cellular structure tends to cause almost totally diffuse scattering. This results in almost equal portions

TABLE 4.1 Mean reflection, transmission and absorption coefficients of green leaves for different radiation wavebands (modified after Ross, 1975).

	PAR† (0·38–0·71 µm) §	NIR‡ (0·71–4·0 µm)	Short-wave (0·35–3·0 µm) §	Long-wave (3·0–100 µm)
Reflection (α)	0·09	0·51	0·30	0·05
Transmission (t)	0·06	0·34	0·20	0·00
Absorption (a)	0·85	0·15	0·50	0·95

† PAR – photosynthetically active radiation
‡ NIR – near infra-red radiation
§ Note these wavelength limits differ slightly from those used elsewhere in this book

of the radiation being reflected back and transmitted on through the leaf and explains why the α and t curves in Figure 4.4 are similar in shape. The pigments are particularly effective absorbers in the blue (0·40 to 0·51 μm) and red (0·61 to 0·70 μm) bands of the visible portion of the electromagnetic spectrum, and are at the core of the photosynthetic process. The waveband between 0·40 and 0·70 μm is therefore designated as *photosynthetically active radiation* (PAR). Within this range there is a small relative peak of reflection and transmission between 5·0 and 5·5 μm. Since this lies in the green portion of the visible it explains the colour of most vegetation as perceived by the human eye. At 0·7 μm absorption decreases sharply and thereafter gradually increases until about 2·5 μm, beyond which absorption is almost total. Therefore in the long-wave region leaves absorb almost all incident radiation and allow no transmission. The primary absorbing agent at these wavelengths is plant water. In this region

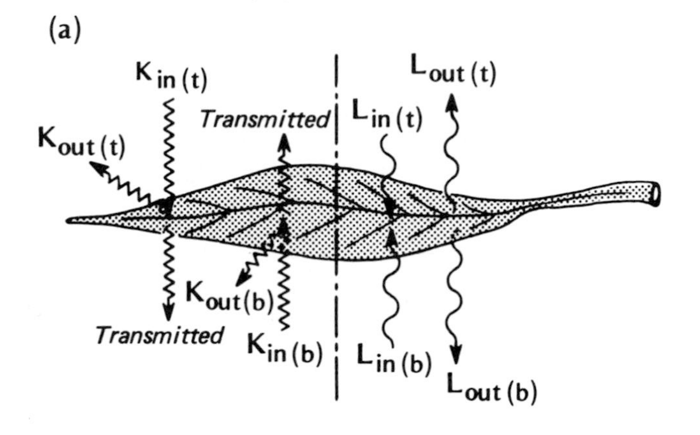

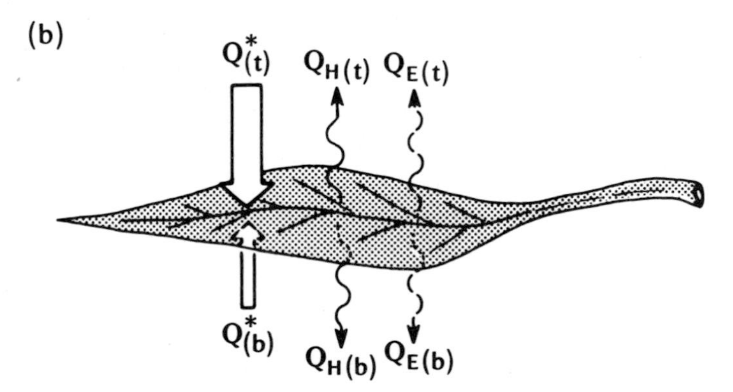

Figure 4.5 Schematic depiction of the fluxes involved in (a) the radiation budget and (b) the energy balance of an isolated leaf.

leaves are classed as almost perfectly full radiators (i.e., $\varepsilon = 0{\cdot}94$ to $0{\cdot}99$) and therefore are also very efficient emitters of long-wave radiation.

The relationships depicted in Figure 4.4 provide an almost ideal radiative environment for leaves. In the red and blue portions of the visible, where light is needed for photosynthesis, absorption is good. On the other hand in the near infra-red (NIR) high reflectivity enables the leaf to reject the bulk of the incident energy. The heat content of this radiation is very high but it is not useful for photosynthesis, therefore its rejection is helpful in offsetting the heat load on the leaf. The high emissivity at longer wavelengths also helps the leaf to shed heat to the environment and keep leaf temperatures moderate.

With these features in mind Figure 4.5a illustrates the fluxes involved in the radiation budget of a leaf. Note that the receipt and loss of radiation takes place on both the upper and lower leaf surfaces, and that the short-wave exchange includes transmission. On this basis we may write the radiation budget of a complete leaf (Q^*_{leaf}) as:

$$\begin{aligned} Q^*_{leaf} &= [(K_{in(t)} + K_{in(b)})(1 - t - \alpha)] + \\ &\quad [(L_{in(t)} - L_{out(t)}) + (L_{in(b)} - L_{out(b)})] \\ &= K^*_{(t)} + K^*_{(b)} + L^*_{(t)} + L^*_{(b)} \\ &= K^*_{leaf} + L^*_{leaf} \qquad (4.6) \end{aligned}$$

In this formulation it has been necessary to discard the use of arrows (which normally refer to the direction of the flux relative to a surface) and replace them with the subscripts (in) and (out), and the further subscripts (t) and (b) indicating the top and bottom surfaces of the leaf. Note also that it has been assumed that reflectivity and transmissivity are equal for the upper and lower sides.

A similar formulation is required to describe the energy balance of a leaf (Figure 4.5b):

$$\begin{aligned} Q^*_{leaf} &= (Q_{H(t)} + Q_{H(b)}) + (Q_{E(t)} + Q_{E(b)}) \\ &= Q_{H(leaf)} + Q_{E(leaf)} \qquad (4.7) \end{aligned}$$

where both the physical and biochemical heat storage has been neglected. This is reasonable since the typical rate of leaf physical storage on a sunny morning is about 6 W m^{-2} (Monteith, 1973), and the biochemical storage rate is usually less than 16 W m^{-2} (p. 94). Therefore their combined effect represents only a small fraction of the net radiation. The poor storage capacity of leaves is a direct outcome of their geometry. They exhibit very large surface areas compared to their mass, therefore they are ideally adapted to heat exchange but are poor heat storers. (For a fuller explanation of the effects of object size see p. 166.)

By day the net short-wave heat gain by the leaf is dissipated by long-wave radiation (L^*) plus the combined convective losses of sensible and latent heat. The latter can be due to the evaporation of water from the leaf exterior (intercepted precipitation, dew, or water exuded from the leaf interior), through the leaf cuticle, or the transpiration of water through stomata. The convective losses are enhanced by the action of leaf fluttering in strong winds, but in general the long-wave radiation emission to cooler surroundings is the most effective means of alleviating excessive solar heat loads.

The convective exchange of heat, water vapour and CO_2 between a leaf and the atmosphere can be usefully viewed as a simple analogue of the flow of current in an electrical circuit. Ohm's Law gives the relationship between the current (amps) in a circuit to the electrical potential (volts) and the resistance of the wire:

$$\text{Current} = \frac{\text{Potential difference}}{\text{Wire resistance}}$$

For entities such as heat, water vapour and CO_2 we may rewrite this to read:

$$\text{Flux rate} = \frac{\text{Concentration difference of property}}{\text{Resistance to flow exerted by the system}}$$

Substituting the appropriate terms we have:

$$\text{Sensible heat} \quad Q_H = -\frac{\rho c_p(\Delta \overline{T})}{r} \tag{4.8a}$$

$$\text{Water vapour} \quad E = -\frac{\rho(\Delta \bar{q})}{r} \tag{4.8b}$$

$$\text{Carbon dioxide} \quad C = -\frac{\Delta \bar{c}}{r} \tag{4.8c}$$

where r represents the appropriate system *resistance* (s m^{-1}). It can be shown (Munn, 1970) that the value of r depends upon the thickness of the layer through which flow must place, and its transfer properties. To some extent r acts as the inverse of the molecular and eddy diffusion coefficients (i.e. the κ's and K's) in the standard flux-gradient equations (Chapter 2). The diffusion coefficients represent the facilitating role of the system in transferring quantities, conversely the resistance represents the degree of hindrance to flow.

Depending on the nature of the pathway for flow the value of r may include a number of consecutive resistances and these should be weighted by their path length in determining the overall resistance. In the case of sensible heat transfer from a leaf the heat must first pass through the laminar boundary layer adjacent to the surface (Figure 4.6) before it is carried away by turbulent eddies in the air above (Figure 4.2b). Therefore

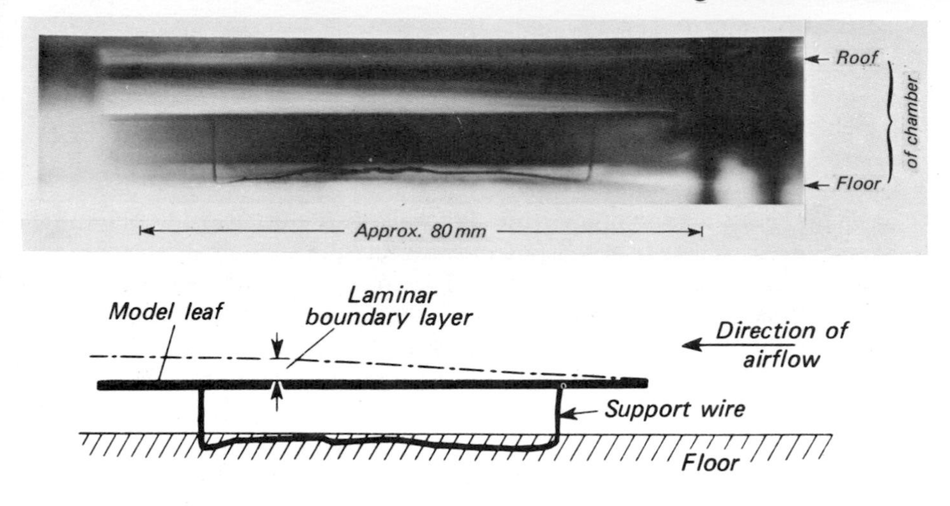

Figure 4.6 The laminar boundary layer above an artificial leaf. The 'leaf' is supported on a wire frame in a chamber through which air is blown from the right-hand side. The layer is made visible by a drop of white smoke – forming liquid applied to the upper right-hand edge. A similar boundary layer also exists on the lower surface but the smoke cannot reach it (after Avery, 1966).

in equation 4.8a the heat flux would depend upon the size of the temperature difference between the leaf surface and the ambient air (i.e. $\Delta\overline{T} = \overline{T}_0 - \overline{T}_a$) and the values of the thermal resistances in the laminar boundary layer (r_b) and the ambient air (r_a) (i.e. $r = r_b + r_a$, analogous to two electrical resistances joined in series). Thus for a given temperature difference the sensible heat flux increases with decreasing r. This can be achieved by an increase in wind speed which reduces the thickness of the laminar layer and increases the turbulent activity in the air above. Therefore r_b and r_a are inversely related to wind speed ($\bar{u}$).

The pathway for water vapour flow is more complex (Figure 4.2b). The source of water vapour is the substomatal cavity where the air is in a saturated state. The outflow to the atmosphere therefore depends on the vapour concentration difference between the cavity and the ambient air (i.e. $\Delta\bar{q} = \bar{q}_0 - \bar{q}_a$), and to the vapour resistance in the path between them. It can be seen from Figure 4.2b that in addition to r_b and r_a the ease of vapour diffusion depends on the stomatal aperture. If the stomata are fully open the stomatal resistance (r_{st}) is at a minimum, but as closure becomes more complete r_{st} becomes increasingly large. Resistance to vapour transfer therefore depends not only upon the wind speed but also upon the stomatal behaviour which in turn depends upon such factors as light intensity, plant turgor and the ambient temperature, humidity and CO_2 concentration. Therefore the total vapour resistance is $r = r_{st} + r_b + r_a$. Evaporation of water from the exterior of a leaf may involve a cuticular resistance to characterize flow through the epidermis, but for intercepted moisture and dew only r_b and r_a are entailed.

The full pathway for CO_2 is even longer. During photosynthesis CO_2 travels to the leaf from the atmosphere (Figure 4.2b). It is transported towards the leaf by turbulence, and diffuses through the laminar boundary layer and the stomate to the walls of the substomatal cavity. In order to reach the site of photosynthesis (the chloroplasts) it must also diffuse into the plant cells. These final resistances constitute the mesophyll resistance (r_m), and are largely chemical in nature. Therefore the CO_2 flux is driven by the CO_2 concentration difference between the air and the chloroplasts and encounters a total resistance r composed of $(r_m + r_{st} + r_b + r_a)$.

It should be appreciated that the resistances identified above are not necessarily equal for all entities (e.g. r_{st} for water vapour may not equal r_{st} for CO_2).

The resistance approach or electrical analogy of fluxes is also of value in analysing other systems. In this book we will use it in connection with the heat flow between the atmosphere and animals (Chapter 6), and buildings (Chapter 7).

Despite radiative and convective heat losses a sunlit leaf is commonly 5–10°C warmer than the surrounding air, and with extreme heat loads some leaves can be as much as 20°C above the ambient temperature. Under these conditions the leaf is in danger of becoming dehydrated so the guard cells diminish the stomatal aperture and reduce the rate of water loss by transpiration. This places an even greater emphasis on the need for passive heat loss by long-wave radiation and sensible heat convection. Continued stress will lead to plant damage. The undersides of sunlit leaves are usually 1–3°C below the upper surface temperature. At night the leaf being an efficient radiator can cool 5–10°C below the surrounding air temperature. The temperature gradient is then directed from the air to the leaf and convection causes a net heat gain.

3 Plant covers and crops

(a) MASS BALANCES

(i) *Water*

The gross water balance of a soil-plant-air volume is illustrated in Figure 4.1b, but this does not show the fact that *inside* the volume there are significant air, soil and plant flows leading to the re-distribution of water, and that there are a number of localized sites where temporary water storage occurs. Figure 4.7 is an attempt to schematize these features in the case of a plant stand where horizontal exchanges (in the air, along the ground, and within the soil) can be neglected. These conditions are met in

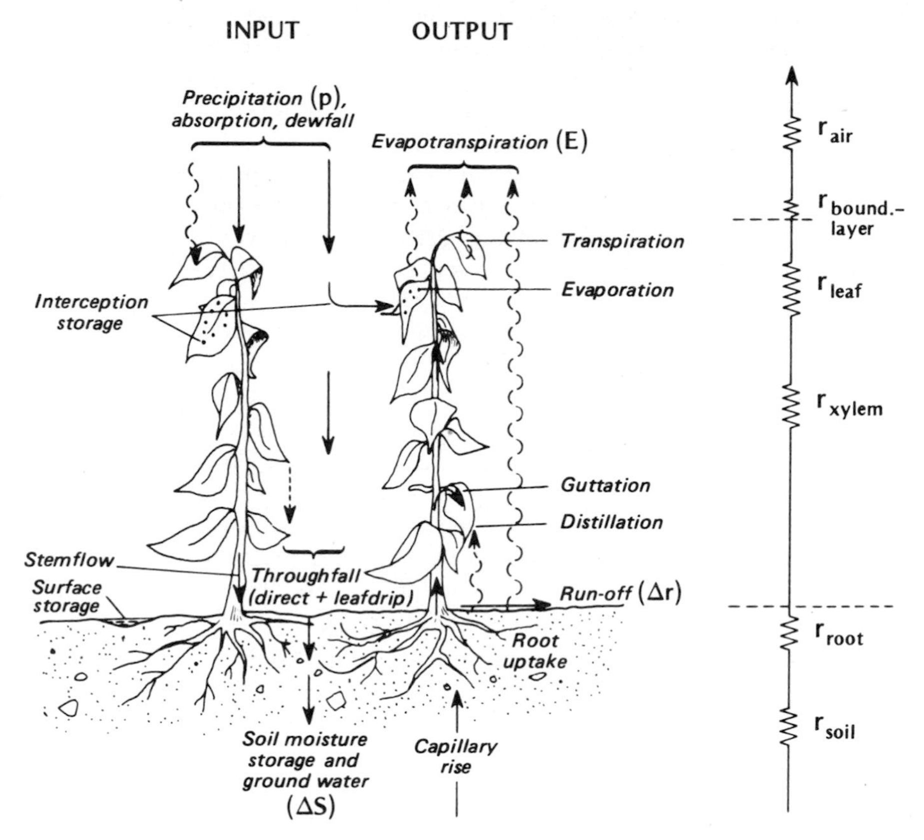

Figure 4.7 The hydrologic cascade in a soil-plant-atmosphere system. At the right is an electrical analogue of the flow of water from the soil moisture store to the atmospheric sink via the plant system.

the middle of an extensive plant community or crop growing on level terrain.

In the absence of irrigation the primary water inputs to the system are rain, snow, fog, dew and frost from the atmosphere, and to a lesser extent soil water rising from below. Rain and snow entering the system is either intercepted by the foliage or falls directly through openings to the ground. The intercepted water contributes to a form of storage, which is further fed by the impaction of fog droplets, condensation (termed dewfall if the water comes from the air above, and distillation if it comes from the soil), and any plant water exuded onto the leaf surface through the leaf cuticle (a process known as *guttation*). The efficiency of precipitation interception depends both upon the nature and amount of the precipitation, and the vegetation characteristics such as the stand architecture, density and the area of foliage. For an initially dry canopy the interception efficiency is high in the early stages of rain or snowfall, or if the amount of input is small. Thus a high proportion of the water is retained by the canopy, but eventually a threshold storage capacity is surpassed and thereafter the efficiency declines. The excess water finds its way to the ground either as a result of leaf drip or by running down the stems.

The water incident upon the ground's surface (direct precipitation + leaf drip + stemflow) either infiltrates into the soil or remains as surface storage in the form of puddles. That which percolates downward enters the soil moisture store along with any deep soil water seeping upward by capillary action. Capillary rise can be substantial if the water table is not too deep and usually occurs as a response to moisture depletion in the upper layers brought about by the vegetative or soil evaporative demand. The root uptake to satisfy this demand can be seen as both the re-distribution of moisture to the plant water store (supplemented by small amounts absorbed through leaves), and as the start of the process leading to transpiration.

Since we have chosen to neglect horizontal outputs, such as net run-off (Δr) and net moisture advection (ΔA), the primary water losses from the system in Figure 4.7 are via deep drainage to the water table, and evapotranspiration to the atmosphere. The evaporative flux is sustained by depleting all of the four water stores (i.e., interception storage, ground surface storage, soil moisture storage and plant water storage). During the growing season the main loss is via the plant system through plant water movement and transpiration. The rate of water flow through the system is a function of the vapour concentration difference between the air and the leaf, and the water potential difference between the leaf and the soil, and is regulated by a series of soil, plant and air resistances (Figure 4.7). In the soil the flow is dependent upon the amount of water present since this determines the hydraulic conductivity of the soil (p. 44) whose inverse is the resistance offered by the soil to moisture extraction (r_{soil}). The plant uptake is governed by the extent of root development (r_{root}) and the ease of internal sap movement by the vascular system of the xylem (r_{xylem}). Diffusion within the leaf (r_{leaf} including the resistances of the mesophyll, the stomate and the cuticle) and to the air ($r_{boundary\ layer}$ and r_{air}) have already been discussed in relation to Figure 4.2b. Water is also lost from the system by evaporation from the soil and from the exterior surfaces of the vegetation. The soil losses may either originate from surface puddles, or from soil layers near the surface. The exterior vegetation losses occur from surfaces wetted by precipitation interception, condensation and guttation.

Except where the canopy is wet, the dominant mode of evaporative water loss from vegetated surfaces is via transpiration, and hence the key resistance to water loss is that of the stomata. Recognizing this, Monteith (1965b) proposed the use of a single *canopy* (*or surface*) *resistance* (r_c) to characterize the physiological control of water loss by a plant community.

In effect r_c considers all the stomata of all the leaves acting in parallel, so that the canopy behaves rather like a 'giant leaf'. In agreement with direct physiological measurements the value of r_c for most crops has been shown to decrease with increasing irradiance (i.e. as the stomata open upon response to light intensity), to increase with greater soil water stress, and to be independent of wind speed (Monteith, 1973). In the case of a wetted leaf $r_c = 0$ because the stomata play no regulatory role. Just as the surface resistance is a bulk physiologic descriptor for the crop, the *aerodynamic resistance* (r_a) of the atmosphere is the bulk meteorologic descriptor of the role of the atmospheric turbulence in the evaporation process. Whereas r_c is closely tied to the stomatal resistance, r_a is dependent upon the wind speed, surface roughness, and atmospheric stability all of which contribute to the level of turbulent activity. Typical values of r_c and r_a are given in Table 4.2.

(ii) *Carbon dioxide*

If we define a soil-plant-air volume similar to that in Figure 4.1 but extending to a soil depth which there is no net CO_2 exchange, and in a situation where there is no net CO_2 advection, then the only net exchange is the vertical CO_2 flux (C) through the top of the volume.

TABLE 4.2 Representative values of the aerodynamic (r_a) and canopy (r_c) resistances for different surfaces types; after Szeicz, 1974.

Surface	r_a† (s m^{-1})	r_c‡ (s m^{-1})	Total r (s m^{-1})
Open water	200	0	200
Short grass (pasture)	70	70	140
Crops	30	40	70
Forests	5	125	130

† Calculated for $\bar{u} = 3$ m s^{-1} at a height of 2 m
‡ Average stomatal aperture, no irrigation

Figure 4.8 shows the diurnal variation of C for a number of days during the growing season of a prairie grassland. During the night the flux is directed away from the vegetation into the atmosphere. The loss of CO_2 from the system is furnished by respiration from the plant tops, plant roots and the soil (soil micro-organisms). The total flux less that from the soil represents the net photosynthetic deficit (ΔP) of the vegetation (equation 4.2). Soon after sunrise the CO_2 flux reverses its direction as the stomata open, and the rate of assimilation exceeds the rate of respiration. The CO_2 supply for this net uptake comes from both the atmosphere and the soil. The flux is usually closely tied to the pattern of solar radiation input and to the stage of crop growth. In the early part of the growing season (May to

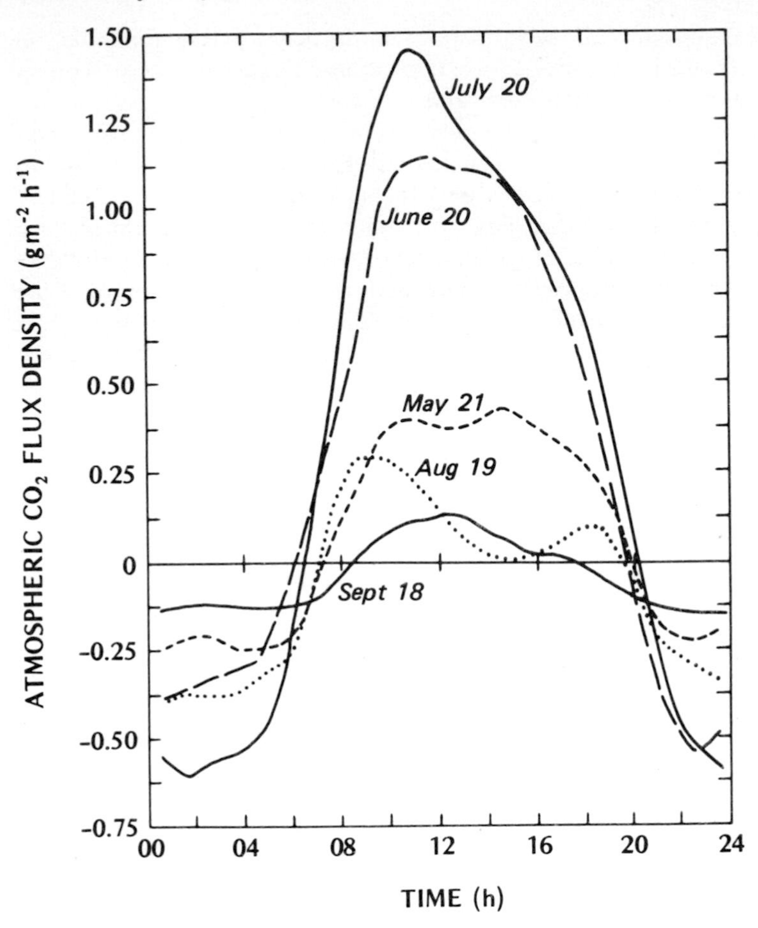

Figure 4.8 Diurnal variation of the vertical flux of carbon dioxide (C) over a prairie grassland at monthly intervals during the growing season. Data are 10-day averages (after Ripley and Saugier, 1974).

July) the magnitude of *C* increases as the vegetation grows and the leaf area increases. But later in the season, even though there is a large leaf area and solar input remains high, *C* drops considerably and the diurnal variation shows a midday minimum. The seasonal decrease of *C* is directly related to soil moisture depletion. Plant water stress and high leaf temperatures restrict photosynthesis and hence the gaseous exchange between the vegetation and the atmosphere. The midday minimum is due to the effects of temperature on the assimilation process, and to a lesser extent to stomatal closure (i.e. an increase in the canopy resistance, r_c). During the winter CO_2 exchanges are very much smaller due to reduced biological activity.

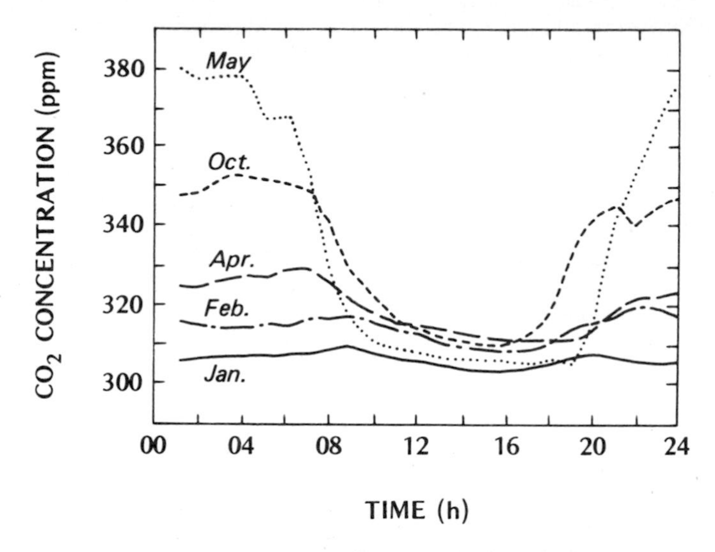

Figure 4.9 Diurnal variation of carbon dioxide concentration at a rural site in Ohio for different months (after Clarke, 1969a).

Table 4.3 shows the typical daily CO_2 balance for a crop of sugar beet. The net assimilation during the daytime is composed of the flux of CO_2 to the plants from both the atmosphere above and the soil beneath, minus the root respiration ($=30{\cdot}2$ g m^{-2}). At night the net balance for the plants is negative due to respiration from the plant tops and roots ($=-7{\cdot}1$ g m^{-2}). Therefore the net daily fixation of CO_2 by the crop is $23{\cdot}1$ g m^{-2}.

Typical diurnal and seasonal patterns of CO_2 concentration[1] in the air are given in Figure 4.9. It shows that since the highest CO_2 release (respiration) rates are found in the growing season, and also since mixing is least at night, the largest concentrations are found on summer nights. On summer days the rate of vegetative assimilation is sufficient to reduce concentrations to those typical of the winter, but the mean daily value is still larger.

(b) RADIATION BUDGET

The radiation budget of plant stands is complex because although the canopy 'surface' is the prime site of radiant energy exchange there is significant internal radiative absorption, reflection, transmission and emission. The transmission of short-wave radiation into a stand shows an almost logarithmic decay with depth of penetration (Figure 4.10a). The height variation is given by a form of Beer's Law (equation 3.1) but in this case the coefficient 'a' relates to the extinction by plant leaves, and z is

[1] Concentrations are given as parts of CO_2 per million parts of air (ppm vol.), to convert to units of mass (μg m^{-3}) see footnote p. 292.

TABLE 4.3 Daily carbon dioxide balance of a crop of sugar beet in Nebraska on a clear day. All values are $g\ m^{-2}/12\ h$ (after Brown and Rosenberg, 1971).

Period of flux *from* the atmosphere (Day, 12 h)		
CO_2 flux from the air (C)	27·8	
CO_2 flux from the soil	4·4	
Net flux to the above-soil portion of the plant (P)	32·2	
Estimated root respiration (R)	−2·0	
Net photosynthesis during the day (ΔP)		30·2
Period of flux *to* the atmosphere (Night, 12 h)		
CO_2 flux to the air (C)	−8·3	
CO_2 flux from the soil	3·1	
Net flux from the above-soil portion of the plant (R)	−5·1	
Estimated root respiration (R)	−2·0	
Net respiration flux during the night (ΔP)		−7·1
Net daily photosynthetic rate (ΔP)		23·1

replaced by the term $A_{1(z)}$ representing the leaf area cumulated from the top of the canopy down to the level z:

$$K\downarrow_{(z)} = K\downarrow_0 e^{-aA_{1(z)}} \tag{4.9}$$

where $K\downarrow_{(z)}$ – average short-wave radiation at the level z; and $K\downarrow_0$ – short-wave received above the canopy. The leaf area modification enables the relationship to apply to vegetation communities with different stand architectures.

The vertical attenuation of solar radiation by foliage affects not only the radiation intensity but also its spectral (wavelength) composition. From Figure 4.4 and Table 4.1 we know that leaves absorb visible (PAR) radiation more strongly than the longer wavelengths of short-wave radiation (i.e. in the near infra-red). Indeed light is reduced to only 5–10% of its above-crop value at the ground beneath a mature plant stand. This selective absorption reduces the photosynthetic value of the radiation as it penetrates.

The albedo of a vegetation stand is lower than the value for its individual leaves because reflection depends not only on the radiative properties of the component surfaces, but also upon the stand architecture and the angle of solar incidence. The latter two factors control the amount of penetration, radiation trapping, and mutual shading within the stand volume. Thus although most leaves have an albedo of about 0·30 (Table 4.1) the albedo of crops and other vegetation communities is less, and to some extent a function of their height (Figure 4.11). For most agricultural crops and natural vegetation less than 1 m in height the surface albedo (α) lies in the

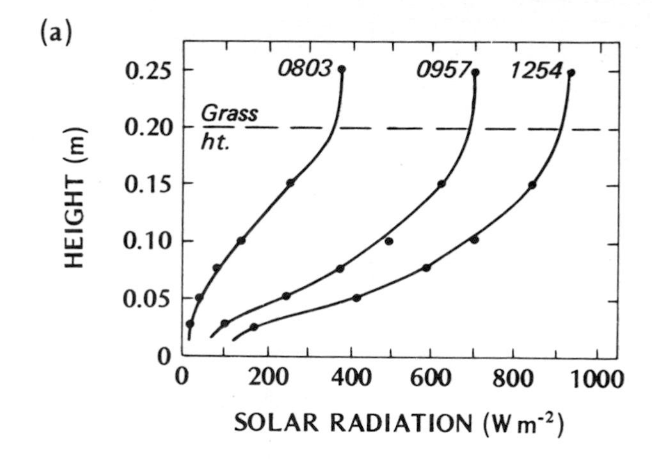

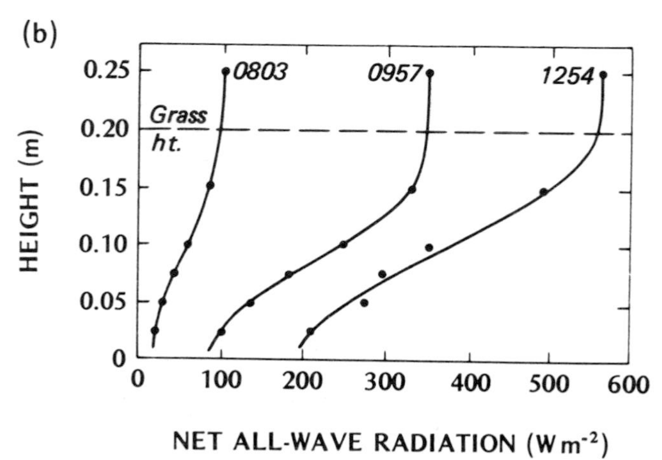

Figure 4.10 Measured profiles of (a) incoming solar ($K\downarrow$), and (b) net all-wave radiation (Q^*) in a 0·2 m stand of native grass at Matador, Sask., on 28 June 1972 (after Ripley and Redmann, 1976).

remarkably narrow range from 0·18 to 0·25 (Table 1.1). There are however two limitations to this simple picture. First, the values only apply to green vegetation with a full surface cover. If the plants are wilted or dead, or the underlying soil is exposed, the generalization does not apply. Second, the values refer to the albedo in the midday period. During the early morning and evening albedos are higher than the range noted above.

Observations over most vegetation surfaces reveal a U-shaped pattern of albedo through the day. Four physical explanations have been forwarded. First, since the spectral composition of short-wave radiation is related to

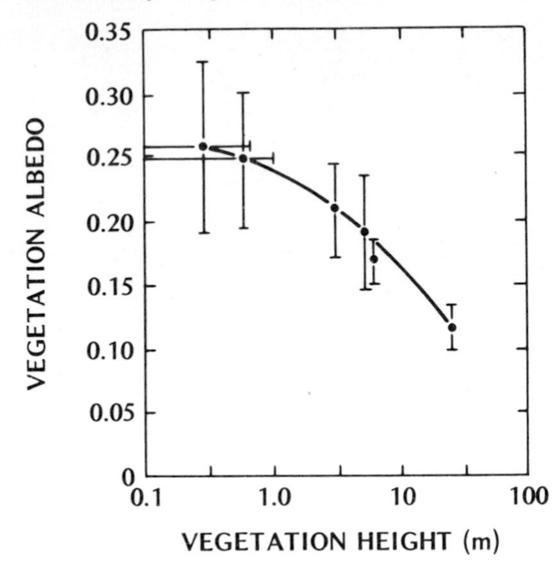

Figure 4.11 Relation between the albedo of vegetation and its height. Vertical lines are two standard deviations, and horizontal lines show seasonal range of canopy height (after Stanhill, 1970).

the length of the atmospheric path of attenuation (and therefore to solar altitude) the daily variation may be a result of the different albedos of leaves at different wavelengths (e.g. Figure 4.4). Second, at higher solar altitudes penetration into the canopy increases thereby causing more solar radiation to be trapped within the stand and less to be reflected. These explanations gain support from the fact that when albedo is plotted against solar altitude a characteristic curve emerges (Figure 4.12). Third, it is possible that leaf wetness in the morning and evening (due to dewfall or guttation) could cause an increase in reflection. Fourth, the variation may be due to changes in crop physiology. For example, under severe heat stress leaves may temporarily wilt thereby losing their horizontal orientation and allowing easier radiation penetration. Some species even have an ability to orient their leaves to maximize or minimize solar heat loading. Since the highest albedos occur at the times of low energy input the effect of this diurnal variation on the total radiation budget is small. The dependence of α upon the solar altitude also explains why tropical albedos are usually less than those for similar surfaces at higher latitudes, and accounts for the observation that the diurnal variation of α is much less with cloudy skies. Its role in explaining seasonal variations of α is obscured by changes in plant phenology, snow, etc.

The long-wave radiation budget (L^*) of vegetation is almost always negative, as with most other surfaces. However within the vegetation it is usual to find that the net loss diminishes toward the ground because of the reduction of the *sky view factor* (SVF). In the case of open, level terrain (or the top of the canopy), long-wave radiation is emitted to the complete sky

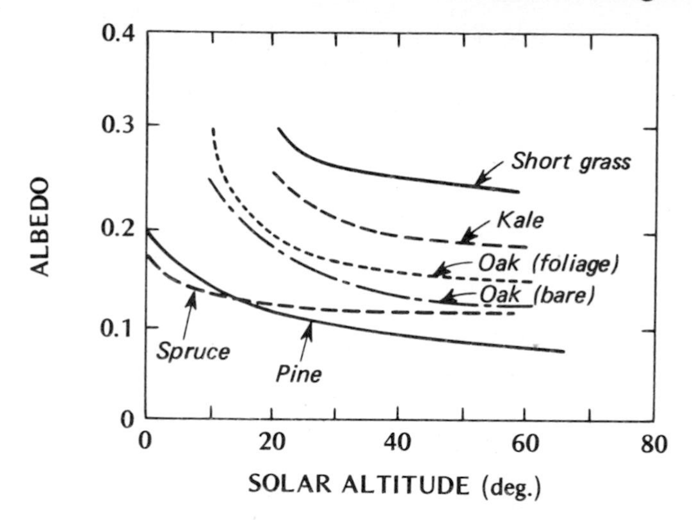

Figure 4.12 Relation between the albedo of vegetation and solar altitude on sunny days. Grass and kale (Monteith and Szeicz, 1961); oak forest (Rauner, 1976); spruce forest (Jarvis *et al.*, 1976); and Scots pine forest (Stewart, 1971).

hemisphere above. Such a site has a perfect SVF of unity. On the other hand for sites where some portion of the hemisphere is obscured by other objects (e.g. overlying leaves) the SVF is proportionately less (Figure 7.2c). At these positions the 'cold' sky radiative sink is increasingly being replaced by relatively warm vegetation surfaces. Deep within the stand the role of the sky is almost eliminated, and the net long-wave exchange simply depends upon the vertical variation of plant surface temperatures. Since these are normally not large the value of L^* is reasonably constant with height. (The SVF concept also explains the greater cooling found on clear compared with cloudy nights (p. 25); and the reduced nocturnal radiative cooling found in furrowed fields (p. 201), near shelterbelts (p. 215), and in urban canyons (p. 248), compared to surfaces where the view of the sky is unobstructed.)

An example of the net all-wave radiation budget (Q^*) above vegetation, and its component fluxes, is given in Figure 1.10. Monteith (1959) suggests that for a given solar input ($K\downarrow$) the net radiation differences between crops should be relatively small provided they have a supply of moisture. He bases this upon the fact that the range of the surface radiative properties (α and ε) is small (Table 1.1) and that evaporative cooling should keep surface temperatures reasonably similar. Within a vegetation canopy the vertical variation of Q^* is much like that of $K\downarrow$ (Figure 4.10b), and in fact Beer's Law is an appropriate means of describing the profile. This is to be expected since as we have already noted L^* does not show large variations with height. The vertical distribution of Q^* is important because it determines the principal sites of heating and transpiration and hence the temperature and humidity structure within the canopy. By day Q^* is positive throughout the depth of the stand, and is especially large near the top where absorption is greatest. In the evening Q^* becomes negative, and the loss is also concentrated near the top of the stand. Below this the budget is almost zero due to internal equilibrating long-wave radiation exchanges.

(c) ENERGY BALANCE

The vertical fluxes of energy between a soil-plant-air system (such as that in Figure 4.1a), and the atmosphere above, are fairly well understood. But the nature of the fluxes *within* such a volume are less clear. Correspondingly here we will concentrate upon the energy balance of the total system (i.e. the vertical fluxes through the top of the canopy, and the gross storage changes inside the volume). Horizontal fluxes will be ignored and are considered separately in Chapter 5.

As an aid to illustrating the linkages between the energy balance and microclimate we will study the results of a single experiment which includes measurements of most of the radiation, energy and microclimatic variables for a period of one day. The data are taken from the work of Long *et al.* (1964) over a barley crop at Rothamsted in England. The day chosen is 23 July 1963 which began with clear skies (except for the 03 to 04 h period), very light winds, and dew on the crop. The daytime was mostly sunny (some cloud between 11 and 14 h) with wind speeds less than 2·5 m s^{-1}. The period finished with cloud and light rain (0·25 mm) from 21 to 24 h.

The radiation budget (Figure 4.13a and Table 4.4) demonstrates features with which we are already familiar (e.g. Figure 1.10 and Table 1.2). The surface albedo exhibited the characteristic U-shape, varying from 0·30 near sunrise and sunset, to 0·21 in the middle of the day. The average albedo based on daily radiation totals was 0·24, indicating that the crop absorbed

TABLE 4.4 Radiation and energy balance components for a barley field at Rothamsted, England (52°N) on 23 July 1963. Data are daily totals (MJ m^{-2} day^{-1}) (based on data in Long *et al.*, 1964).

Radiation budget		Energy balance		Derived terms	
$K\downarrow$	20·6	Q_H	−1·1	α	0·24
$K\uparrow$	4·9	Q_E	12·2	E (mm)	4·88
K^*	15·7	ΔQ_S†	1·2	Q_E/Q^*	0·99
L^*	−3·4				
Q^*	12·3				

†ΔQ_S is approximated by the soil heat flux (Q_G) because physical and biochemical heat storage in the canopy-air volume were not measured.

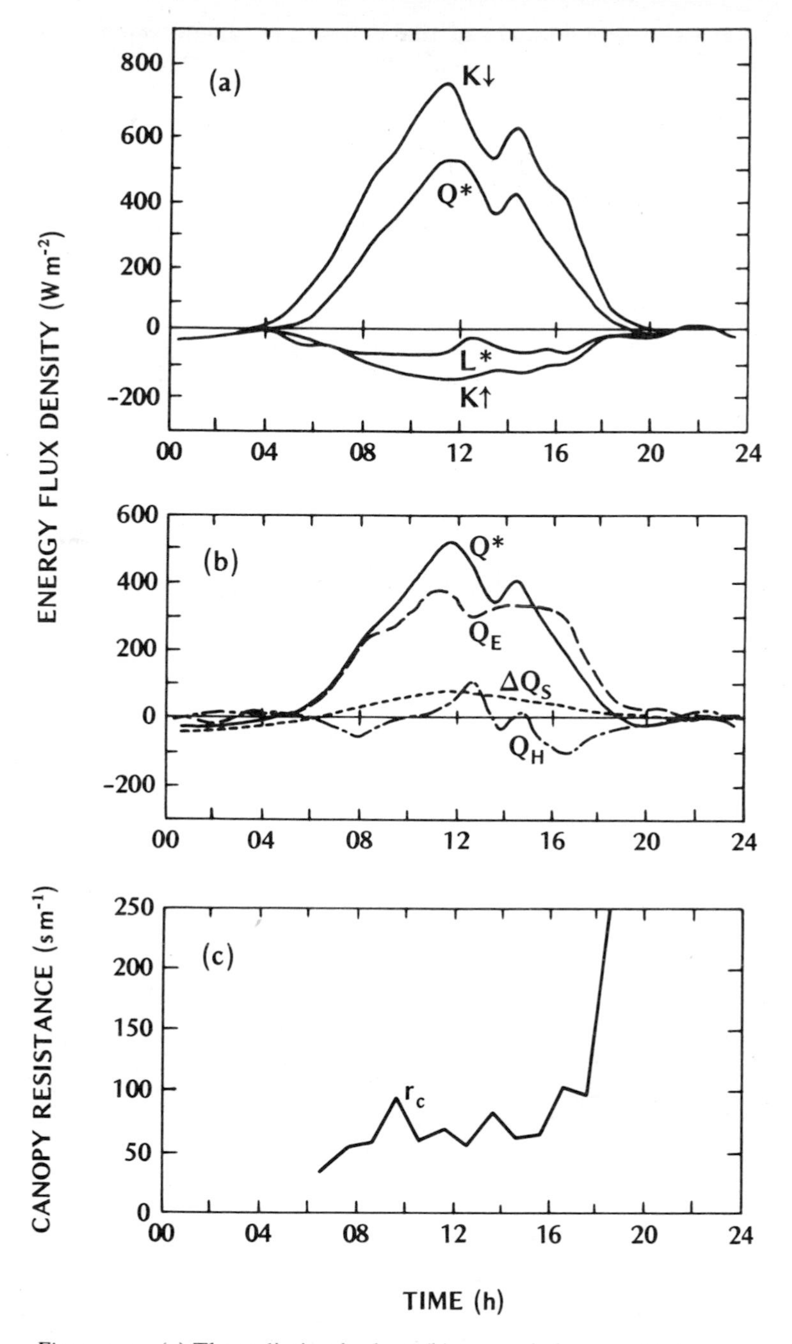

Figure 4.13 (a) The radiation budget, (b) energy balance and (c) canopy resistance of a barley field at Rothamsted, England on 23 July 1963 (modified after Long *et al.*, 1964, and Monteith *et al.*, 1965).

approximately three-quarters of the short-wave radiant input. The net long-wave budget was always negative but never large. This may have been due to the moderating effect of evaporative cooling on leaf surface temperatures and/or a relatively high atmospheric emissivity due to high vapour content. The long-wave loss clearly diminished at the times of increased cloud cover. The net all-wave radiation was in-phase with the incoming solar radiation during the day, and of course equal to the long-wave loss at night.

The energy partitioning (Figure 4.13b and Table 4.4) shows that evaporation was the dominant means of dissipating the daytime radiative surplus. On some occasions Q_E even exceeded Q^*. This was made possible by the flow of heat *from* the atmosphere to the crop in the morning and afternoon. With so much energy being used to evaporate water, little was left to heat the crop above the air temperature and thereby create the necessary temperature gradient for upward sensible heat flux. On a daily basis the crop was in receipt of sensible heat. Heat storage in the soil was relatively small because of the shade provided by the canopy. On a daily basis a small amount of heat entered storage (ΔQ_S) as is normal for the summer.

The dominant role of Q_E in the balance (note on a daily basis $Q_E \simeq Q^*$) was made possible by an abundant supply of soil moisture. It allowed transpiration to proceed without the need for undue physiological control by the vegetation. This is corroborated by the diurnal pattern of the canopy resistance (r_c) shown in Figure 4.13c. In the early morning r_c was very small because the crop was covered with dew. Evaporation of this water is not controlled by the stomata and the canopy resistance approached that for open water (Table 4.2). After the dew dried r_c increased slightly and remained reasonably constant for most of the day. During this period the plant water supply was keeping pace with the transpiration demand and the stomata remained open. The increase of r_c in the late afternoon was probably due to decreasing light intensity and possibly to increasing water stress, both of which act to close the stomata. The pattern in Figure 4.13c is typical of that for crops with adequate water availability except that if dew is absent the early morning values are larger due to the lack of light. The total pattern then takes on a U-shape.

We will interrupt consideration of the barley crop here to point out that significantly different conditions prevail when soil moisture becomes a limiting factor, and that vegetation can play a more active role in regulating water loss to the atmosphere. Figure 4.14 gives the results of a study over alfalfa near Phoenix, Arizona. The crop was flooded by irrigation on 28 May. Following this the crop was allowed to deplete its moisture store throughout the month of June which was characterized by a constantly high radiant input (Figure 5.5), a dry windy atmosphere, and no rain. For the first 20 days following irrigation evaporation proceeded at close to the

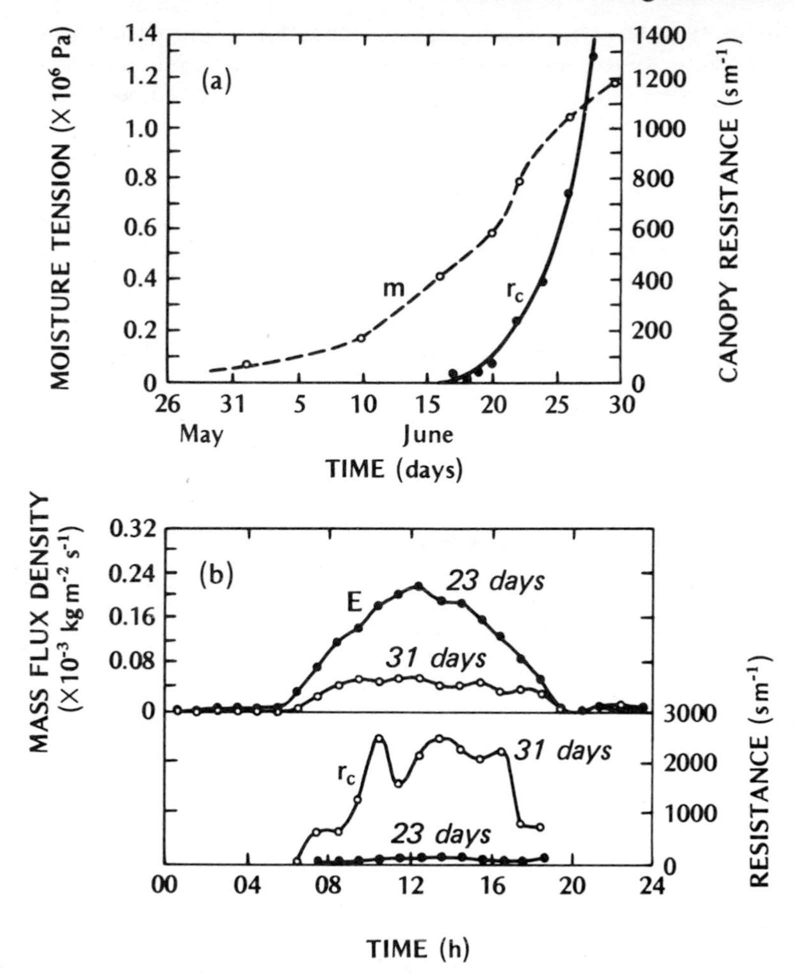

Figure 4.14 Water relations of a field of alfalfa at Phoenix, Arizona (33°N). The field was irrigated on 28 May and this was followed by a drought throughout June. (a) Daily average soil moisture tension and canopy resistance, and (b) diurnal variation of the evaporation rate and canopy resistance at periods of 23 and 31 days after irrigation (modified after van Bavel, 1967).

maximum possible rate. Soil moisture tension increased with time but did not reach a point where the evaporation demand could not be freely supplied (Figure 4.14a). By the 23rd day after irrigation (20 June) the role of the canopy resistance began to become evident, and thereafter assumed increasing importance. Eight days later (28 June) soil moisture was held at a tension of approximately $1 \cdot 1 \times 10^6$ Pa (near the limit for availability to plants, see Figure 2.7) and r_c became very large during the middle of the day. Obviously water stress and high leaf temperatures induced stomatal closure so that the system resisted water loss. The net effect was to eliminate the normal curve of evaporation and to replace it with a smaller more constant rate.

(d) CLIMATE

Returning to the barley crop, Figure 4.15 gives the hourly mean profiles of wind speed, temperature, vapour pressure and carbon dioxide during

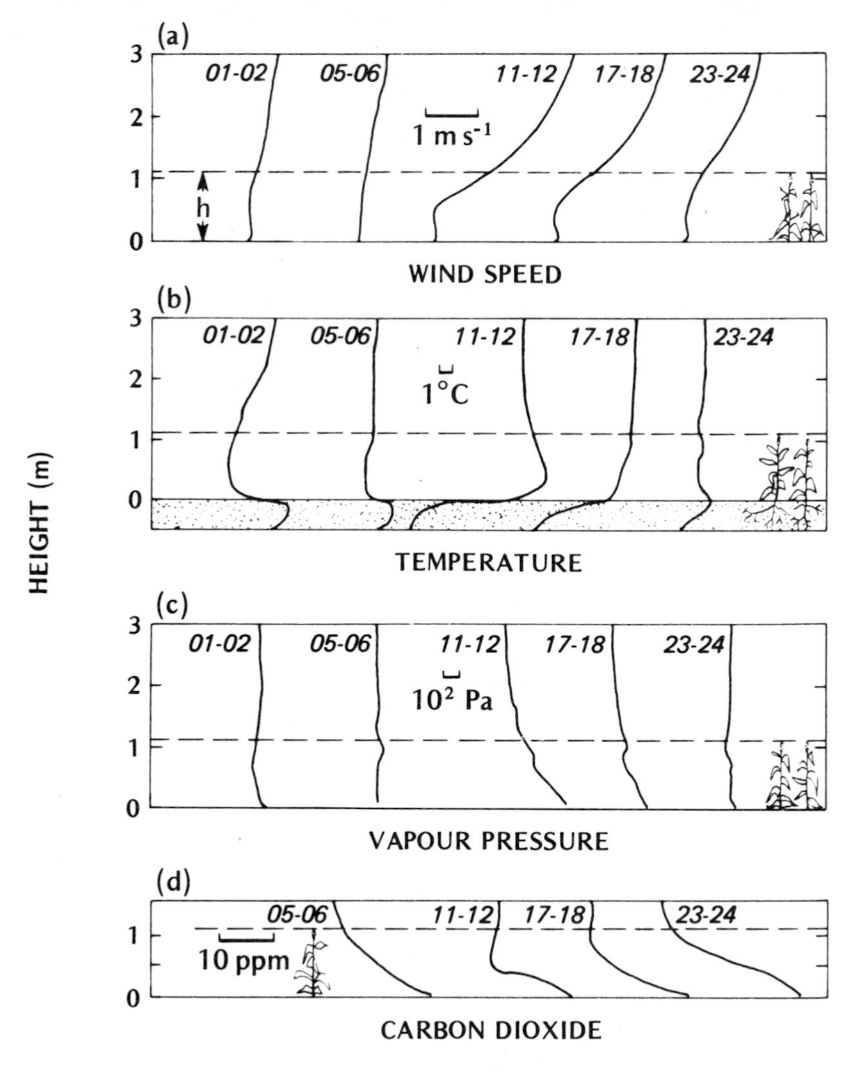

Figure 4.15 Profiles of (a) wind speed, (b) temperature, (c) vapour pressure and (d) carbon dioxide concentration in and above a barley field at Rothamsted, England on 23 July 1963 (modified after Long *et al.*, 1964).

the same day as the data in Figure 4.13. Note that in each case the form of the profile above the crop corresponds to those for simple non-vegetated surfaces as outlined in Chapter 2. The profiles appear to respond to an active surface located just below the top of the canopy.

The wind speed profiles (Figure 4.15a) above the crop exhibit the logarithmic shape already discussed in relation to Figure 4.3 and equation 4.5. During the middle of the day when wind speeds were highest the wind speed gradient (slope of the profile) increased and the turbulent transfer of momentum would have been correspondingly greater leading to deeper penetration into the canopy. Within the crop the profile depends upon the internal stand architecture, but it is common to find a wind speed minimum in the mid to upper canopy where foliage density is greatest; then a zone of slightly higher speeds in the more open stem layer; finally decreasing again to zero at the ground.

Accurate description of the wind profile above a vegetation surface depends upon the specification of the parameters z_0 and d (equation 4.5). Typical values are given in Table 2.2 but it will be noted that their range is large. This is for two reasons: first, both parameters depend upon the vegetation height, and second, both depend upon the wind speed. The relation of d to h is given by equation 4.4, and that for z_0 can be approximated by:

$$\log z_0 = \log h - 0{\cdot}98 \tag{4.10}$$

for tall, dense stands (Szeicz *et al.*, 1969). The relation of z_0 and d to wind speed is however less clear. Many crops show a decrease of both parameters with increasing wind speed. This is commonly attributed to the flexibility of the plants so that at high speeds the surface elements become more streamlined (reduce z_0) and their mean height lowered (reduce d) by bending. Monteith (1973) however points out a number of crops where this does not apply, and that there may be a number of wind speed 'regimes' within which even the same crop may behave differently to increasing wind speed.

The air temperature profiles (Figure 4.15b) show that the main heat exchange was centred just below the top of the canopy. At night long-wave radiation emission from the crop gave rise to a temperature minimum just below the crown so that temperatures increased upwards into the atmosphere, and downwards within the vegetation. Therefore in accordance with the flux-gradient relationship (Chapter 2) sensible heat will converge upon this active layer from both directions. However the upward flux is likely to be weak because the temperature gradient is slight, and turbulent transport is dampened within the stand. Notice that the strength of the above-canopy inversion was reduced during cloudy periods because of the curtailed net long-wave radiation loss (e.g. compare the 01 to 02 h profile with that for 23 to 24 h). By day the principal site of net radiation absorption was near the canopy crown and hence this was the level of maximum heating. Temperatures usually decrease both upwards and downwards from this level so that sensible heat is carried up into the air and down into the crop. The lapse profile above the canopy was not very pronounced because of the evaporative cooling at the surface of the foliage. Indeed, as the sensible heat flux results (Figure 4.13b) show, evaporation was often sufficient to induce an inversion above the crop and a downward sensible heat flux from the air. The net effect of the diurnal heating and cooling is to raise the level of the maximum diurnal temperature range (Figure 2.2) from the soil surface to near the top of the vegetation.

The soil temperature regime beneath crops is similar to that for non-vegetated surfaces, but the amplitude of the temperature wave is dampened due to the radiation shading afforded by the canopy. At night a weak cooling wave travels downwards. The daytime heating wave is somewhat stronger and dependent upon the amount of radiation penetration as well as the nature of the soil.

The shape of the nocturnal vapour pressure profiles (Figure 4.15c) are more complicated. During periods with light winds and clear skies canopy cooling induced dewfall and the vapour profile was inverted (see 01 to 02 h profile in Figure 4.15c, and the Q_E flux in Figure 4.13b). At other times weak evaporation continued and the vapour profile lapsed with height (see 05 to 06 h profile, and the corresponding flux). Leaf wetness due to dewfall and guttation is important in the early morning because it lowers the canopy resistance and the relative enhancement of evaporation delays crop heating. The daytime profiles are more straightforward. Since both the soil and the canopy were moisture sources (Figure 4.7) the vapour profile decreased with height all the way from the soil up through the vegetation and into the atmosphere. Reduced turbulent transfer in the crop near the soil permitted the vapour to accumulate. This cool and humid environment is the ideal habitat for many insects and small rodents.

In the case of carbon dioxide (Figure 4.15d) both the soil and the vegetation are sources at night. The soil releases CO_2 as a result of bacterial action on decaying organic material, and the vegetation releases are due to respiration. Therefore the CO_2 concentration decreased with height all the way from the soil to the atmosphere. By day the soil continues to release CO_2 but since photosynthesis far outweighs plant respiration, the canopy is a net CO_2 sink. Concentrations were therefore a minimum in the middle to upper canopy layer, and increased in both directions away from this position.

4 Orchards and forests

The special features of vegetation systems outlined at the beginning of this chapter apply to orchards and forests as well as to plant and crop covers. In fact in many respects it is acceptable to view tree stands as being merely the larger-scale counterparts of plant stands. They do however have some special features that have climatic importance. First, the stand architecture is often more clearly demarcated. The forest canopy is commonly quite dense, whereas the trunk zone is devoid of foliage (except where there are multiple tree layers or dense underbrush). Second, the increased biomass and the sheer size of the stand volume contribute to the possibility that heat and mass storage may not be negligible over short periods. Third, the height and shape of the trees leads to greater radiation trapping and a very much rougher surface. Fourth, especially in the case of coniferous forests the stomatal control may be different.

(a) MASS BALANCES

The general features of the hydrologic cascade in a vegetation system (Figure 4.7) require little or no modification in the case of tree stands. The inputs to the system consists of the same terms but it is worth noting that forests are able to retain a larger proportion of their precipitation as interception storage. The canopy of a deciduous forest can intercept 10–25% of the total annual precipitation, and the range for coniferous forests

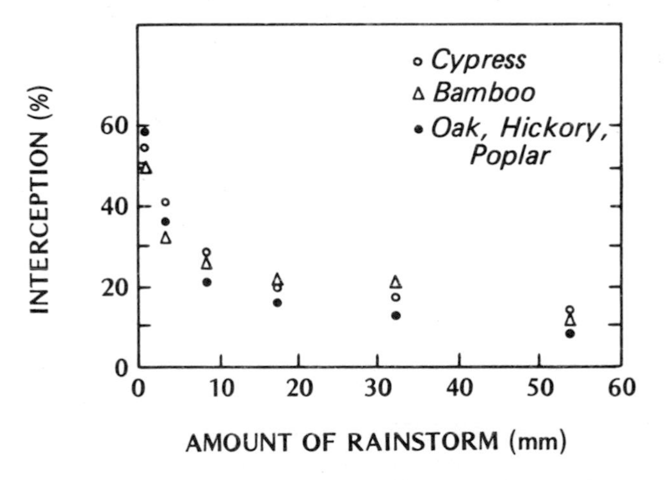

Figure 4.16 The relation between the rainfall interception efficiency of tropical and temperate forests and the amount of rain precipitated by a storm (after Pereira, 1973).

is 15–40%. In general snowfall interception efficiency is similar to that for rainfall. For an individual storm the efficiency can be almost 100% in the early stages but when the maximum storage capacity is exceeded the efficiency drops. Figure 4.16 illustrates this fact for a range of tropical and temperate forests. It also shows that interception depends more upon the nature of the rainstorm than upon the tree species. The maximum storage capacity for rain is from 0·5 to 2 mm for all forest types, and from 2 to 6 mm for snow (Rutter, 1975).

As with plant stands the main water output from most forests is by evapotranspiration either through the evaporation of water from the soil or wetted foliage, or by transpiration via the stomata. The relative roles of these losses for a coniferous forest in summer are given in Table 4.5. It can be seen that transpiration is usually the main pathway to the atmosphere but that in wet years this is rivalled by evaporation of intercepted water. The importance of wetted-leaf losses can also be observed on a short term basis, and the high evaporation rates associated with this are explained in the energy balance section (p. 132).

TABLE 4.5 Relative losses of water from a coniferous forest by evaporation in wet and dry years. Data for May to September (after Federov, 1965).

	Precipitation (mm)	Evaporation (mm)			
		Total	Transpiration	Evap. beneath canopy	Evap. of canopy storage
Wet summers†					
Average	548	415	153	112	150
Percent		100	37	27	36
Dry summers†					
Average	259	357	151	119	87
Percent		100	43	33	24

† Wet years – 1953, 1957, 1962; Dry years – 1951, 1959, 1960.

The principal features of the carbon dioxide balance of forests are essentially the same as for other vegetation. The daytime CO_2 flux from the atmosphere (C) is related to the assimilative capacity of the canopy and this is governed by stomatal activity. Jarvis *et al.* (1976) working in a spruce forest near Aberdeen, Scotland, showed that C increases with short-wave irradiance ($K\downarrow$) and decreases as the vapour pressure deficit (vpd) becomes larger (i.e., at high light levels and with low evaporative demand the stomata are encouraged to open fully and to capture CO_2 for photosynthesis). They state that for typical conditions ($K\downarrow = 300$ W m^{-2} and vpd $= 300$ Pa) $C = 2{\cdot}9$ g m^{-2} h^{-1}. If the deficit increases to about 1400 Pa the flux of CO_2 is eliminated, probably because of stomatal closure. The reverse flow at night due to respiration from the soil, wood and the leaves is typically about 0·5 g m^{-2} h^{-1}. In terms of biochemical heat storage these rates correspond to flux densities of 9 and 2 W m^{-2} respectively, and are therefore climatically negligible.

(b) RADIATION BUDGET

The principal radiative exchanges in forests and orchards are associated with the canopy layer, and the trunk zone is less important. Schematically the canopy can be represented as an isolated layer possessing an upper and lower boundary as shown in Figure 4.17. This treats the canopy rather like a 'giant leaf' (Figure 4.5a), except that it allows some transmission of long-wave radiation through gaps in the canopy. It illustrates the possibility of radiative trapping within the system, and the need to consider the canopy layer as an elevated long-wave radiative source which is capable of emitting energy both upwards to the sky and downwards to the forest floor. We will use Figure 4.17 as the basis for discussing radiative exchange above and within tree stands. The results in Table 4.6 give an idea of the relative magnitudes of many of the fluxes numbered in Figure 4.17. These numbers will also be used to identify fluxes in the following discussion.

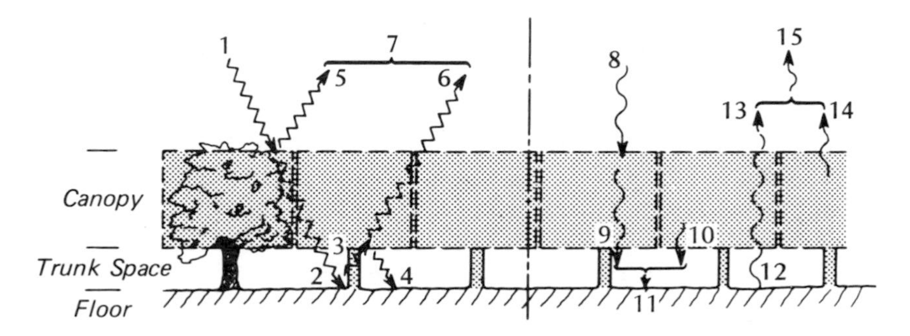

Figure 4.17 Schematic model of radiation exchanges above and within a forest (after Gay and Knoerr, 1970). (Flux numbers are referred to in the text and Table 4.6. Dashed lines indicate partial transmission through the canopy.)

The short-wave radiation incident upon a tree community (flux 1) is either reflected (flux 5), absorbed by the canopy, or transmitted through to the floor (flux 2). In agreement with the situation in plant stands the form of the attenuation with height is approximately given by Beer's law in the form of equation 4.9. The amount penetrating to the floor depends upon many factors including the height, density and species of the stand and the angle of solar incidence, but in general less than 20% of flux 1 reaches the floor of a mature stand, and it may be as little as 5%. For example, Figure 4.18 shows the relation between $K\downarrow$ above a canopy and for a point on the floor. The record for the floor is an indication of the highly variable radiation environment found there due to shading and sunflecking. During

TABLE 4.6 Radiant energy cascade within and above a forest system. Data are daily totals (MJ m^{-2} day^{-1}) for a pine forest in Connecticut on 30 September 1964; data after Reifsnyder, 1967.

Radiation component	Forest system† (canopy + floor)	Relation to numbered fluxes in Figure 4.17	Floor	Relation to numbered fluxes in Figure 4.17
$K\downarrow$	18·2	1	1·0	2 + 4
$K\uparrow$	1·8	5 + 6 = 7	0·2	3
α^+	0·10	7/1	0·17	3/(2 + 4)
K^*	16·4	1 − 7	0·9	(2 + 4) − 3
$L\downarrow$	24·6	8	29·5	9 + 10 = 11
$L\uparrow$	35·3	13 + 14 = 15	30·1	12
L^*	−10·7	8 − 15	−0·6	11 − 12
Q^*	5·7	(1 − 7) + (8 − 15)	0·3	(2 + 4) − 3 + (11 − 12)

† Measured or calculated above the canopy
+ Non-dimensional

the 8 h period of measurement the total $K\downarrow$ above the canopy was 16·1 MJ m^{-2}, and on the floor was 2·8 MJ m^{-2}, giving an average transmission of 18%. In the case of the orchard in Figure 4.19a the transmission is 21%, but for the forest in Table 4.6 it is only 6%.

The canopy not only reduces the magnitude of short-wave radiation reaching the floor but also affects both the proportion arriving as diffuse-beam (D), and its spectral composition. Close inspection of the $K\downarrow$ trace for the floor in Figure 4.18 reveals a 'background' level of radiation, most evident in the early morning and late afternoon, but also seen as the base of the midday 'spikes'. This 'background' is D. At low solar altitudes it is the only short-wave radiation penetrating to the floor because the direct-beam has a long path length through the canopy and a very low probability of finding an unobstructed route. At midday D is discernible as the lowest flux level found in shade. Passing a smooth curve through this 'background' level it emerges that D represents 46% of $K\downarrow$ on the floor, whereas above the canopy it is only 15% of the incident short-wave. Thus the canopy acts to diffuse the solar beam.

In general tree canopies lead to a depletion of the blue region of the spectrum (0·40 to 0·45 μm) and an enrichment of the red and near infra-red (0·65 to 0·75 μm). The light on the floor is therefore photosynthetically less active and taken in combination with the reduced magnitude of radiation, this tends to limit the variety and productivity of plants growing on the floor. Light penetration also varies with the age of a stand. Initially as the trees grow there is a progressive decrease in the light penetration until a

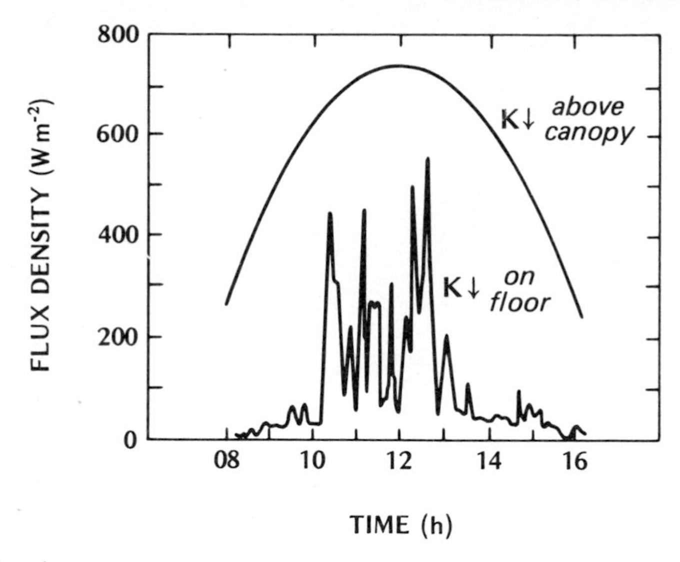

Figure 4.18 Short-wave radiation measured above the canopy, and at one point on the floor of a 23 m stand of Loblolly pine near Durham, N/C (36°N) on 30 October 1965 (after Gay *et al.*, 1971).

minimum value is achieved, thereafter increasing competition for light allows only the strongest trees to prosper and a natural thinning and increased penetration results.

The total short-wave radiation reflected from orchards and trees (flux 7) includes that from both the canopy (flux 5), and from the floor (flux 6). The latter flux being that emerging from the canopy after the initial floor reflection (flux 3) has been depleted by multiple reflection between the floor and the underside of the canopy (flux 4), and by absorption whilst passing through the canopy layer. Forest albedos are low by comparison with most other natural vegetation (Table 1.1 and Figure 4.12). They are lowest for conifers (especially spruce), higher for bare deciduous and greatest for deciduous trees in full leaf. Since the spectral reflectivity of coniferous needles (Jarvis *et al.*, 1976) and deciduous leaves is similar to that of plants (Figure 4.4), the low albedos are probably due to greater trapping. This is illustrated by the results from an orange orchard given in Figure 4.19. For a single layer of orange leaves 32% is reflected, 19% transmitted and 49% absorbed, but for the orchard canopy only 15% is reflected, 21% transmitted and 64% absorbed. Hence the natural orientation of leaves and the depth of the canopy greatly enhance absorption due to trapping as a result of multiple internal reflection. On a much smaller scale, enhanced scattering may explain why the diurnal albedo variation of coniferous trees is reduced in comparison with deciduous trees and low plant covers (Figure

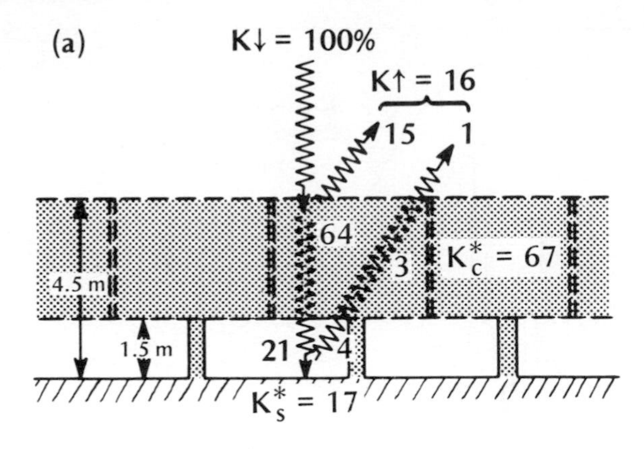

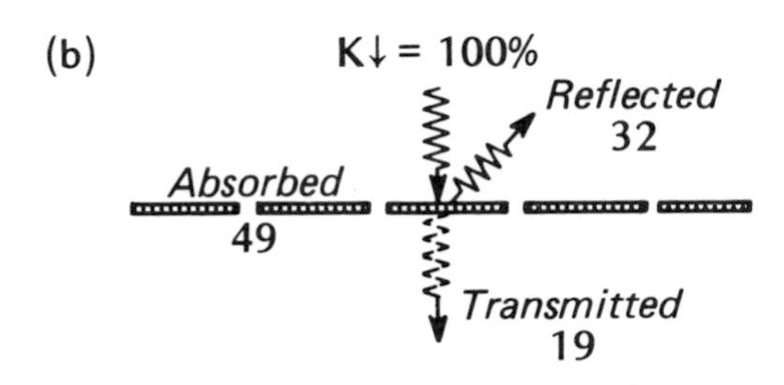

Figure 4.19 Short-wave radiation budget of (a) an orange orchard, and (b) a single-layer mosaic of fresh orange leaves. All values expressed as percentages of the incident radiation (after Kalma, 1970).

4.12). Even at low solar altitudes the roughness of needle clusters may be sufficient to increase scattering and thereby to trap radiation more efficiently than the 'smoother' surfaces of other vegetation covers.

The high absorptivity of evergreen trees makes them extremely important in the radiation budget of a high latitude landscape, especially in the spring in areas where conifers are scattered through otherwise snow-covered terrain with a high albedo. This contrast is accented at low solar altitudes because then the trees present their maximum surface area for irradiance, and their receiving surfaces are almost normal to the solar beam. Their relative warmth makes them sources of long-wave radiation which is readily absorbed by the surrounding snow thus hastening the local melt and the exposure of surfaces with lower albedos.

There is little information from which to generalize the long-wave exchanges above and within tree communities, but Figure 4.17 and Table 4.6 serve to illustrate the anticipated characteristics. The budget above the canopy is composed of the normal atmospheric input ($L\downarrow$, flux 8) and the forest system output ($L\uparrow$, flux 15) which consists of that portion of the floor emission which penetrates the canopy (flux 13), plus the canopy emission (flux 14). As with most surfaces $L\uparrow$ usually exceeds $L\downarrow$ and the budget is negative both on an hourly and a daily basis. Beneath the canopy the situation is more complex. The budget of the floor for example must include an input (flux 11) due to that portion of sky radiation which penetrates the canopy (flux 9), plus the emission from the underside of the canopy (flux 10), and an output due to the temperature and emissivity of the floor. With full foliage the role of the sky is likely to be small and the budget becomes simplified to considerations of the difference in temperature and emissivity between the bottom of the canopy and the floor. Usually these differences are small, and therefore L^* on the floor is a small term (Table 4.6).

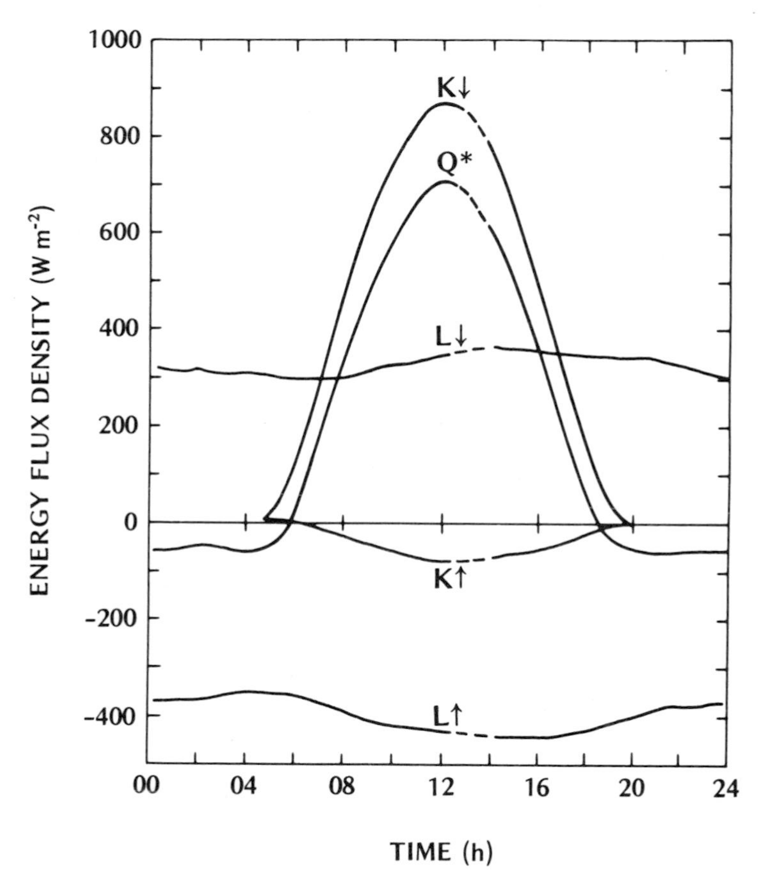

Figure 4.20 Component fluxes of the radiation budget of a 28 m stand of Douglas fir at Cedar River, Washington (47°N) on 10 August 1972 (after Gay and Stewart, 1974).

The diurnal variation of the fluxes comprising the radiation budget of a coniferous forest is given in Figure 4.20, and their daily totals are compared with those from other sites in Table 4.7. The course followed by the component fluxes, and by the resultant net all-wave radiation is similar to that for other vegetation except that the low albedo (0·09) gives a smaller reflection term and a slightly greater Q^* absorption by day. However some caution should be exercised in this interpretation, because Q^* depends on L^* which includes a term ($L\uparrow$) which varies with the canopy 'surface' temperature. Temperatures relate to the total energy balance and the thermal properties of the materials, and not to the radiation budget alone.

The profile of Q^* within a tree stand tends to approximately follow that within plant stands (e.g. Figure 4.10b). The percentage of the above canopy Q^* found on the floor varies with stand height, density, species and solar altitude, but as with $K\downarrow$ it is common to find a reduction of 80% or more. In the case of the pine forest in Table 4.6 the daily Q^* total for the floor was only 5% of that in the open, and the corresponding figure for the orange

TABLE 4.7 Radiation budgets and energy balances of three forests during the summer. Data are daily totals (MJ m^{-2} day^{-1}) for Scots and Corsican pine at Thetford, England on 7 July 1971†; Douglas fir at Cedar River, Washington on 10 August 1972†; and Douglas fir at Haney, British Columbia on 19 July 1970.‡

	Thetford	Cedar River	Haney
Radiation budget			
$K\downarrow$	27·7	26·0	23·6
$K\uparrow$	2·3	2·4	2·2
K^*	25·4	23·6	21·4
$L\downarrow$	29·3	27·7	N.A.
$L\uparrow$	35·0	34·5	N.A.
L^*	−5·7	−6·8	−6·9
Q^*	19·7	16·8	14·5
Energy balance			
Q_H	11·7	7·0	4·8
Q_E	7·0	10·0	9·9
ΔQ_S	1·0	−0·2	−0·2
Derived terms			
α	0·08	0·09	0·09
β	1·67	0·70	0·48
E (mm)	2·80	4·00	3·96
Q_E/Q^*	0·36	0·60	0·68

Sources: † Gay and Stewart, 1974; ‡ McNaughton and Black, 1973.

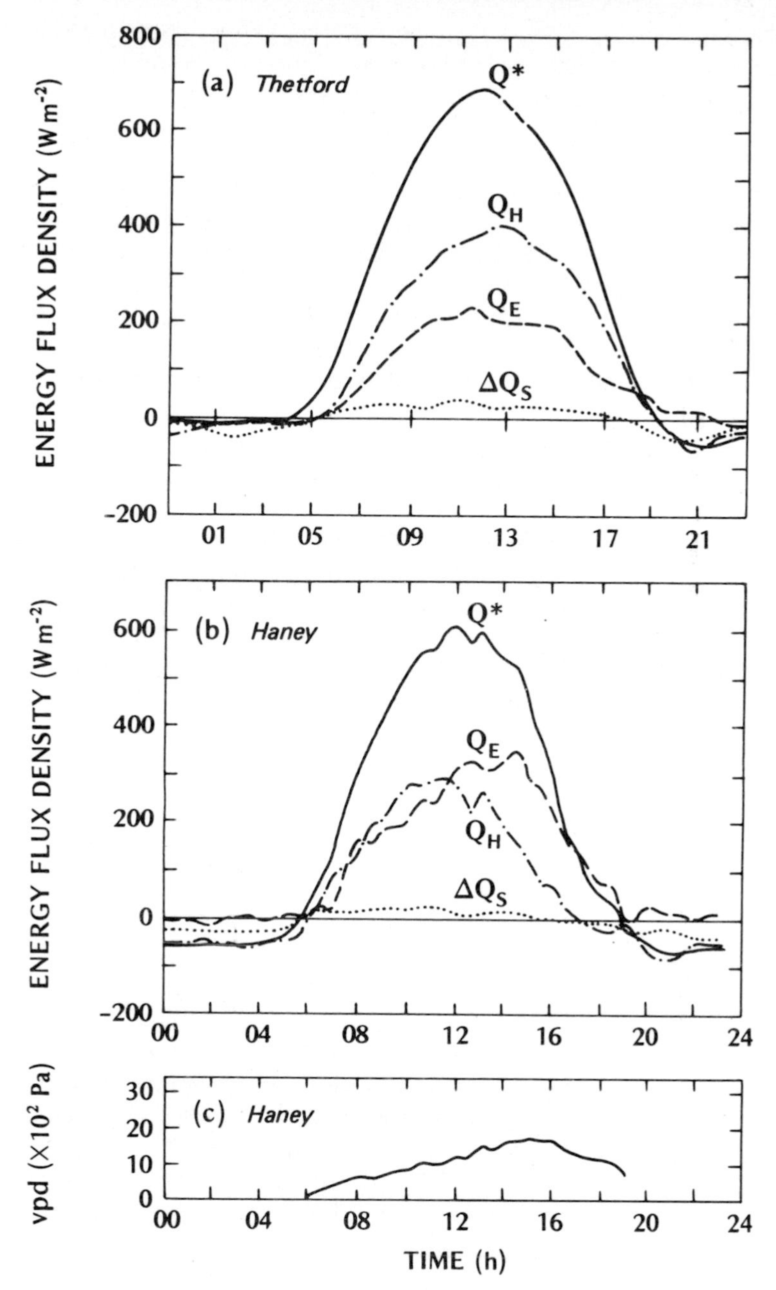

Figure 4.21 Diurnal energy balance of (a) a Scots and Corsican pine forest at Thetford, England (52°N) on 7 July 1971, and (b) a Douglas fir forest at Haney, B.C. (49°N) on 10 July 1970, including (c) the atmospheric vapour pressure deficit. Thetford data from Gay and Stewart, 1974 and Haney data from McNaughton and Black, 1973.

orchard in Figure 4.19 was 11% (varying from 16% at noon to 6% in the late afternoon). At night the level of maximum radiation emission is found within, rather than at the top of, the canopy.

(c) ENERGY BALANCE

The energy balance of a soil-forest-air volume (equation 4.1, Figure 4.1a) is illustrated by the results in Figure 4.21 and summarized by the daily energy totals in Table 4.7. Both the Thetford and Haney sites are characterized by extensive coniferous forests, one pine the other fir, with no obvious restriction to water availability. The albedo of the Thetford site is slightly lower, and the daylength longer than at Haney. These features combined with some nocturnal cloud at Thetford (00 to 05 h) and some daytime cloud at Haney (11 to 20 h) gave the former a larger net radiative surplus for the day. In most respects the Cedar River site (Table 4.7, Figure 4.20) is similar to that at Haney and the diurnal energy balance curves have been omitted for brevity.

Apart from these differences the Q^* and physical heat storage (ΔQ_S) terms are similar for the Thetford and Haney sites. (Biochemical heat storage, ΔQ_P, has been neglected on the assumption that it only represents 2–5% of Q^*, see p. 123.) Although ΔQ_S is relatively small it cannot be ignored on an hourly basis since at night it is of the same order of magnitude as Q^*. For example the storage release from the biomass alone can amount to 40 W m^{-2}, and to this must be added the releases from the soil and the air. Soil heat storage can be important where radiation is able to reach the floor relatively unhindered. Storage in the air becomes significant if the stand volume is large. On a *daily basis* however, net heat storage in a forest system is negligible (Table 4.7) since the daytime uptake and the nocturnal release compensate each other.

On the other hand the partitioning of the available energy into the turbulent sensible (Q_H) and latent (Q_E) heat fluxes was quite different at the two sites. At Thetford only one-third of Q^* was used to evaporate water, whereas at Haney and Cedar River approximately two-thirds was used for this purpose. The Q_E differences are particularly evident in the afternoon period. This is brought about by a phase-shift in the relationship of Q_H and Q_E relative to Q^* between Thetford and the other two sites (i.e. at Thetford Q^*, Q_H and Q_E all peak at midday, but at Haney (and Cedar River, not shown) Q_H peaks before, and Q_E after, the time of maximum Q^*). To inquire into this apparent anomaly we will consider the factors controlling evaporation from a vegetated surface.

In Chapter 2 it was noted that the evaporation rate depends upon the availability of water and energy, the strength of the surface-to-air vapour concentration gradient, and the intensity of turbulent motion. In this chapter we have added to this list the physiologic role of stomatal activity for vegetation surfaces. In the terminology of the electrical analogy (p. 102)

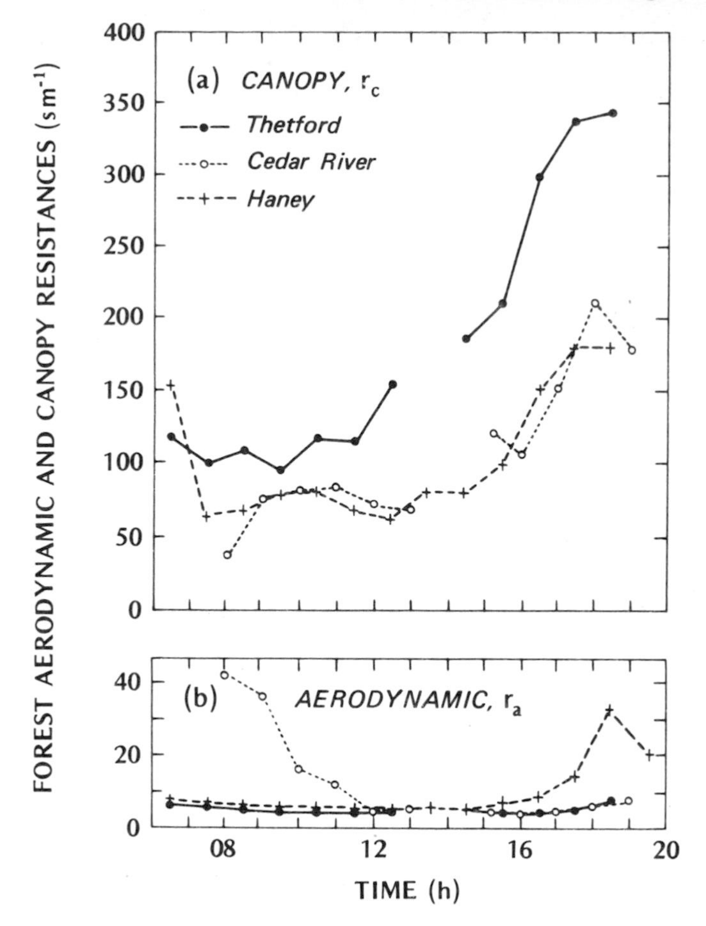

Figure 4.22 Diurnal variation of (a) the canopy and (b) aerodynamic resistance of coniferous forests (compiled from Gay and Stewart, 1974, and McNaughton and Black, 1973).

the role of atmospheric turbulence is characterized by the aerodynamic resistance (r_a), and that of the stomata by the canopy resistance (r_c).

In the case of the forests being compared here, differences in the availability of water or energy are not considered significant. At both sites r_a was small, and during the interesting afternoon period any r_a differences between sites were minor (Figure 4.22b). The higher r_a values at Cedar River in the morning, and at Haney in the evening, were due to local wind stagnation. Forests show low r_a values in general because of their large roughness (Tables 2.2 and 4.2), thus for a given wind speed, the atmosphere is more turbulent over a forest than any other natural surface (excluding topographic effects). In fact turbulent diffusivities (K_M, K_H, K_E etc.) are typically two orders of magnitude greater over forests compared to crop surfaces. When taken in conjunction with the diurnal pattern of vpd (Figure 4.21c) this gives a very efficient system for transporting vapour into the air, with the potential for a peak loss in the afternoon (i.e. near the time of maximum air temperature). This potential seems to have been realized at Haney but not at Thetford.

This leads us to consider whether the phase differences in Figure 4.21 could be explained by stomatal activity. Figure 4.22a confirms that distinct r_c differences exist, with values at Thetford being approximately twice those at the other sites. Whether this is related to species or site factors is not known. Since r_a and r_c act in series to restrict vapour transfer it is clear that r_c is the dominant limitation (r_c values are at least one order of magnitude greater than r_a) at all sites. Most significantly, at Thetford this control was particularly large in the afternoon. It is therefore suggested that at Thetford the increased stomatal control is sufficient to completely offset the vpd-driven tendency for high afternoon evapotranspiration, and thereby keeps Q_E in phase with Q^* and Q_H (Gay and Stewart, 1974).

The importance of biological control is further illustrated by the situation when a forest canopy is wet following rain. Under these conditions evaporation takes place from the wet leaf surfaces, stomatal control is virtually absent, and r_c approach zero. As long as there is sufficient wind to keep r_a small the evaporation rate under cloudy skies can approach that for a 'dry' canopy with clear skies. If the sky clears the rate can be as much as five times greater (Rutter, 1967). The energy required to attain this high evaporation rate commonly exceeds the available radiant energy (Q^*), but the same turbulent conditions which cause the rapid vapour transport also allow a ready flow of sensible heat from the air to the cooler canopy and thereby to supplement the radiation supply.

There is less detailed work available on the behaviour of deciduous forests. It appears however that with full foliage in spring and summer the energy balance is approximately in accord with that for conifers. The higher albedo reduces radiant absorption slightly but the dominant role of evaporation in energy partitioning remains. In the autumn sensible heat becomes increasingly important as transpiration decreases and the leaves die. In the winter if the ground is frozen or snow-covered virtually all of Q^* is converted into sensible heat to warm the trees by conduction, or to warm the atmosphere by convection.

Considerable research has been concerned with the question as to whether forests 'waste' more precious water to the air than low plant covers (grassland or crops). These comparisons are often motivated from a need to know the effects of forest clearance upon the local water balance, especially with respect to run-off. Tentatively it appears that in rainy areas where water is readily available the evaporation from forests is 10–20% greater than in nearby areas with low plant covers. In some forested areas as much as 100% of Q^* is used as Q_E on an annual basis. This may in large measure be due to the greater precipitation interception efficiency of forests (p. 121), and the subsequent rapid evaporation of this water due to their greater radiation absorptivity and their lower aerodynamic resistances. In drier climates most studies again show greater E from forests, which may in part be due to the ability of deep-rooted trees to tap water supplies not readily available to crops and grasses (Rutter, 1972).

(d) CLIMATE

As one enters a forest on a sunny summer day the climate changes noticeably. Air motion is weak, and it is cooler and more humid. This is because we are walking in a layer well below the level of the active surface where the primary site of drag on airflow, radiation absorption, and evapotranspiration is located. The typical climatic profiles in Figure 4.23 illustrate this fact, and their general form shows almost complete agreement with those from crops (Figure 4.15). The only major difference is that above-forest gradients are much weaker. This is directly attributable to the greater mixing over the much rougher forest surface which readily diffuses atmospheric properties throughout a deep surface layer.

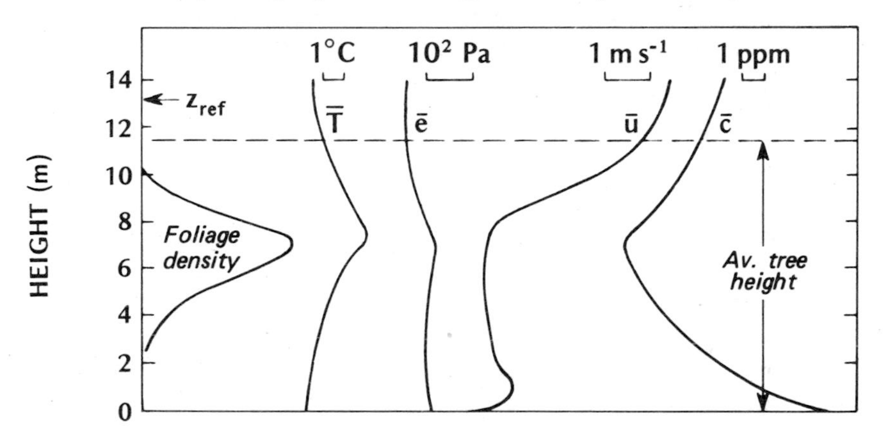

Figure 4.23 Typical mean hourly profiles of climatological properties in a Sitka spruce forest at Fetteresso near Aberdeen (57°N) on a sunny day in July 1970 at midday. Also included is the profile of the leaf area showing the vertical distribution of foliage density. The following characterize conditions above the canopy at a reference height of 13 m above the ground: $K\downarrow = 605$ W m^{-2}, $Q^* = 524$ W m^{-2}, $\overline{T} = 11{\cdot}8$°C, $\bar{e} = 11{\cdot}1 \times 10^2$ Pa, $\bar{c} = 315{\cdot}4$ ppm, $\bar{u} = 3{\cdot}9$ m s^{-1} (after Jarvis *et al.*, 1976).

The wind profile above a forest in neutral stability conforms to the logarithmic law with appropriate modification to include a zero plane displacement (equation 4.5). There is a sharp decrease of wind speed in the upper canopy down to the level of maximum leaf area (Figure 4.23).

Beneath this winds are very weak in the lower canopy and there is a mini-jet in the trunk space before becoming zero at the floor. In the case of a leafless deciduous forest there would be less drag exerted and a correspondingly greater penetration into the canopy, resulting in higher mean wind speeds throughout the stand depth. Comparison between wind speeds above forests with those above fields show the former to be slower (Oliver, 1974) if the height scale is based at the zero plane displacement level. This effect is explained by the increased drag and mixing over the forest and is illustrated by the two profiles on the left-hand side of Figure 2.9a.

The daytime air temperature and humidity profiles exhibit maxima at the level of maximum leaf area where radiative absorption and transpiration provide the most heat and water vapour (Figure 4.23). Beneath this there is a temperature inversion because the canopy is warmer than the floor where only weak radiative absorption occurs. Within the soil (not shown) there is usually very little temperature variation with depth or time. The humidity profile is less easily generalized within the stand because of the possibility of evaporation from the soil, or plants on the floor. At night the temperature profile is reversed with a minimum in the upper portion of the canopy, an inversion above, and lapse below. At this time the reduced sky view factor for positions beneath the canopy reduces long-wave radiative losses and helps to maintain relatively mild conditions within the stand compared with those in the open or above the trees. The greatest frost and dewfall are therefore found just beneath the crown of the canopy. If dewfall is absent the nocturnal humidity profile is a weakened version of that during the day.

The form of the CO_2 profile by day reflects the fact that the canopy is a CO_2 sink (photosynthesis) and the soil a CO_2 source (respiration). The gradient within the stand is relatively strong because of the lack of vigorous mixing to diffuse the CO_2 released by the soil. At night when the canopy also respires, concentrations decrease all the way from the soil up into the air above the forest.

CHAPTER 5

Climates of non-uniform terrain

In the preceding sections we have considered the energy and water balances of a range of relatively simple surfaces. The balance relationships we have been able to use have assumed that the surface in question is extensive and flat, and is virtually uniform in character (ideally it is an infinite homogeneous plane). If this is so, then the response of the surface to the energy and mass cascades is everywhere equal. This means that all gradients of climatological properties will be perpendicular to the active surface, and all fluxes are in the vertical direction.

In the real world relatively few surfaces are flat, few could be considered homogeneous, and none are infinite. On the contrary, the Earth's surface is a patchwork quilt of surface slopes and types. The scale of the units composing this quilt extend all the way from the oceans and continents down to individual leaves and even smaller. As an example consider the case of a group of agricultural fields, some ploughed and bare, some fallow, some in crops. Each of these fields possesses its own combination of radiative, thermal, moisture and aerodynamic properties, such as albedo, soil conductivity, soil moisture, surface roughness, etc. Each field therefore will tend to regulate and partition the available energy and water in a different manner, giving each a unique energy and water balance. These differences will be manifested as different surface, sub-surface and atmospheric climates in terms of temperature, humidity, and wind speed profiles. Thus there will be spatial discontinuity of climates, and horizontal gradients will exist. Near the surface at the boundaries between the fields these gradients will be greatest, and horizontal interactions will occur.

The climatic response of the above array of fields would be further complicated if they were situated in an area of varied topography. Solar loading differences would arise because of differences of slope and aspect, moisture availability would vary because of areal precipitation and drainage characteristics, and the wind field would be affected by channelling and shelter effects. One of the greatest challenges in modern atmospheric science is to understand the way in which these interactions take place. Only then will we be able to approach a fully dimensional climatology which accounts for the space and time domains.

In our further study of the climates of non-uniform terrain it is convenient to deal first with the effects of spatial variation in surface character, and second with the effects of topography.

1 Effects of spatial inhomogeneity

Two different cases will be considered: first, we will note the modification of atmospheric properties as air, already in horizontal motion, moves from one distinct surface type to a different one – these are *advective effects*; second, we will consider the circulation of air induced by contrasting surface properties when regional winds are weak – these are *thermal circulation systems.*

(a) ADVECTIVE EFFECTS

Three different advective effects are generally recognized: the '*clothesline-effect*', the '*leading-edge or fetch effect*', and the '*oasis effect*'.

(i) '*Clothesline-effect*'

This effect is normally restricted to the flow of air *through* a vegetative canopy. The ideal conditions exist at a vegetative stand border, as for example at the edge of a crop surrounded by warmer, drier ground; or at the edge of a forest bordered by fields. The situation is well illustrated by the horizontal flow (ΔQ_A) depicted in Figure 4.1a. If the air entering the crop from the right is warm and dry it will increase both the heat supply, and the vapour pressure gradient between the transpiring leaves and the crop air. The net result is to enhance the evaporation rate and hence to more rapidly deplete the soil moisture close to the stand border. Further into the crop the air cools down and acquires moisture, thus causing it to adjust to the more typical conditions within the stand.

At the edge of crops the soil moisture depletion, crop desiccation, greater wind buffeting, and greater exposure to new plant pests and disease commonly combine to stunt crop growth for a few metres in from the border. The thermal contrast between the interior and exterior of a forest can even be sufficient to induce a thermal breeze at the stand border. By

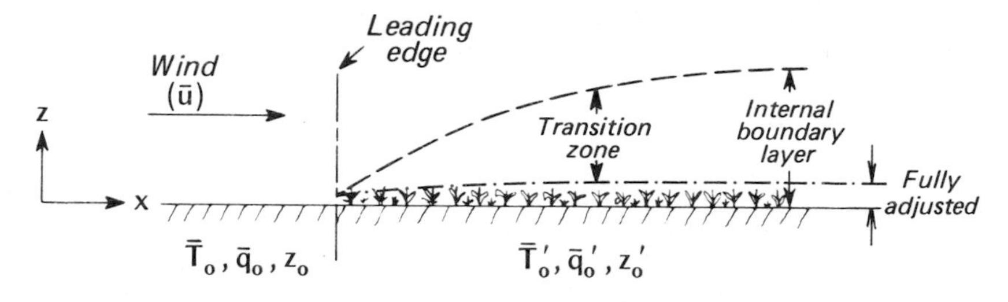

Figure 5.1 The development of an internal boundary layer as air flows from a smooth, hot, dry, bare soil surface (with surface values $\overline{T}_0$, $\bar{q}_0$, z_0) to a rougher, cooler and more moist vegetation surface (with new surface values $\overline{T}'_0$, $\bar{q}'_0$, z'_0).

day, the flow is from the cooler forest to the surrounding fields or grassland, and may extend for a few hundred metres horizontally away from the forest. By night, the reverse flow should occur but the greater frictional retardation of the forest restricts its effects to a few metres in from the edge.

(ii) '*Leading-edge or fetch effect*'

As air passes from one surface-type to a new and climatically different surface, it must adjust to a new set of boundary conditions. As shown in Figure 5.1 the line of discontinuity is called the *leading-edge*. The adjustment is not immediate throughout the depth of the air layer, it is generated at the surface and diffuses upward. The layer of air whose properties have been affected by the new surface is referred to as an *internal boundary layer*, and its depth grows with increasing distance, or *fetch*, downwind from the leading-edge. It is only in the lower 10% of this layer that the conditions are fully adjusted to the properties of the new surface. The remainder of the layer is a transition zone wherein the air is modified by the new surface but is not adjusted to it. The properties of the air above the internal boundary layer remain determined by upwind influences and not those of the surface immediately beneath.

To illustrate the manner in which adjustment takes place consider the case of air flowing from a dry bare soil surface to a fully moist low vegetation cover. From our previous discussion of the climates of these two surface-types (Chapters 3 and 4) we may anticipate that on a summer day the air has to adjust from a relatively smooth, hot and dry surface to one which is rougher, cooler and wetter. Initially we will restrict consideration to moisture changes in the new internal boundary layer. If we assume there is no major across-wind exchange (called the y direction) then we may conveniently analyse the vertical and along-wind (z and x directions respectively) in terms of the two-dimensional 'boxes' shown in Figures 5.2a. Air with a vapour content $\rho\bar{q}^{\dagger}$ ($\mathrm{kg\,m^{-3}}$) moves from the dry to the wet

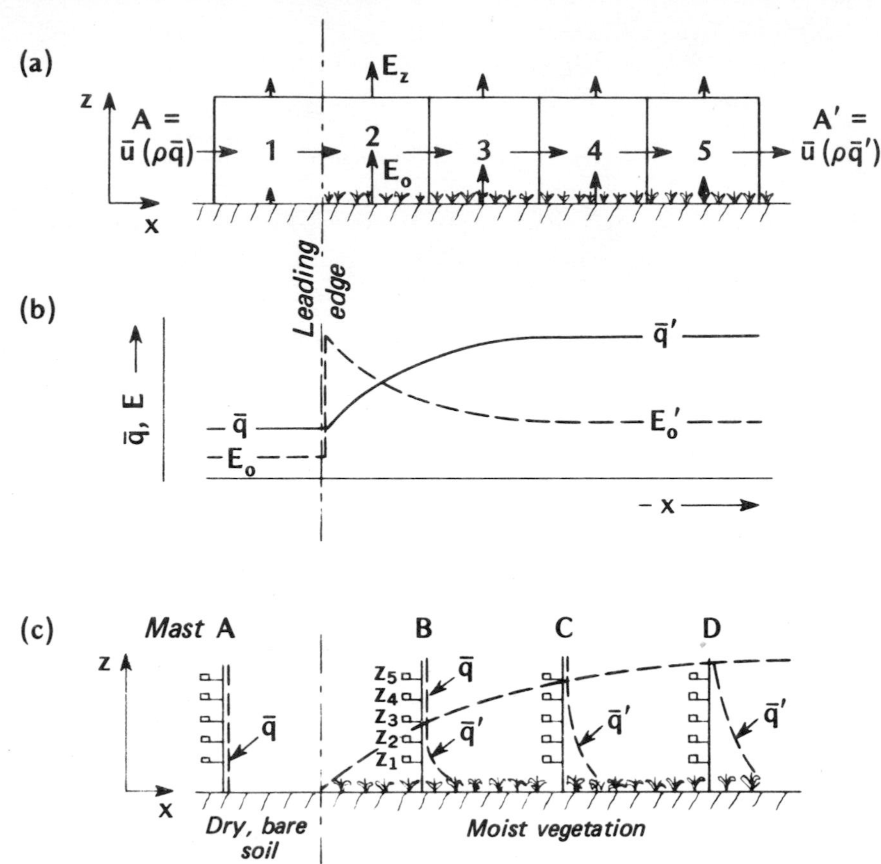

Figure 5.2 Moisture advection from a dry to a wet surface. (a) Evaporation rates and the vapour balance of a surface air layer (fluxes of vapour are proportional to the length of the arrows). (b) Surface evaporation rate (E_0), and mean water vapour concentration of the air layer. (c) Vertical profile of water vapour in relation to the developing boundary layer.

surface at a speed $\bar{u}^{\dagger}$ (m s^{-1}). Hence the rate of horizontal moisture transport, A (kg m^{-2} s^{-1}) is:

$$A = \bar{u}(\rho\bar{q}) \qquad (5.1)$$

In Figure 5.2a the vertical fluxes of water vapour into and out of the boxes are shown by vector arrows the lengths of which are proportional to the strength of the flux. The vertical arrows represent the evaporation fluxes, and the horizontal arrows the advective fluxes. We will assume there to be no net condensation or evaporation of vapour *within* the boxes, and hence assuming conservation of vapour the sum of the lengths of the input arrows

† Averaged quantity over the depth of the layer.

must equal that of the output arrows. If the vertical evapoıation arrows are of different lengths this indicates divergence or convergence of the vertical flux (Figures 2.4a, b) and a non-constant flux layer. If the horizontal arrows are dissimilar it indicates divergence or convergence of the horizontal flux (Figure 2.4c) and that advection is in operation, and the *difference* in their length is a measure of the net moisture advection (ΔA).

The vapour flow over the 'dry' upwind surface (box 1 in Figure 5.2a) is essentially non-advective. The horizontal arrows are equal and small, and the vertical arrows are similarly in balance (i.e. evaporation at the surface, E_0 equals that passing through the reference level z at the top of the box, E_z). As the air crosses the leading-edge into box 2 the surface evaporation rate rises sharply because there is an extremely strong gradient in vapour concentration between the moist surface and the 'dry' air. An increase in E_0 results in an increase of vapour $\bar{q}$ in the box (Figure 5.2b), and hence a larger horizontal transport $A = \bar{u}(\rho\bar{q})$ out of the downwind side of box 2 (Figure 5.2a). To conserve the vapour balance of the box it follows that E_z will be less than E_0. This process continues in boxes 3 and 4 but because of the accumulation of vapour with distance the surface-to-air vapour gradient weakens and with it the surface evaporation input. Finally at box 5 a new equilibrium situation arises where the mean content reaches a value ($\rho\bar{q}'$) which is more typical of the moist surface, and which does not overstimulate the surface evaporation regime. Downwind of box 5 both the horizontal and vertical fluxes are constant with distance and the air layer has become fully adjusted to the properties of the new surface.

The corresponding adjustment of the vertical profile of water vapour is given in Figure 5.2c. The upwind profile is very weak as befits a dry surface. The increased vapour content occurs first near the surface, which is the vapour source, and is diffused upwards to affect a deeper layer with increasing distance downwind.

The height of the fully adjusted boundary layer can be viewed as the level at which the vertical vapour flux equals the local surface value. This layer has been found to grow rather slowly requiring a fetch distance of 100 to 300 m for every 1 metre increase in the vertical. It is important to take account of this when setting-up instrumentation masts. For example in Figure 5.2c the levels z_4 and z_5 at mast B should not be used if the data are intended to characterize the conditions of the vegetated surface. The depth of the total internal boundary layer develops much more rapidly, commonly requiring a fetch of from 10 to 30 m for every 1 metre increase in the vertical, but the rate depends on the relative change of roughness between the two surfaces and the state of atmospheric stability. For example the rate will be greater if air moves from a smooth to a much rougher surface during unstable conditions because turbulence would be well developed and able to diffuse properties easily.

In addition to moisture changes the contrast of surface temperature and

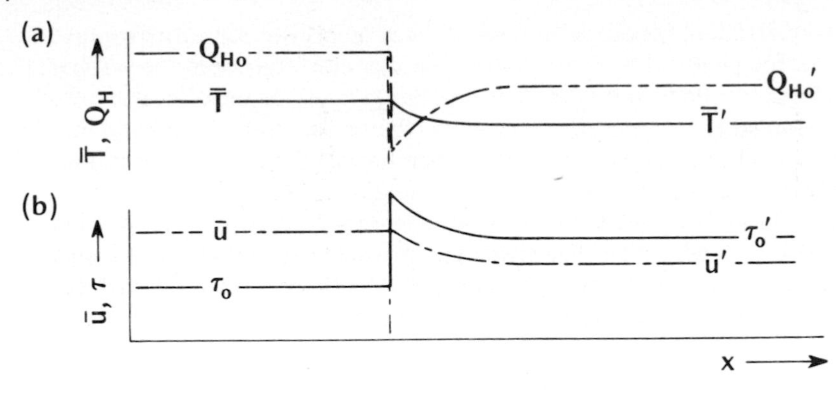

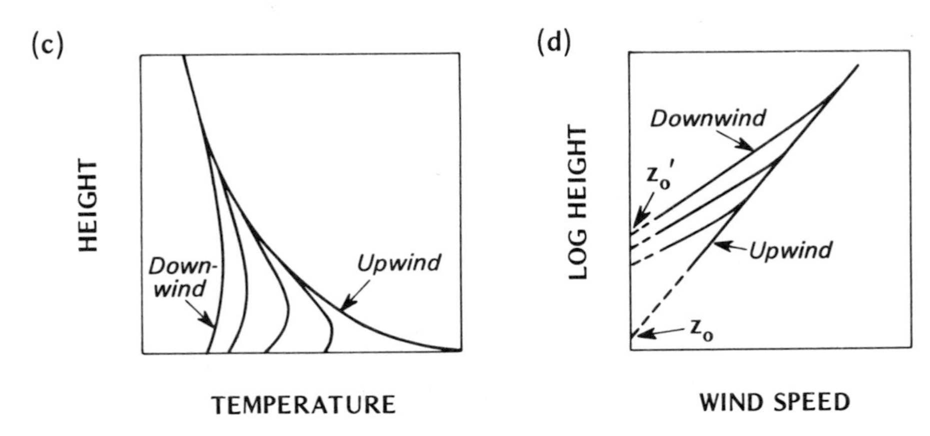

Figure 5.3 (a) Adjustment of surface sensible heat flux (Q_{Ho}) and mean air temperature ($\overline{T}$) as air passes from a hot to a cooler surface. (b) Change in surface shearing stress (τ_0) and mean wind speed ($\bar{u}$) as air flows from a smooth to a rougher surface. Associated modification of the vertical profiles of (c) air temperature, and (d) wind speed at different distances downwind of the leading edge.

roughness between the bare soil and vegetated surfaces would give rise to adjustments in the sensible heat flux and momentum exchange downwind of the discontinuity, and these would in turn influence the mean air temperature and wind speed. The anticipated form of these changes are shown in Figure 5.3. The increased evaporative cooling of the vegetated surface would considerably reduce the value of the surface heat flux compared with that upwind. In fact the cooling may produce a surface-based inversion (Figure 5.3c). This means that the sensible heat flux will be directed *towards* the surface. The heat involved in this flow will be drawn from the advecting hot air and therefore contributes to its cooling (Figure

5.3a) until a new equilibrium temperature is established. The greater roughness of the vegetation will exert a greater drag on the air. This increases the surface shearing stress and decreases the mean wind speed (Figure 5.3b). The wind profile in the upwind region (Figure 5.3d) is logarithmic, and its extrapolation would intersect the height axis at a small value of z_0 (see p. 47). Downwind of the leading edge the profiles have a 'kink' where the steeper slope of the new profile intersects that of the upwind profile. The point of intersection becomes higher as the boundary-layer adjustment deepens downwind. The greater slope reflects the increased shearing, and extrapolates to a much larger roughness length, z_0' (Figure 5.3d).

In reality for a case such as we have been discussing it is not possible to separate entirely the changes in energy, mass and momentum. For example the momentum changes will increase the turbulent diffusivities (K's) over the rougher surface, and hence enhance Q_E and Q_H even if the vapour and temperature gradients did not change.

Leading-edge effects occur wherever there is a marked discontinuity in surface properties. Figure 5.4 shows a surface temperature cross-section

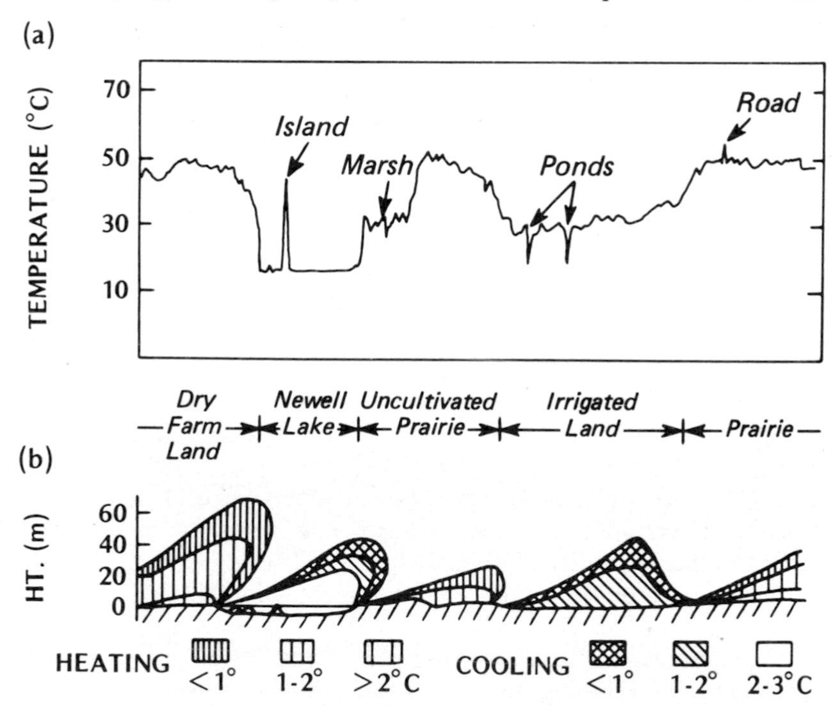

Figure 5.4 (a) Horizontal profile of surface radiation temperature, and (b) hot and cold 'plumes' over a diverse prairie landscape. Based on aircraft observations on the afternoon of 6 August 1968 near Brooks, Alberta (after Holmes, 1969).

obtained from a radiation thermometer mounted on an aircraft. The instrument senses $L\uparrow$ and this can be related to surface temperature from equation 1.4 if the emissivity of the radiating surfaces is known (p. 9). This midday transect across the prairies shows sharp changes in temperature (up to 35°C) due to the different climatic properties of cultivated/uncultivated, dry/irrigated farm land, and water bodies. The vertical structure shows the development of a series of boundary layers, some warmer and some cooler than the bulk air temperature. At the upwind edge of each new surface an internal boundary layer forms, and hence the structures appear as hot or cool 'plumes'. The heated plumes being unstable have more vigorous vertical development than those from the cooler, more stable lake or irrigated-land. The former were found up to 1000 m, the latter never above 60 m.

There are two types of fog which illustrate the effect of advection of air across water of a very different temperature. Paradoxically one occurs with warm air flowing over cold water, the other with cold air over warm water. Near coastlines winds or ocean currents can cause cold water to well up to the surface from below. Air flowing towards the land across the warmer surface waters offshore encounters the leading edge of this band of cold water along the coastline and the lowest air layers are cooled. If air temperatures are depressed to the dew-point a band of *cold-water advection fog* is formed. Another leading edge is crossed at the coastline and the lowest layer becomes re-heated over the warmer land surface. The fog therefore thins out from below as it travels inland and evaporates.

On the other hand *warm-water advection fog* occurs when very cold air is transported across a much warmer water body. The saturation vapour pressure at the water surface therefore exceeds the vapour pressure of the air, whether it is saturated or not. The warmer, moister surface air is unstable and convects moisture easily up into the cooler air. Some of the super-saturated mixture then condenses to give a fog which visually appears like rising steam (the same effect is evident with exhaled breath on a cold day, or when hot water is run into a bath in a cool room). These conditions can occur over arctic oceans (where it is called arctic sea 'smoke'), or over open bodies of water in otherwise cold continental climates, and over some industrial cooling ponds. It can also be seen over lakes on summer or autumn mornings, especially if during the night cold air has drained down onto the lake from the surrounding hills.

(iii) *'Oasis-effect'*

Due to evaporation cooling, an isolated moisture source always finds itself cooler than its surroundings in an otherwise rather arid region. The desert oasis is the most obvious example of this situation. Table 5.1 shows the energy and water balances of oases in comparison with their surrounding terrain. The semi-desert area evaporates all of its precipitation leaving

TABLE 5.1 Annual energy and water balances of Tunisian Oases (after Flohn, 1971).

Surface type	Area (km^2)	Albedo α	Q^*	Q_H	Q_E	β	Q_E/Q^*	E	p
			($MJ\ m^{-2}\ day^{-1}$)					($mm\ yr^{-1}$)	
Semi-desert	35,000	0·20	6·9	5·8	1·0	5·8	0·14	150	150
Oases (av.)	150	0·15	8·6	−3·1	11·8	−0·26	1·37	1680†	150

† To support this evaporation rate the precipitation supply is supplemented by irrigation from artesian wells.
Note: On an annual basis $Q_G \simeq 0$.

virtually nothing for run-off. Even so this consumes only a small proportion of the available radiant energy because precipitation is so limited. The surplus heat is therefore dissipated as sensible heat to warm the air, resulting in a large Bowen ratio (β).

In the oases on the other hand, the free availability of water permits evaporation to exceed precipitation by one order of magnitude, and the energy necessary to accomplish this is more than that supplied by radiation (i.e. Q_E is greater than Q^*). This apparently anomalous situation is explained by the fact that the *atmosphere* supplies sensible heat to the surface because the oasis is cooler than the regional air in which it is embedded. Therefore there is a continual air-to-oasis inversion temperature gradient driving a downward directed heat flux, and the process is aided by air mass subsidence over the oasis. (Note that the reversal of Q_H gives the oasis a negative Bowen ratio.)

In this example the ratio $Q_E/Q^* = 1{\cdot}37$, in an extreme case the 'oasis-effect' has been observed to produce a ratio of $Q_E/Q^* = 2{\cdot}5$ for a shorter period over an irrigated field of cotton (Lemon *et al.*, 1957). The irrigated field of alfalfa near Phoenix, Arizona considered previously in relation to Figure 4.14, also exhibits the 'oasis-effect'. Figure 5.5 shows the average daily energy balance components in June following the irrigation in late May. For the first half of June an 'oasis-effect' is clearly evident; evapotranspiration exceeds the net radiation, and Q_H is directed towards the crop. During this period the average Bowen ratio is about $-0{\cdot}25$, and the ratio Q_E/Q^* is approximately 1·5. However, by the middle of June soil moisture starts to become restricted, the surface resistance (r_c) increases, and evaporation rates decrease. Since Q^* remains relatively constant this requires an adjustment by Q_H. By 25 June the Bowen ratio is $+1{\cdot}0$, and Q_E/Q^* is about 0·5. In the absence of further irrigation or precipitation the energy balance would return to a semi-arid type.

Many other 'oasis-effect' advective situations can be envisaged. In each of the following examples a cool, moist surface is dominated by larger-scale warmer, drier surroundings: a lake in an area with a dry summer climate; a

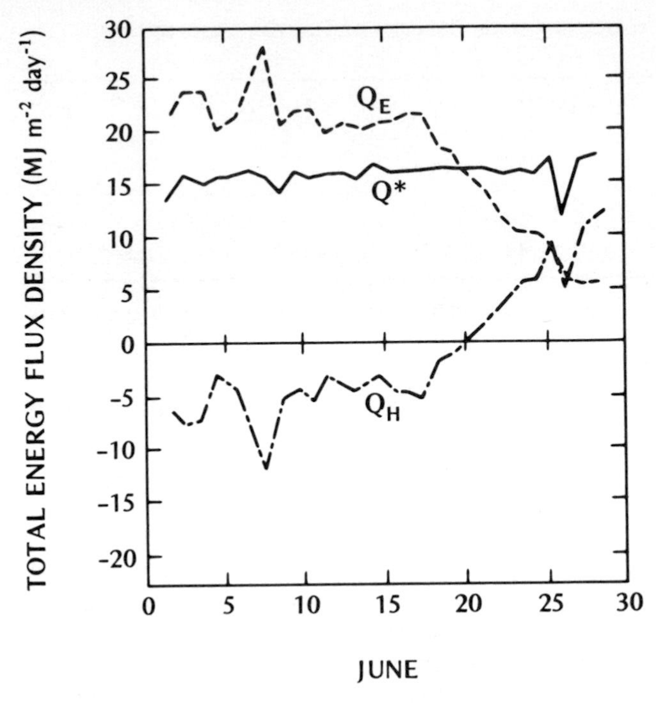

Figure 5.5 Average daily energy balance of an alfalfa crop in June 1964 near Phoenix, Arizona (33°N). The crop was irrigated by flooding in late May and this was followed by drought throughout June (see Figure 4.14) (after van Bavel, 1967).

glacier in a mountain valley; an isolated snow patch; an urban park; an isolated tree in a street or on open, bare ground. In each case we may expect evaporation to proceed at an increased rate compared with that from an extensive area of the same composition, and it is quite possible that the energy required to accomplish this will exceed the local radiative surplus.

(b) THERMAL CIRCULATION SYSTEMS

The juxtaposition of contrasting thermal environments results in the development of horizontal pressure gradient forces, which if sufficient to overcome the retarding influence of friction will cause air motion across the boundary between the surfaces. Land and water surfaces possess contrasting thermal responses because of their different properties and energy balances, and this is the driving force behind the *land and sea* (*lake*) *breeze* circulation system encountered near ocean or lake shorelines. Compared with most land surfaces a water body exhibits very little diurnal change in surface temperature. The four main reasons for this were outlined on p. 88, but in summary water is different because it (i) allows transmission of short-wave radiation to considerable depths, (ii) is able to transfer heat by convection and mixing, (iii) converts much of its energy surplus into latent rather than sensible heat, (iv) has a large thermal inertia due to its higher heat capacity. Thus although Q^* may be greater over water (because of its low albedo, p. 81), the effectiveness of Q_E and ΔQ_S as thermal sinks means that Q_H is small (Table 3.5). By day Q_H is small because most of the energy is channelled into storage or latent heat; at night it is small because the long-wave radiative cooling is largely offset from the same water store (Figure 3.13b). The reduced convective heat flux (Q_H) to and from the air means that atmospheric warming and cooling rates ($\Delta \overline{T}/\Delta t$) are relatively small over water bodies. In contrast the convective fluxes and rates of temperature change over land are large and show a marked diurnal variation.

These land-water temperature differences and their diurnal reversal (by day – land warmer than water; at night – land cooler than water) produce corresponding land-water air pressure differences. These in turn result in a system of breezes across the shoreline which reverse their direction between day and night (Figure 5.6). Thermal breeze systems of this type are best developed in anticyclonic summer weather because almost cloudless skies

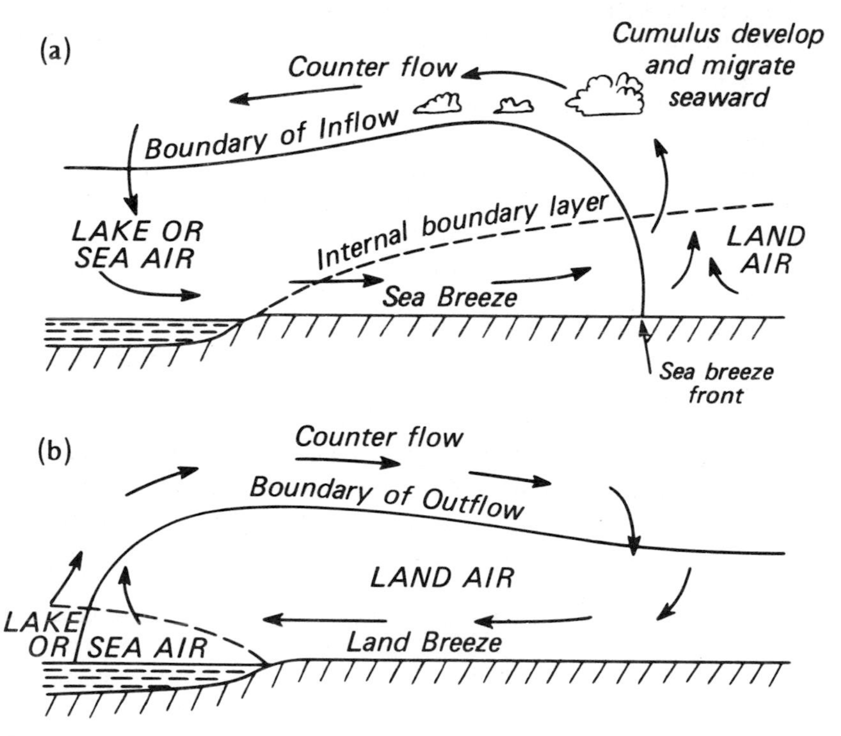

Figure 5.6 Land and sea (lake) breeze circulations across a shore-line (a) by day and (b) at night during anticyclonic weather.

and weak synoptic-scale winds permit the maximum differentiation between surface climates. Increased cloud or stronger winds modify or obliterate these local winds. Note that both the land and sea breezes are really the low-level portion of a complete circulation cell.

The daytime sea breeze circulation (Figure 5.6a) has a greater vertical and horizontal extent, and its wind speeds are higher, than in the nocturnal land breeze (Figure 5.6b). This is because by day the solar forcing function is in operation and instability is greatest. Commonly the sea breeze blows at 2 to 5 m s^{-1}, extends inland as far as 30 km, and affects the air flow up to a height of 1 to 2 km (Figure 9.6). On the other hand the land breeze is usually about 1 to 2 m s^{-1} in strength and smaller in both horizontal and vertical extent. During the sea breeze the cooler more humid sea or lake air advects across the coast and wedges under the warmer land air. The advancing *sea breeze front* produces uplift in what is already an unstable atmosphere over the land. The front is therefore commonly associated with the development of sea breeze cumulus clouds which are caught up in the counter flow aloft and are carried seaward where they dissipate because they have been removed from their moisture source, and because over the water the air subsides, and warms adiabatically (Appendix A1).

The lowest layer of sea or lake air which is advected across the coast is modified by the 'leading-edge effect'. The stable marine air is warmed over the land producing a more unstable internal boundary layer which grows in depth with distance from the shore (see also Figure 9.3d and discussion). If the coastal water is upwelling and therefore relatively cold it may remain cooler than the land even at night. The lack of temperature reversal may then encourage the sea breeze to continue through the night.

There are a number of practical implications associated with the land and sea breeze circulation. For example Mukammal (1965) studied the injurious flecking of tobacco leaves near the shoreline of Lake Erie. It was known that the damage was caused by ozone (O_3), but there was no obvious local source of this pollutant. The study showed that ozone formed by photochemical action over Lake Erie (due to the action of sunlight on emissions from lakeshore cities) was brought down to lake-level in the offshore subsidence of the lake breeze (Figure 5.6a), and advected inland in pulses by the lake breeze. In another study van Arsdel (1965) solved an equally perplexing problem near Lake Superior. Blister rust disease was found to be affecting pine trees located about 15–20 km inland from the lakeshore. The diseased trees however lay at least 8 km further inland than the currant bushes which produced the disease spores, and the intervening trees were not similarly infected. It was discovered that the spores were only released at night, and that they were probably carried out over the lake by the land breeze, then aloft and landward again in the counter flow, finally being deposited on the pine trees at the base of the descending portion of the cell (Figure 5.6b). Although the spores were not tracked along this route their probable trajectory was verified by tracing smoke and balloons. Further examples of the importance of sea breeze circulations to air pollution transport are given in Chapter 9, and an analogous city breeze system is outlined in Chapter 8. Of course not all effects of the sea breeze are harmful, in particular we should note that the inhabitants of coastal settlements often find the cool sea breeze to be beneficial in offsetting a hot climate.

2 Effects of topography

It is convenient here to divide topographic effects into those due to the varying input of solar radiation, and those related to the generation, or modification, of airflow.

(a) RADIATION LOADING EFFECTS

The radiation received by a surface is usually the major determinant of its climate. This radiant input is composed of the components S, D and $L\downarrow$, of which only the direct-beam receipt (S) is dependent upon the angle at which it strikes the receiving surface. The relationship between the radiation received by a surface and the incident beam is given by the *cosine law of illumination* (Figure 5.7a):

$$S_{slope} = S_i \cos \Theta \tag{5.2}$$

where, S_{slope} – the radiant flux density incident on the surface (AB in Figure 5.7a), S_i – the radiant flux density perpendicular to the incident beam (i.e. on the imaginary surface CD), and Θ – the angle between the direct-beam and a normal to the surface. (Note that in order to calculate Θ for any surface at any time it is necessary to know not only the angle of the slope

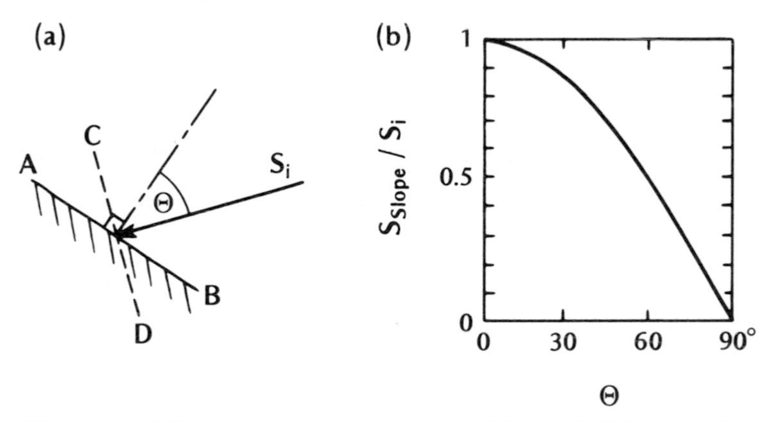

Figure 5.7 (a) Diagrammatic representation of the angle Θ between the surface and the incident direct-beam short-wave radiation, S. (b) The form of the cosine law of illumination.

relative to a horizontal plane but also both the azimuth angle and the zenith angle of the Sun.) At a given time and location S_i is unlikely to vary very much spatially (depending upon atmospheric conditions) and hence variations in the slope and azimuth angles presented by topography determine the radiant loading differences across the landscape. Clearly the slope that most directly faces the Sun (i.e. where Θ approaches zero, and therefore cos Θ approaches unity) will receive the most radiation; whereas if the Sun is almost grazing the surface (i.e. Θ approaches 90°, and cos Θ approaches zero) minimal radiation is received. Because of the cosine form (Figure 5.7b) it can be seen that the direct-beam solar input is almost uniformly high for angles of Θ less than 30°, but above this the receipt drops at an increasing rate.

Figure 5.8 shows the daily input of direct-beam radiation to various slope angles and aspects at latitude 40°N at the times of the solstices (when the Sun is overhead at the Tropics of Cancer and Capricorn); and equinoxes (when the Sun is overhead at the Equator). In the northern hemisphere, south-facing and horizontal surfaces show symmetrical energy receipt centred on midday.

At the equinoxes the maximum direct-beam input will be upon a south-facing slope of 40° at midday (i.e. Θ = zero because the Sun is overhead and $S_{slope} = S_i$). The closest slope to this in Figure 5.8a is south 45° (Θ = 5°), next is horizontal (Θ = 40°) and then south vertical (Θ = 50°). East-facing slopes receive the early morning solar beam more effectively than the south-facing slopes. Hence the curves for east-facing slopes rise more sharply after sunrise, with the east 45° slope receiving more than the east vertical. However as the Sun's azimuth changes through the day, east-facing slopes soon achieve their 'local solar noon', and the incident radiation decreases rapidly towards their 'local sunset' (12 h for east vertical, and 15 h for east 45°). Although not shown in Figure 5.8a the situation for west-facing slopes would be symmetrical with that of the east-facing slopes, showing higher receipts in the late afternoon than the south-facing slopes. At the times of the equinoxes vertical north-facing slopes receive no direct-beam short-wave radiation input, their radiant receipt being limited to diffuse-beam short-wave and long-wave radiation from the atmosphere. The input to a north 45° slope is very slight at all times.

At the summer solstice (Figure 5.8b) the horizontal, east- and north-facing slopes experience sunrise before the south- and west-facing slopes (west is not shown but is symmetrical with east). East-facing slopes are illuminated as for the equinoxes but with an earlier sunrise, and a larger maximum input. Horizontal surfaces receive direct-beam radiation throughout the day, and of the slopes shown they obtain the maximum intensity at the midday peak (for a horizontal surface Θ = 16·5°, for south 45° Θ = 28·5°, and for south vertical Θ = 73·5°). The vertical north- and south-facing slopes show mutually exclusive illumination, the north being in

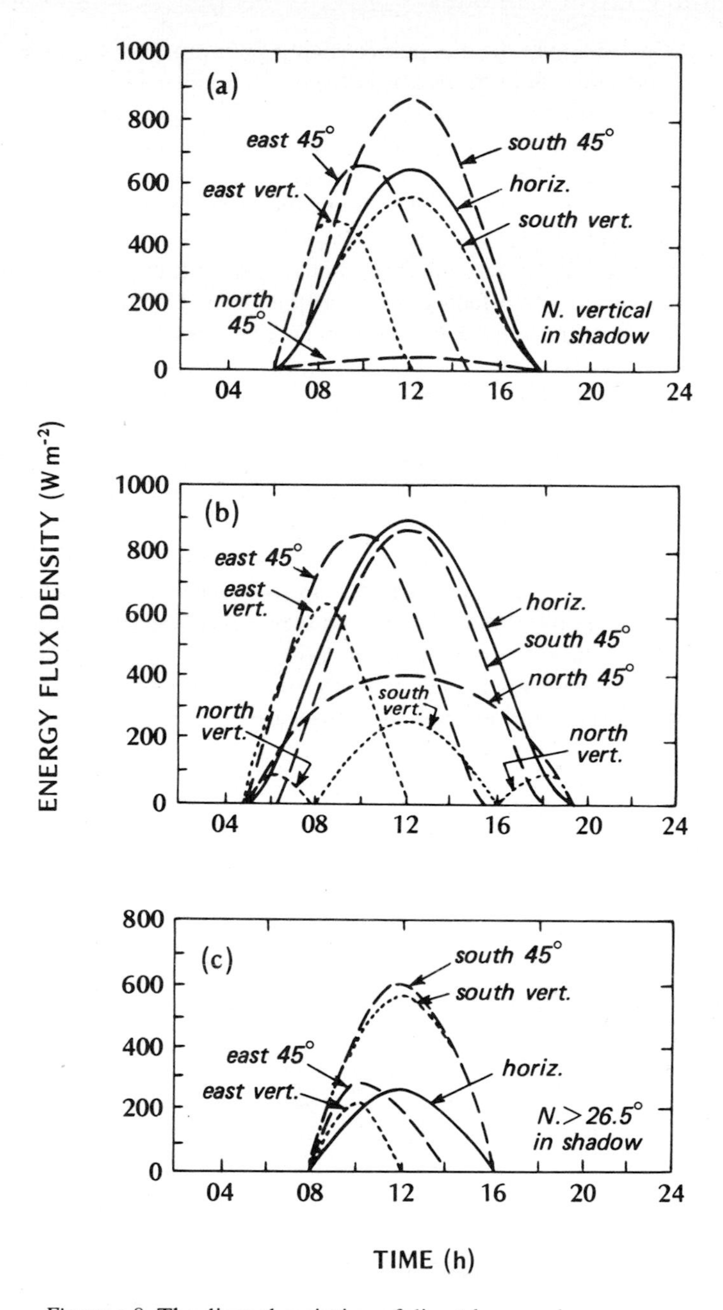

Figure 5.8 The diurnal variation of direct-beam solar radiation upon surfaces with different angles of slope and aspect at latitude 40°N for (a) the equinoxes (21 March, 21 September), (b) summer solstice (22 June), and (c) winter solstice (22 December) (after Gates, 1965).

receipt early and late in the day (experiencing two sunrises and sunsets per day), and the south only in receipt between 08 h and 16 h.

By the winter solstice (Figure 5.8c), north-facing slopes of greater than 26·5° receive no direct-beam at all, whereas south-facing slopes are most favourably placed. The length of day is considerably shorter, which combined with the generally lower intensities give relatively small daily radiation incomes.

In general the effect of moving from high to low latitude is to increase the illumination of the north-facing slopes at the expense of the south-facing slopes. Also, since the Sun is never more than 47° from the zenith at midday in the tropics, the effect of topographic slope changes is reduced compared to the high latitudes where small slope or aspect changes may be of considerable practical importance.

Figure 5.9 gives the direct-beam short-wave radiation totals at the time of the equinoxes at latitude 45°N (approximately equivalent to Figure 5.8a). It shows marked differences in the receipt of direct-beam radiation on slopes of different aspect. Note that the maximum load would be on a south 45° slope ($\Theta = 0°$), whereas no direct-beam would reach north-facing slopes of greater than 45° angle. The same authors produced maps of direct-beam radiation by placing a grid over a topographic map, determining the slope angle and azimuth for each grid position, and then computing the radiant input. Spatial energy distributions of this type form

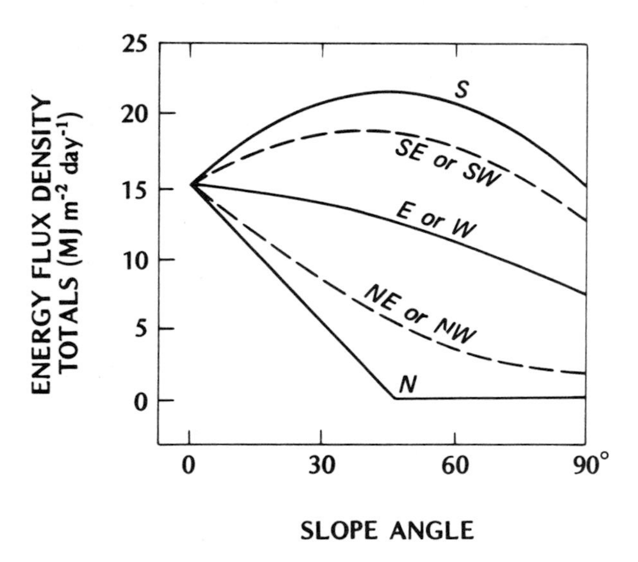

Figure 5.9 Total daily direct-beam solar radiation (S) incident upon slopes of differing angle and aspect at latitude 45°N at the times of the equinoxes (diagram constructed by Monteith, 1973, using data from Garnier and Ohmura, 1968).

an excellent base for the understanding of micro-climatic variations in regions of complex topography.

Although only an approximation, we may assume that diffuse-beam input from cloudless and cloudy skies is equal for all positions of the sky hemisphere, and therefore does not contribute to spatial variability of solar receipt at the surface. But the receipt of direct-beam under partial cloud cover clearly does give marked differences between areas in direct illumination and those in shadow from cloud. If the cloud is in motion and not occupying preferred positions then over a sufficient period of time such differences will average out spatially. On the other hand cloud may be directly related to surface features such as sea breeze (p. 146) or anabatic cloud (p. 152). This would lead to different solar loading across the landscape which may re-inforce or dampen the cause of cloud development (i.e. can be a positive or negative feedback).

TABLE 5.2 Effects of topography on the surface energy balance of bare ground in the Turkestan Mountains (41°N). Data are daily energy totals (MJ m^{-2} day^{-1}) based on monthly average for September (from Aisenshtat, 1966).

Site	Energy balance				Dimensionless quantities		
	Q^*	Q_H	Q_E	Q_G	α	β	Q_E/Q^*
Horizontal	14·4	9·4	2·1	2·1	0·14	4·5	0·15
North-facing (33°)	6·0	3·5	1·7	0·7	0·20	2·0	0·28
South-facing (31°)	17·6	12·6	3·1	1·9	0·15	4·1	0·18

Topographically-induced radiation variations lead to energy balance differences across the landscape. An example of this is given in Table 5.2 where it can be seen that the south-facing slope receives almost three times more net radiation than the north-facing slope. Some of this increase may be attributed to the lower albedo of the south-facing slope, but the primary effect is probably due to its more favourable aspect. The slope surfaces were bare debris, and the local climate was semi-arid, which accounts for the energy balance partitioning in favour of sensible heat (Q_H), and the high Bowen's ratio (β) values. However the south-facing slope pumps more than three times as much sensible heat into its lower atmosphere and such strong differential heating is likely to produce local slope winds (p. 152). The lower β value for the north-facing slope may also indicate a greater availability of moisture for evaporation but there were no soil moisture data to test this possibility.

It is evident therefore that orientation of a surface with respect to the solar beam is a very powerful variable in determining its energy income. It also follows that the naturally uneven configuration of the landscape produces a wide spectrum of microclimates, and these have implications for other aspects of the physical environment. Plant and animal habitats are affected, thereby leading to distinctly different assemblages of flora and fauna on slopes of different angle and aspect. Similarly these differences are often reflected in the type of land-use especially in agriculture and forestry. Hydrologic activity is also likely to vary as a result of different rates of evaporation, lengths of snow cover retention, and probabilities of avalanching on different slopes. Equally geomorphic processes such as frost-shattering, mud slumping, soil creep, and rock exfoliation are directly or indirectly related to thermal and/or moisture conditions.

(b) TOPOGRAPHIC WIND EFFECTS

(i) *Topographically-generated winds*

Valleys, especially those in mountainous regions, produce their own local wind systems as a result of thermal differences. As with the land and sea breeze thermal circulation the local winds of valleys are best developed in anticyclonic weather in summer. Under such conditions, with almost cloudless skies and weak large-scale motion, differential warming or cooling of different facets of the landscape gives rise to horizontal temperature and pressure gradients, which cause winds. The exact nature of these wind systems depends on the orientation and geometry of the valley. The best developed and most symmetric wind system might be anticipated in a deep, straight valley with a north-south axis. In valleys with other orientations or possessing complex geometries (e.g. bends, constrictions, etc.) the flow pattern may be asymmetric or incomplete. For convenience we will consider the case of a simple north-south valley, but even then there must be some asymmetry with time due to the diurnal variation of solar radiation input to west- and east-facing slopes.

By day the air above the slopes and floor of the valley will be heated by the underlying surface to a temperature well above that over the centre of the valley. As a result shallow, unstable upslope (or *anabatic*) flow arises, and to maintain continuity a closed circulation develops across the valley involving air sinking in the valley centre (Figure 5.10a). Commonly the uplift along the slopes leads to the formation of convective *anabatic clouds* along the valley ridges. In tropical valleys this may lead to greater precipitation along ridges in comparison with the valley floor. The cross-valley circulation also effectively transports sensible heat (Q_H) from the surrounding active surfaces to warm the whole valley atmosphere. Therefore when compared to the atmosphere at the same level over an adjacent plain (or further downstream) the valley air is much warmer, and in a manner analogous to the sea breeze, a plain-to-mountain flow

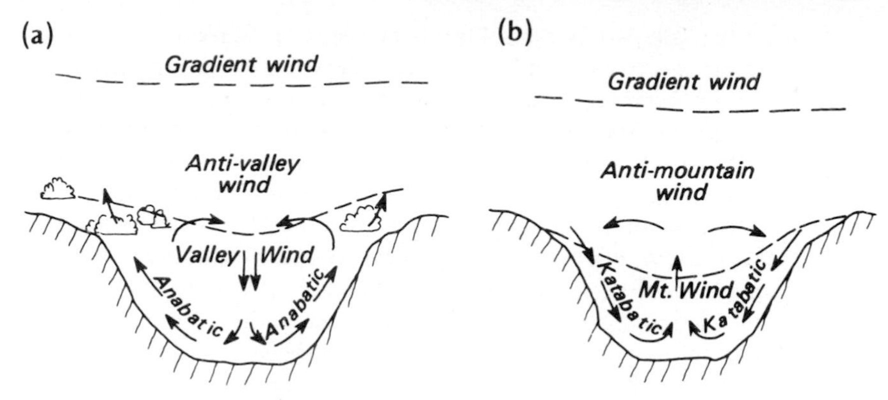

Figure 5.10 Mountain and valley wind system viewed at a valley cross-section with the reader looking up-valley. (a) By day slope winds are anabatic, and the valley wind fills the valley and moves upstream (into the page away from the reader), with the anti-valley wind coming downstream (out of the page). (b) At night the slope winds are katabatic and reinforce the mountain wind which flows downstream, with the anti-mountain wind flowing in the opposite direction above.

develops. The up-valley flow is termed the *valley wind*, and fills the entire valley. The maximum pressure gradient is near the surface, and hence the maximum wind speed is as close to the ground as the retarding influence of the surface allows. Above the ridges there is a counter flow (the *anti-valley wind*) which flows down-valley by day (Figure 5.11). Again the similarity with the sea breeze circulation cell is evident. Above the anti-valley wind there will be yet another wind associated with the large-scale synoptic flow pattern.

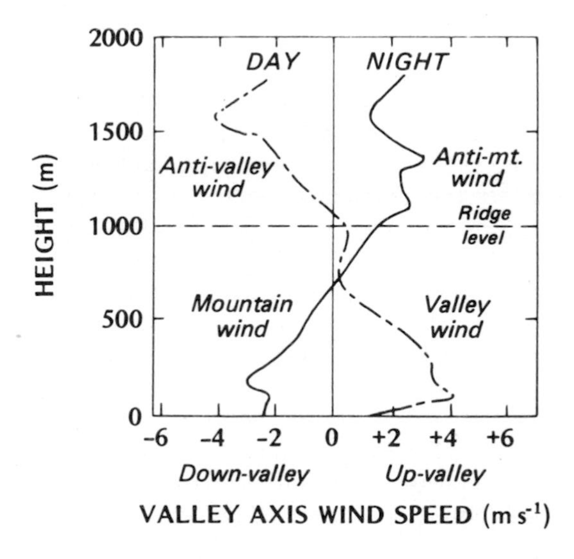

Figure 5.11 The vertical distribution of along-valley winds in a 1 km deep valley on Mt. Rainer, Washington. Horizontal scale is graduated in units of wind speed and separated into two wind directions (up and down valley) (after Buettner, 1967).

At night the valley surfaces cool by the emission of long-wave radiation. The lower air layers cool and slide down-slope under the influence of gravity. These *katabatic* winds usually flow gently downhill at about 1 m s^{-1}, but greater speeds are observed where the cold layer is thicker and the slope steeper. The convergence of these slope winds at the valley centre results in a weak lifting motion (Figure 5.10b). All of the downslope flows combine into a down-valley flow known as the *mountain wind* which seeps out of the mountain valleys onto the adjacent lowlands. A counter flow (the *anti-mountain wind*) flows up-valley aloft (Figure 5.11). The drainage of cold air down-slope or down-valley often occurs as intermittent surges rather than a continuous flow. The reason for this behaviour is not certain, but it appears that the stable cold air may become retarded or blocked by obstacles in its path, until a threshold value is reached beyond which it overcomes the restraining influences and plunges forward.

Katabatic flows are commonly found over ice and snow surfaces. If the head of the valley considered above was occupied by a glacier or snowfield the additional cold air would have augmented the nocturnal down-valley wind. In fact katabatic winds are also found by day over glaciers and ice-caps. The cold ice surface gives rise to a semi-permanent temperature inversion, and cools the overlying air layer because Q_H is directed towards the surface (Figure 3.6). The cool skin of air slides down the glacier, over its snout, and cuts under the less dense valley wind which is moving up-valley. The *glacier wind* soon dies out however because the air is slowed, by friction with the valley floor and the opposing force of the valley wind, and because it is thermally modified at its lower boundary as it advects over the warmer valley floor. Similar katabatic flows exist on a larger scale on the surface and at the margins of the continental ice-caps. In this case the cold layer is deeper, and the wind speeds are greater than for glacier and valley winds. Winds as strong as 20 m s^{-1} are often encountered.

From a practical standpoint the less spectacular gentle flow of cold air on good radiation nights is equally important. Height differences of less than one metre may allow cold air to drain to the lowest lying portions of the landscape (e.g. hollows, basins, valleys). The coldest (and densest) air settles to the lowest levels and therefore temperature increases with height above the valley floor producing a *valley inversion*. In this stable stratification temperature varies directly with elevation (Figure 5.12). Should cooling be sufficient to depress temperatures below the dew-point the stratification is made visible by the presence of radiation fog in the lowest-lying spots. If temperatures fall below freezing these same areas experience the greatest frost risk. Such *frost 'pockets'* should be avoided when planting frost-susceptible plants and trees (p. 207).

Under these conditions as one moves up the valley slopes from the floor the thermal conditions ameliorate (Figure 5.12), until the top of the pool of cold air is reached. Above this point the normal adiabatic decrease of

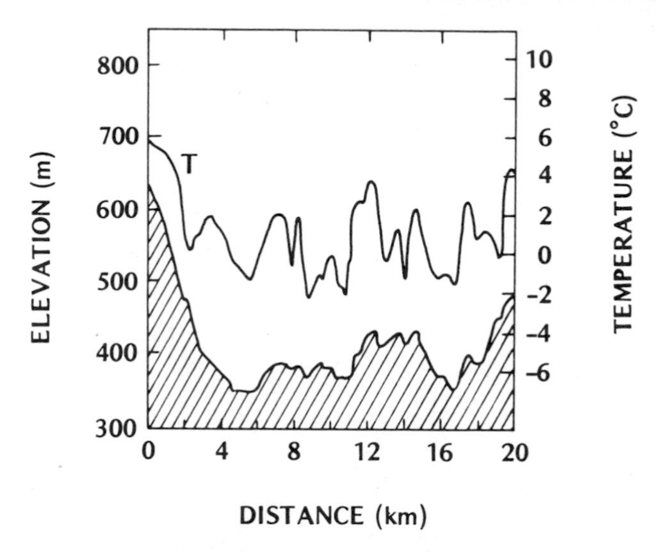

Figure 5.12 Variation of air temperature with distance along a traverse route over hilly terrain in the early morning following a good radiation night. Note the correspondence of elevation and temperature. The vertical distance scale is exaggerated to aid comparison (after Hocevar and Martsolf, 1971).

temperature with height (Γ) usually prevails. Thus the most favourable location on the valley sides is just above the level to which the cold pool builds up. This is known as the '*thermal belt*' and its height depends upon the geometry of the valley and the area of cold air sources which feed the cold pool. The belt usually corresponds to a contour band along the valley sides which is therefore favoured for the siting of thermally sensitive crops (e.g. fruit orchards, vineyards) and native dwellings.

(ii) *Topographically modified winds*

The airflow over non-uniform terrain is not easy to generalize. Every hill, valley, depression, tree, rock, hedge, etc. creates a perturbation in the pattern of flow, so that the detailed wind climate of every landscape is unique. It is however possible to isolate some typical flow patterns around specific features and we will consider three such adjustments here, but if the integrated wind field is required it is probably only possible to model the situation by building a scale model and subjecting it to flow simulations in a wind tunnel.

Flow around an isolated obstacle Figure 5.13 shows the pattern of airflow around a cylinder. If the flow is laminar it is accelerated over the upstream

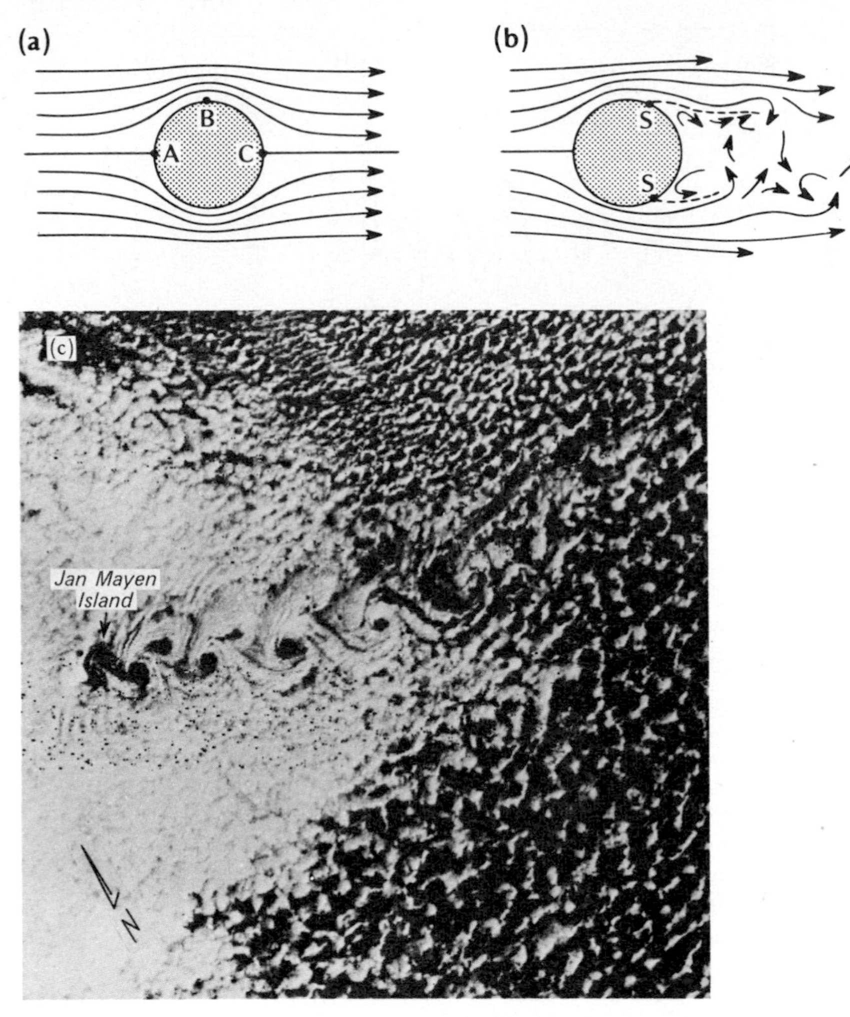

Figure 5.13 Plan view of the pattern of flow around an upright cylindrical object with (a) laminar flow, and (b) fully turbulent flow with separation. (c) Satellite picture of Kármán vortex street downwind from Jan Mayen Island on 18 September 1974, with winds of 15·5 m s^{-1} in the cloud layer (after Bayliss, 1976).

portion of the cylinder (i.e. from A to B in Figure 5.13a) and decelerated thereafter. Intuitively this must occur because air near the cylinder has to travel further around it than air further away. The reverse situation applies as air passes from B to C so that the flow slows down. On the other hand for turbulent flow such as is normal for the real atmosphere, the pattern is as in Figure 5.13b. Under these conditions the air is retarded by friction with the obstacle on its upstream side so that on continuing past its speed is insufficient to overcome the downstream slowing effect noted for Figure 5.13a. The flow is brought to a standstill at point S where *flow separation* takes place. Behind the obstacle within the separated flow the motion is considerably more turbulent. This zone is called the *wake* and includes a *lee eddy or vortex* immediately behind the obstacle where some of the air becomes trapped in a zone of relatively low pressure. At regular intervals a vortex detaches itself and moves downstream. The vortices are shed alternately with clockwise and counterclockwise rotation. The downstream series of these is known as a *Kármán vortex street* and a good example is provided by the cloud patterns often seen behind islands (Figure 5.13c).

The characteristics of separated flow closely correspond to the distribution of winds around an isolated hill (Figure 5.14a). As the air 'squeezes' over and around the hill there is an acceleration in wind speed over its upwind side. With strong winds there is separation from the top and sides and a turbulent wake develops in the lee. Further downwind the wake dissipates and the flow resumes its initial character. Light rain or snow suspended in the air tends to be preferentially deposited in the zone of least wind speed behind the hill. Similar patterns of airflow and precipitation

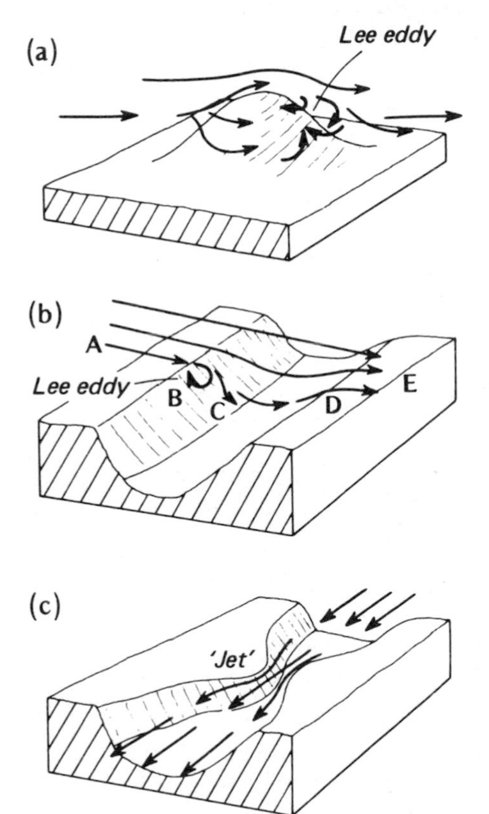

Figure 5.14 Modification of existing airflow due to (a) an isolated hill, (b) a valley, and (c) a topographic constriction.

accumulation occur around other isolated obstacles such as bushes, trees, tussocks and boulders. It should be noted that this is the reverse distribution to most mountain ranges where the windward slopes receive the most precipitation and the leeward slopes the least. The mechanisms involved in the two cases are completely different.

Flow over a break of slope Elements of the perturbations shown in Figure 5.13b are also evident in the flow over a valley (Figure 5.14b) or any other feature which involves a sudden drop or rise in the underlying surface such as a cliff or an escarpment. However, if the break is especially abrupt the pattern conforms more to that around a sharp-edged cube (e.g. Figure 8.1) than around a cylinder. Consider an airstream at A flowing across a deep valley lying normal to its path. After crossing the windward ridge the flow has suddenly to adjust to occupy a much deeper layer. The airstream cannot expand immediately thereby causing separation at the ridgeline, and a lee eddy in the low pressure zone at B. At this location the flow is more turbulent and ground-level winds are typically *opposite* to the direction at A. Divergence (expansion) of the airstream causes the average wind speeds to drop and the minimum wind at ground-level is usually found at point C, on the downslope between B and the valley floor.

On the opposite side of the valley the airstream is forced to converge again resulting in a relative increase of wind speed to a maximum at D near the top of the slope. Thereafter it readjusts to its original condition so that at some distance E the speed returns to a value similar to that at A.

The strength of the lee eddy circulation, and the degree of shelter within the valley are both decreased if the valley slopes are gentle and if the regional wind strikes the valley at an angle-of-attack more parallel to the valley axis.

The top of the windward slope is to be avoided for most applications. For example a pollution source placed within the lee eddy will find it difficult to disperse effluent because it may become trapped in the closed circulatory cell (Figure 5.15a). With very high wind speeds the strength of

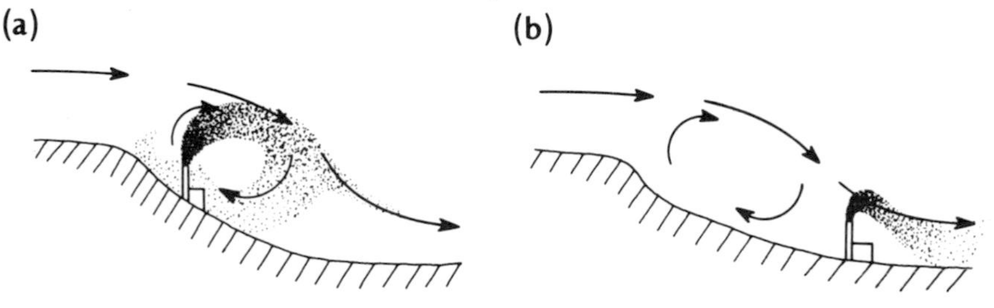

Figure 5.15 Problems of pollution dispersal on the windward slope of a steep-sided valley. In (a) the plume contents are trapped in the lee eddy, and in (b) are forced to ground level by 'downwash'.

the up-slope counter flow can be sufficient to rip the roof off a chalet or barn. During snow drift conditions the lee eddy is responsible for the formation of overhangs (cornices) along the ridgeline. If a pollution source is placed slightly downslope of the lee eddy it may still encounter problems due to '*downwash*' (Figure 5.15b); that is the plume may be caught in the descending flow and brought to ground-level near the valley floor.

Flow through a topographic constriction When air is forced to pass through a constriction it forms a 'jet' of high wind speeds. This occurs in valleys that narrow (Figure 5.14c) or through mountain passes, and on a smaller scale through gaps in hedges, lines of trees etc. Upon leaving the constriction the flow diverges and decelerates.

CHAPTER 6

Climates of animals

1 Special features

The interactions between the atmosphere and animals represents one of the highest levels of complexity in the boundary layer. In attempting to extrapolate the principles gained in preceding chapters to the case of animals four special characteristics of animal-atmosphere systems become evident:

(a) ENERGY AND WATER BALANCES OF ANIMALS

The energy balance of an animal (Figure 6.1a) may be written:

$$Q^* + Q_M = Q_H + Q_E + Q_G + \Delta Q_S \qquad (6.1)$$

where, Q_M – rate of heat production by metabolic processes (see p. 164), and ΔQ_S – net change of body heat storage. In this balance Q_M is *always* a heat source, Q^* is a heat source during periods of strong radiant heat loading, and Q_H and Q_G can become heat sources if the air surrounding the animal, or the ground it is in contact with, is warmer than the body temperature of the animal. Otherwise Q^*, Q_H, Q_E and Q_G all represent channels of heat loss to dissipate the animal's metabolic heat output. Net heat storage can be an energy gain or loss, but in many animals it must remain close to zero because the range of tolerable body temperatures is small. Equally unless the animal is prone, in a burrow, or immersed in water, heat gain or loss by conduction is usually negligible by comparison with the other exchanges.

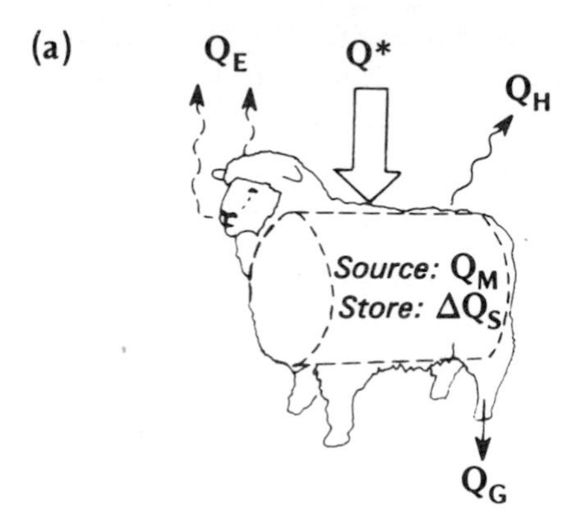

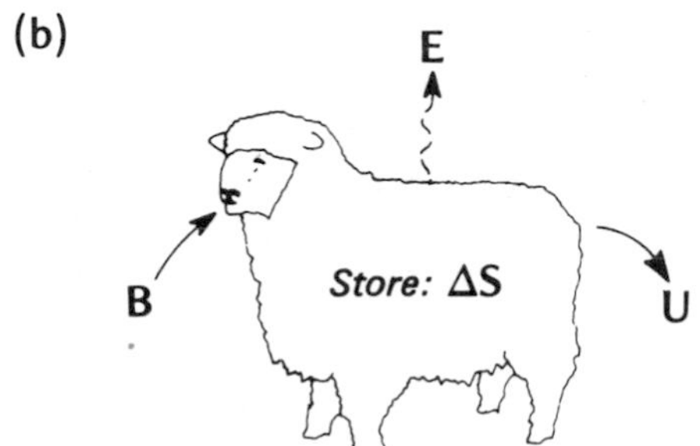

Figure 6.1 Schematic depiction of fluxes involved in (a) the energy, and (b) water balances of an animal.

Equation 6.1 refers to the complete three-dimensional balance of the animal's body volume. Therefore the terms are fluxes spatially-averaged over the complete surface area of the animal regardless of whether it is involved in the exchange of that property or not. For example if we consider the daytime short-wave radiation exchange we know that some areas will be sunlit (i.e. receiving $S + D$) whereas others will be in shade (i.e. receiving only D). In addition, the sunlit area assumes very different shapes depending upon the geometry of the animal and of the incident beam (see Figure 6.16). Further we know that the underside of the animal is likely to receive more of the energy reflected back up from the surrounding ground, and that the albedo of the animal is unlikely to be spatially uniform, or even the same for direct as against diffuse radiation. If we add the fact that the animal is capable of moving to different radiation environments (e.g. seeking shade) the situation becomes exceedingly complicated, and therefore difficult to monitor or model. This explains the tendency of research workers to simplify the radiation exchange to that of a cylinder (horizontal for four-legged animals, vertical for erect humans) or a sphere (for birds). This approach recognizes the animal's body, rather than its appendages (i.e. arms, legs, tail, etc.) to be the primary site of exchange (see the horizontal cylinder in Figure 6.1a). Unfortunately spatial averaging also masks the variability of heat exchange over an animal's body. For example long-wave radiation and convective sensible heat losses depend upon the animal's surface temperature and this may be higher where the pelage is thin or absent (e.g. ears, abdomen, tail, scrotum, etc.). Similarly in animals unable to sweat, the site of evaporative cooling may be entirely localized to the respiratory tract.

The corresponding volumetric water balance of an animal (Figure 6.1b) is given by:

$$B = U + E + \Delta S \qquad (6.2)$$

where, B – net water intake (usually via eating and drinking, although some animals are able to absorb water through their exterior covering), U – water loss in urine and faeces, and ΔS – net change of body water storage (for small animals this may be directly equated to weight change). The evaporative loss (E) may take place via three pathways. First, water may be evaporated and diffused directly through the skin or other permeable integument. Second, it may be ejected through pores in the skin by the sweat glands and be evaporated from the skin surface. Third, evaporation may occur from the moist surfaces of the respiratory tract.

(b) THERMOREGULATION

Successful functioning of a living organism depends upon the relationship between itself and its surrounding environment. There are two basic classes of organisms distinguished by the degree to which they internally control

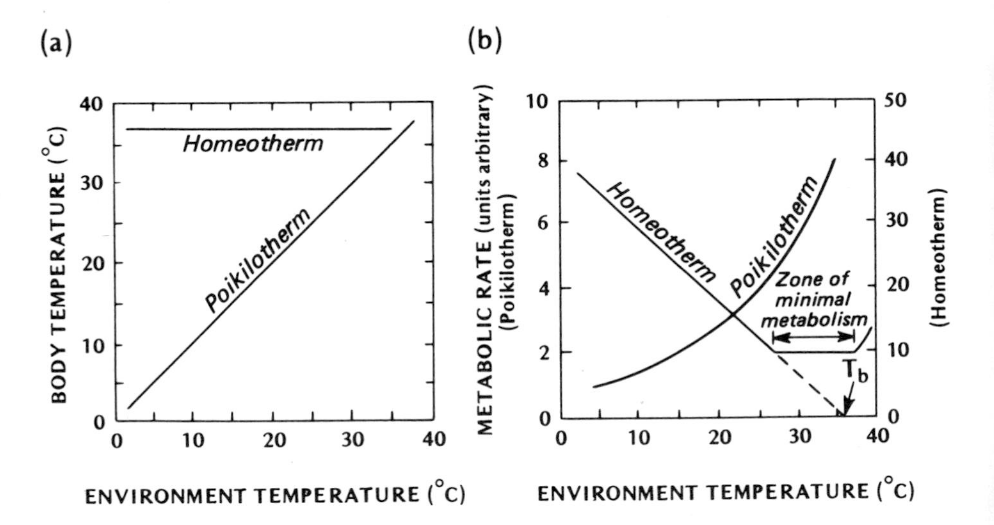

Figure 6.2 The effect of environmental temperature upon (a) body temperature, and (b) the rate of metabolic heat production for typical poikilotherms and homeotherms (adapted after Bartholomew, 1968).

their thermal balance with the external environment. *Poikilotherms* are organisms whose temperature is almost totally dictated by that of the surrounding thermal environment. Plants and 'cold-blooded' animals (most insects, reptiles and fish) are poikilothermic. *Homeotherms* on the other hand are able to maintain a relatively constant deep-body temperature by means of physiological mechanisms which vary the production of metabolic heat or the loss of heat by radiation, conduction and convection. 'Warm-blooded' animals (humans, most mammals and birds) are homeothermic, and the precision to which this thermoregulation is achieved is termed *homeostasis*. Wide variation away from this condition (i.e. when ΔQ_S becomes significant and deep-body temperatures rise or fall) is deleterious to the health of the animal.

These relationships are illustrated in Figure 6.2. Over a range of normally encountered environmental temperatures[1] (between 0 and 40°C) an hypothetical poikilotherm has a body temperature approximately equal to that of the environment (Figure 6.2a). The poikilotherm therefore expends no energy on thermoregulation. Its metabolic rate is governed solely by its temperature, roughly according to van't Hoff's rule which states that the rates of most biochemical reactions doubles with each 10°C increase in temperature. This doubling for each increment in temperature produces the exponential curve in Figure 6.2b. Over the same range an hypothetical homeotherm (Figure 6.2a) holds its body temperature virtually constant (usually between 34 and 42°C). To maintain this independence from environmental temperature however it has to pay a large energy cost. The metabolic rate of the homeotherm must increase linearly with decreasing environmental temperature (Figure 6.2b), because the lower the external temperature the more work it must do to maintain its constant body temperature. Above a threshold temperature, however, the animal can remain comfortable with a relatively constant minimum metabolic output, in what is therefore called the *zone of minimal metabolism*. Beyond this zone at higher temperatures the metabolic rate increases non-linearly. It is interesting to note that extrapolation of the linear portion to zero metabolic output (dashed line in Figure 6.2b) intersects the horizontal axis at the animal's deep-body temperature (T_b).

Thermoregulation in animals may be achieved by behavioural (voluntary) or physiological (involuntary) responses to the external thermal environment. The following are the common behavioural responses:

Movement – unlike most plants, animals are capable of physically moving into an environment which for the moment places the least stress upon them. Such a location is termed their *preferendum*. It provides the most equable set of conditions so that the result of the energy exchanges

[1] Environmental temperature refers to the combination of the temperature of the air and of surrounding radiating surfaces.

given in equation 6.1 is suitable. For example a poikilotherm may find its body temperature becoming too hot at a sunny, calm and dry site because of strong radiative heat loading and the lack of heat dissipation via turbulence. Therefore it finds it advantageous to seek out a shaded, and/or windy, and/or more moist site.

Posture – an animal may control the size and nature of its surface areas involved in energy exchange by orienting the body appropriately, by curling-up or stretching-out, and other posture changes. For example a husky dog in a blizzard will lie down, curl-up and put its back towards the wind. This arrangement minimizes the surface area capable of losing heat via radiant and turbulent processes; conserves metabolic heat inside a shell of poor conductivity (its fur); and minimizes heat loss through its nose, mouth and ears by tucking its nose inside its fur and flattening its ears.

Ingestion – intake of relatively warm or cold fluids can affect the heat content of the body. For example, the ingestion of cool water by a bird causes it to lose some body heat in warming the water up to body temperature.

Construction of shelter – some animals construct special shelters against the environment, including burrows in the soil and in trees, nests, and in the case of humans very elaborate shelter in the form of buildings and clothing.

In addition, homeotherms in particular have built-in physiological responses to environmental stress. These include changes in metabolism, dilation and contraction of blood vessels, increased and decreased pulse rate, sweating and panting, erection of an insulating layer of hair or feathers. Fuller consideration is given to these features in the section on the climates of homeotherms (p. 176).

(c) ANIMAL METABOLISM

Metabolism refers to the process in living organisms whereby substances are transformed into tissue with an attendant release of energy and waste. Plant photosynthesis is a metabolic process in which sunlight, carbon dioxide and water produce dry matter, oxygen and heat (p. 94). In the case of humans and other animals food (carbohydrate) is the energy source rather than light, and oxygen (inhaled via the lungs and carried by haemoglobin in the blood stream) is used to convert this food into heat, and through muscular activity, into work.

The total amount of metabolic heat produced depends on the state of the surrounding physical environment, and also upon the diet, body size, age and level of activity of the animal. The total production (Q_M) can be conveniently broken down into two components, viz:

$$Q_M = Q_{Mb} + Q_{Mm} \qquad (6.3)$$

The basal metabolic rate (Q_{Mb}) is the baseline level of heat production by an animal, at rest, in a thermally pleasant environment. Thus, having eliminated the influences of muscular activity or thermally stressful surroundings, we may expect Q_{Mb} to be related to simple physiologic features such as size and age. Indeed, as will be shown, there is a general tendency for Q_{Mb} to increase with the size of the animal, but to decrease with age. The Q_{Mm} term in equation 6.3 is the heat released by the muscular activity of the animal in doing work (i.e. moving external objects). The contribution of Q_{Mm} to the total production increases markedly as the level

TABLE 6.1 Adult human metabolic heat production (Q_M), at different levels of activity (after Fanger, 1970).

Activity	Approximate metabolic rate	
	W	W m^{-2}†
(a) *Resting*		
Sleeping	70	40
Seated, quiet‡	100	60
Standing, relaxed	120	70
(b) *Walking*		
Level, 3·2 km h^{-1}	200	120
Level, 5·6 km h^{-1}	320	190
Level, 8·0 km h^{-1}	570	340
5% grade, 3·2 km h^{-1}	300	170
15% grade, 3·2 km h^{-1}	450	270
25% grade, 3·2 km h^{-1}	660	390
(c) *Occupational*		
Office work (typing, filing, etc.)	90–120	50–70
Driving	100–200	60–120
Domestic work (cooking, washing)	160–350	90–200
Moving 50 kg bags	400	230
Digging trench	600	350
(d) *Sports*		
Gymnastics	350	200
Tennis	450	270
Squash	710	420
Wrestling	860	500

† Flux density calculations based on the surface area of a nude adult (known as the DuBois area) $\simeq$ 1·7 m^2.
‡ This rate approximates the basal rate (Q_{Mb}) $\simeq$ 50 k cal m^{-2} h^{-1}, sometimes called 1 metabolic unit (or MET).

of the animal's activity escalates (e.g. from standing, to walking, to running, etc.). This is shown in Table 6.1 for a wide range of activities by humans. As a unit of comparison note that the basal rate for an adult human is approximately equivalent to the heat (radiant and thermal) given off by a household light bulb (approximately 100 W). Note also that muscular activity can increase Q_M by almost an order of magnitude in comparison with the basal rate.

(d) EFFECTS OF ANIMAL SIZE

The basal metabolism of animals (including humans) is related to their weight (mass) as shown in Figure 6.3. Poikilotherms kept at a body temperature of 20°C metabolize at approximately 5% of the value for a homeotherm with a deep-body temperature of 39°C, and the same is true for hibernating mammals.

To maintain a constant core temperature in relatively hot (or cold) surroundings a homeotherm must lose (or gain) energy through one or more of the flux channels in equation 6.1. It can be shown (e.g. Lowry, 1967) that this energy requirement is proportional to the surface area of the animal, but only to the two-thirds power of its mass. Thus a large animal, with considerable mass, requires a relatively small energy requirement to keep its deep-body temperature stable, whereas a smaller animal requires more energy per unit of its mass. As an example, in a situation where an elephant requires 33 J kg^{-1} of body weight, a mouse requires 837 J kg^{-1}. The ratio of the mass of an animal to its surface area is therefore very important. Those with small mass:surface ratios (slender, long limbs) are

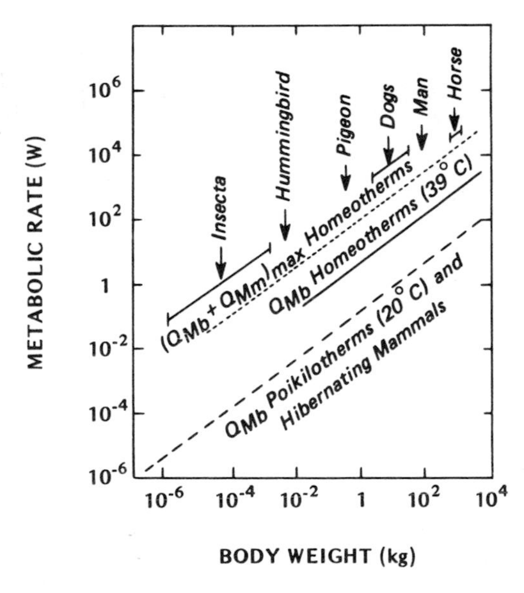

Figure 6.3 Relation between the rate of metabolic heat production and body weight for a wide range of animal species. Solid line – basal rate of homeotherms; dotted line – maximum rate for sustained work by a homeotherm; dashed line – basal rate of poikilotherms and also hibernating mammals (adapted from Hemmingsen, 1960).

Figure 6.4 Response of new-born piglets to contrasting thermal environments. In (a) at 15°C the piglets huddle together, draw in their limbs, and cover their noses in order to conserve their combined metabolic heat output and minimize the surface area available for heat loss. In (b) at 35°C the piglets are spread apart and relaxed in posture (after Mount, 1968).

good heat exchangers and well adapted to heat dissipation and warm environments. Those with large mass: surface ratios (sturdy body, short limbs) are best suited to heat conservation and cold environments.

As mentioned previously, changes in an animal's posture enable thermoregulation by altering the surface area (and therefore the mass: surface ratio) involved in energy exchange. Figure 6.4 clearly shows this principle in the case of a group of new-born pigs both as individuals and as a group. In the case of rabbits their ears, when extended in warm conditions, form a very effective means of losing heat to the environment. In cold weather, and with strong winds, the ears are laid flat and streamlined along the back presenting a much smaller surface area, and less aerodynamic resistance. The appendages of animals usually have small mass: surface ratios in comparison with the rest of the body and are hence sites of good heat exchange.

Similarly with regard to the water balance of animals the amount of water available depends upon their mass but the rate of evaporation depends upon their surface area. Thus for small animals with small mass: surface ratios evaporative cooling is an impossible luxury which if indulged leads to rapid dehydration. Large animals can use water for thermoregulation and can survive longer without recourse to a water supply.

2 Climates of poikilotherms

(a) FISH

Most fish are observed to be almost ideally poikilothermic. The relation between body temperature and environment (water) temperature is very close to a perfect 1:1 relation as depicted in Figure 6.2a. The energy exchange between a fish swimming well below the surface and its surrounding water is a simple balance between metabolic heat production, and heat loss by conduction to the water, i.e. $Q_M = Q_G$. If the fish were close to the surface this balance may have to contain Q^* due to the downward transmission of short-wave radiation.

The inability of most fish to elevate their body temperature over that of the surrounding water is rooted in their method of respiration. To survive the fish must extract oxygen from the water, and this is achieved by passing the water over the fish's respiratory surfaces (gills). However, the dissolved oxygen content of water is very small compared to the oxygen content of air, and consequently the water must be continuously drawn through the gills. The water being flushed through the gills has a high heat capacity (Table 2.1) and forms a very effective energy sink for the warm arterial blood flowing from the heart to the gills for oxygenation. The oxygenated blood flows to the rest of the circulatory system, is utilized by tissues, and releases Q_M. This raises the blood temperature slightly but the heat gain is dissipated as the blood returns to the gills again. Thus the fish is unable to elevate its body temperature by more than a few tenths of a degree. Even an increased output of Q_M through greater muscular activity does not result in body warmth because greater activity means greater oxygen consumption, and this in turn means greater heat loss.

Although the temperature of most fish is intimately linked to the ambient water temperature to within 1°C, different species are found over a wide range of water temperatures from just below 0°C to almost 40°C. Species at either end of this range could not survive if immersed in water at the other end of the range, indeed sudden changes of less than 10°C may be fatal. To avoid such changes fish commonly migrate seasonally or even diurnally to seek out their preferendum. This may be achieved by swimming considerable distances, or to a different depth at the same location.

In contrast to this picture there are a few large, fast-swimming species (e.g. tuna, marlin, swordfish and mackerel shark) which can maintain significantly elevated body temperatures. They are able to do this because they possess a different circulatory system to that just discussed. They illustrate the principle of *countercurrent heat exchange*, and since this also applies to some homeotherms it will be discussed here. The crucial feature of a countercurrent heat exchange system is that it allows the flow of heat, and the flow of blood, to be uncoupled. Normally the arterial blood warmed in the body core flows out to the appendages, is cooled and flows back to the core via the veins. In a countercurrent system the arteries and the veins are closely packed together so that the cool venous blood is warmed by heat conduction from the warm arterial blood. Thus the *heat* flow is short-circuited, and shunted back into the core without reaching the periphery or appendages. When the arterial blood reaches the periphery it is cooler and closer to the ambient temperature, and therefore loses less heat to the environment. As a result the peripheral and appendage temperatures are much lower than those of the core. This efficient heat conservation technique is employed by engineers in industrial process design.

In fish with countercurrent circulation systems the main blood vessels are

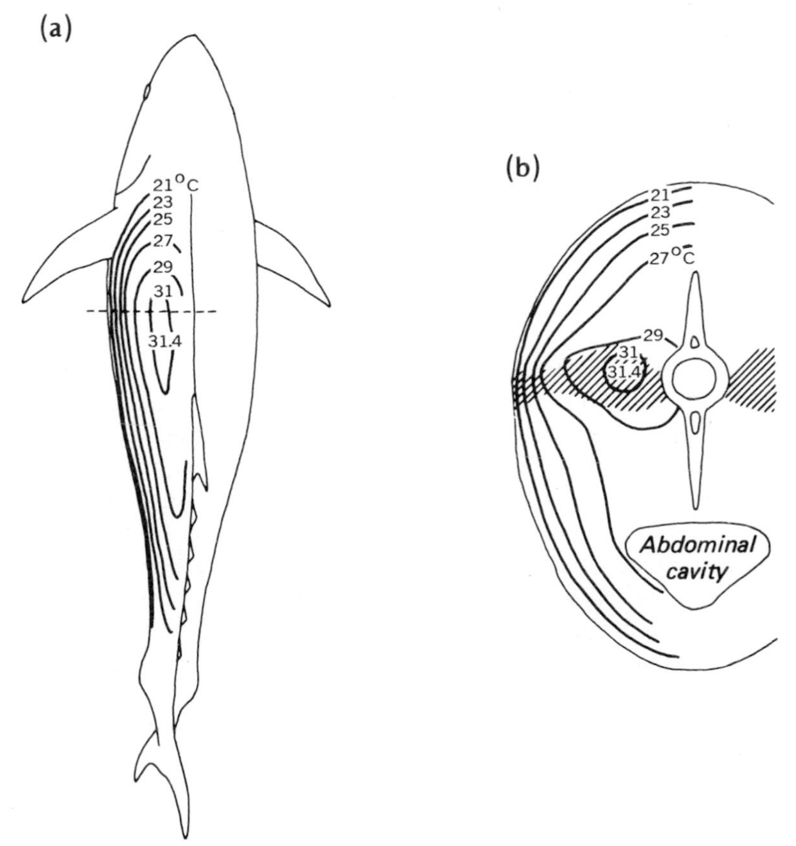

Figure 6.5 Temperature distribution *within* the body of a blue fin tuna swimming in 20°C water (a) in plan view, and (b) in cross-section through the position marked by a dashed line in (a) (after Carey, 1973).

smaller and more numerous, and their main arteries and veins are arranged in pairs. Even more important are the countercurrent nets called *retia* (tissue composed of closely intermingled veins and arteries) that lie adjacent to the main swimming muscles. Through the operation of this system these fish are able to maintain parts of their bodies at temperatures of up to 20°C above that of the surrounding water. An example of the temperature distribution in a blue fin tuna is given in Figure 6.5. Note that the body temperature is as much as 12°C above the water temperature, and that the warmth is concentrated in the muscles (shaded) used for propulsion. Elevated temperature greatly enhances muscle power and swimming speed. Birds, bats and some large insects need similarly high temperatures in order to gain muscle power for flight. The ability of tuna and similar fish to conserve Q_M and thermoregulate through countercurrent retia, gives them great mobility and less dependence on water temperature.

(b) AMPHIBIANS

Amphibians such as frogs, toads and salamanders behave thermally like fish when in water (i.e. remaining very close to water temperature), but act very differently when exposed to the complex energy environment on land. The key to their existence on land is their water balance. Because of the free flow of water through their skin they are continually in fear of desiccation, and they act very much like a wet-bulb thermometer. Evaporation is greatest if the amphibian is exposed to strong radiation in a warm, dry atmosphere with air movement. Under these conditions a true land-based reptile such as a python will lose about 0·1% of its body weight by evaporation in one day. An amphibian, like a salamander, on the other hand will lose water at a rate equivalent to 950% of its weight in one day!

Not surprisingly therefore amphibians employ behavioural thermoregulation, and seek out shaded, cool, moist and calm sites on land. Most have relatively low metabolic rates and can remain inactive without food for long periods, only stirring when conditions are amenable. Amphibians also tend to maintain relatively low body temperatures (5 to 25°C). Some frogs do allow elevation of their body temperatures over that in the environment by basking, but this is only possible if they make frequent trips for water, or sit partially submerged in water. This posture enables them to maintain a favourable water balance by acting like a pump; water is drawn in through the submerged portion of the skin, and evaporated from the area open to the atmosphere.

(c) REPTILES

The rate of metabolic heat production by reptiles, as with all poikilotherms, is rather low. They are also characterized by having relatively poor body insulation against heat loss. Therefore if a reptile is to survive, let alone be active, it must depend upon external energy sources. The prime such source is the Sun, and reptiles are especially adept at basking so as to maximize the positive aspects of their radiation budget. They are also able to gain lesser amounts of heat by conduction from warm surfaces. The main channels of heat loss are by conduction (to cooler surfaces), and by convection (to the air) of sensible heat. Evaporative cooling is not possible through the skin of reptiles, but in any case it would not be a feasible means of thermoregulation because of the danger of desiccation (p. 167).

As an illustration of the thermal climate of reptiles consider the example of a lizard in a hot desert environment. Lizards are most active when their body temperatures are between 30 and 40°C, although some can survive with temperatures as low as 3°C and others up to 45°C. Notice that the upper lethal limit is only a few degrees above the normal active range. Lizards are therefore susceptible to being killed by overheating because they have no rapid means of cooling physiologically, such as the process of sweating employed by some mammals. It follows that in order to maintain an acceptable range of body temperature they must alter their behaviour to exploit the thermal properties of their environment.

In the morning the lizard slowly emerges from its burrow, climbs onto a rock (above the cold surface air) and basks in the Sun. Commonly it optimizes the receipt of solar radiation by orienting its body to expose maximum surface area to the direct-beam at the optimum receiving angle (i.e. it is exploiting the cosine law of illumination, Figure 5.7). If its body becomes too hot the lizard seeks shade under or in a bush, behind a stone, or in a burrow. After sufficient cooling it reappears and seeks out another basking position utilizing solar radiation or heat conduction from warm surfaces. There is a subtle array of basking, shading, partial burrowing and other methods employed. The lizard is also capable of using a few physiological controls at high temperatures. These include the elevation of its heart rate to enhance dissipation of heat via the blood system to the exterior; increased Q_M output by muscular activity in large lizards; a slight increase in Q_E from the mouth as a result of increased breathing rate; and in certain species some ability to alter body colour and thus the absorption of solar radiation. In the evening when cooling becomes a problem the lizard retreats to its burrow and occupies a position at a depth where the nocturnal cooling wave is least felt. It becomes inactive, assumes a body temperature similar to that of the soil, and awaits the warming cycle of the next day.

The net result of these thermoregulatory processes may be a body temperature almost as stable as that of some homeotherms. The combination of behavioural and physiological responses enables reptiles to survive otherwise severe diurnal environmental fluctuations. In the cold season many reptiles hibernate in burrows, thereby eliminating activity, and maintaining Q_M as a weak energy source. In this mode the reptile strikes a simple poikilothermic balance with the surrounding soil by conduction (much like a fish immersed in water).

(d) INSECTS

Although there is a tremendous number of insect species they all have the general characteristic of being small relative to the other animals considered here. Therefore they have small mass : surface ratios, so that like leaves they are good energy exchangers but poor energy storers. This dictates that insects are generally poikilothermic and strongly subject to the surrounding thermal environment. Unlike leaves, however, they do not have a continuous external source of water to sustain significant rates of evaporation. Body water is obtained via food, drink and in some cases by absorption of water vapour. It is lost by evaporation through the insect's outer covering, its respiratory system, and by excretion. As noted previously (p. 167) small animals cannot allow free evaporation without rapid desiccation so it follows that insects are characterized by relatively impermeable exterior coverings, and Q_E is negligible in their energy balance.

During the daytime when an insect is exposed to the Sun and is active, the net heat gain via radiation (Q^*), metabolic heat production (Q_M) and conduction (Q_G) can become excessive so that ΔQ_S increases, and the insect approaches its upper temperature limit for activity, or even survival. With evaporation (Q_E) not performing an efficient role in dumping heat, the only heat loss channel remaining is sensible heat convection (Q_H). In many insects at least 80% of energy loss is via this process, especially if the insect is capable of flying.

To maintain its body temperature within a range suitable for activity (such as food foraging, predator avoidance, etc.) an insect exhibits a number of behavioural and, in some cases, physiological responses. Behavioural avoidance of overheating includes simple seeking-out of shade, shading of the body by the wings (by certain butterflies), or the encouragement of convective losses by flight. Some ants and beetles have long stilt-like legs which enable them to remain out of contact with the hot surface, other insects climb up plant stalks to enhance convective losses at a height where the wind speed is higher (Figure 2.9), and air temperature cooler (Figure 2.2), yet others burrow to avoid radiant loading entirely. Behavioural avoidance of cold includes a wide range of basking attitudes. For example, arctic butterflies orient their wings and bodies to obtain maximum solar radiation intensity (cf. lizards), or press themselves close to warm surfaces. They seek out sheltered locations for activity (e.g. gullies, creeks, etc.), and restrict flight to the lowest one metre of the atmosphere to minimize convective losses. These techniques enable some species to elevate their body temperature by as much as 17°C above that of the air, and to remain active even though air temperatures are low.

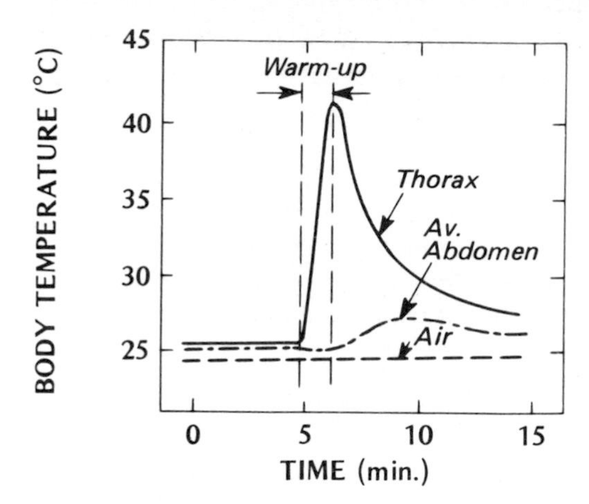

Figure 6.6 Thoracic, abdominal and air temperatures during the warm-up and cool-down phases of a queen bumble bee readying for flight. Warming is produced by shivering of flight musculature in the thorax and the resulting temperature rise of 15·5°C in 1·3 min is restricted to this same region. The slight drop in abdominal temperature during warm-up is probably due to evaporative cooling caused by increased respiratory movement (adapted after Heinrich, 1974).

Physiologic thermoregulation in insects is particularly well displayed in certain of the larger flying species (e.g. bees, moths, locusts, dragonflies, etc.). These insects have a restricted range of body temperature within which flight is possible. Commonly when at rest or walking, the insect's body temperature is close to that of the air, and thus in cool conditions is not always within the range required for flight. They are however able to elevate their body temperature by very rapid generation of metabolic heat within their flight musculature. In this activity the flight muscles are operated in a form of vigorous shivering which does not involve moving the wings themselves. This releases Q_{Mm} sufficient to produce a total metabolic output many times greater than when at rest, and this heat is conserved for use in flight by the following means. The heat is generated in the thorax (middle portion of the body where the wings are located), and not 'wasted' in heating the head or abdomen (Figure 6.6); the heat is trapped within the thorax by insulating air sacs which lie between the sites of heat generation (muscles), and heat loss (body wall); and radiation and convection losses from the body exterior are restricted by means of a coat of hairs or scales which forms an insulating layer of air around the thorax. These physiologic resistances to heat flow restrict the loss of Q_M, and as a result some large flying insects can elevate their thorax temperature by 20–30°C over that of the surrounding air! This amazing mechanism allows these insects to remain ready for flight, or actually fly, when conditions would otherwise dictate that they remain grounded.

These same physiologic features can however become a problem for large insects in flight because the heat build-up can result in overheating. Some cicadas simply have to cease flight in very warm weather and must seek out shade. This is acceptable because cicadas only eat when stationary. On the other hand many large moths eat whilst in flight, and overheating could become a problem (e.g. body temperatures exceeding 40°C) because their insulation restricts passive heat loss. This is overcome by pumping blood from the warmer thorax to the cooler uninsulated abdomen where the large surface area allows more efficient radiant and convective heat loss. Small, uninsulated flying insects such as fruit flies and midges permit rapid heat loss because of their very small mass:surface ratios, and thus heat build-up in flight is not a problem. Such insects are never more than 1°C above the ambient temperature.

The incubating queen bumble bee (Figure 6.7) presents a fascinating heat

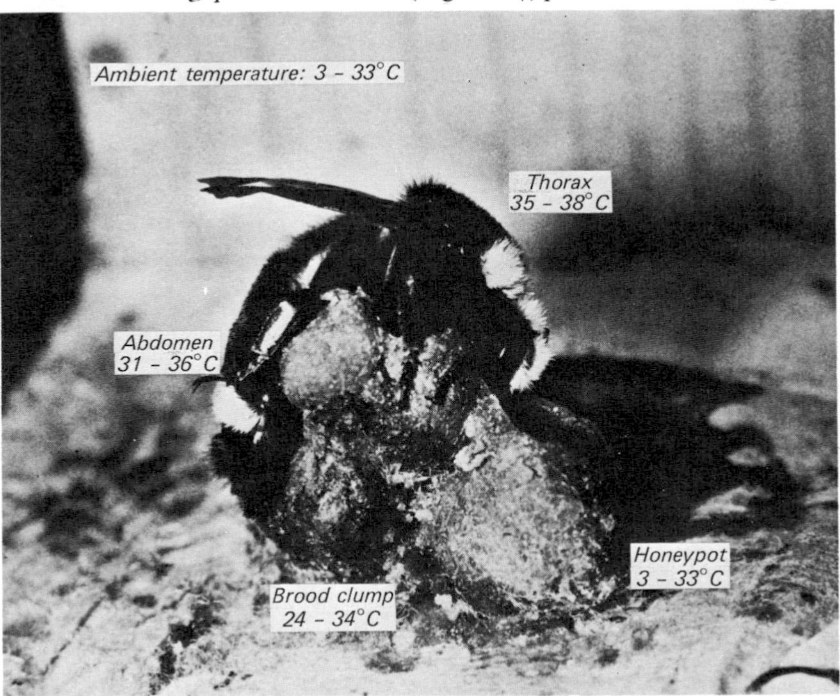

Figure 6.7 A queen bumble bee incubating her brood clump. She presses her abdomen onto the clump and faces the honeypot. Heat generated in the insulated thorax is pumped into the abdomen, the underside of which is not well insulated and is in good thermal contact with the clump (after Heinrich, 1974). Copyright 1974 by the American Association for the Advancement of Science.

balance model, utilizing both behavioural and physiologic means to heat her brood clump. The energy for the heating is provided by the sugar in the honeypot, which is consumed by the bee and converted into metabolic heat by shivering in the thorax. The heat is pumped via the blood to the un-insulated abdomen, which is in direct contact with the incubating pupae and eggs, and therefore is conducted to them. Convective heat loss is largely prevented from the thorax and the top and sides of the abdomen by an insulating pile of hairs. In the example illustrated this almost ideal arrangement maintained the brood clump in the range from 24 to 34°C, whilst the ambient air temperature ranged from 3 to 33°C.

There is also a larger energy balance problem to consider. The queen bee mentioned above was supplied with energy in the form of sugar collected by her workers. But for most other bees (and many other insects) it is necessary to weigh the energy used in activity (walking, flying, keeping the nest warm, maintaining body temperature, etc.), against that available as food. For the bee, nectar is the only source of food energy, therefore it must visit flowers to gain energy. However the process of foraging requires considerable energy expenditure. Therefore the bee must ensure that the energy receipts to be gained exceed the losses if it is to remain active. Factors which weigh heavily in this balance include the air temperature, the distance between flowers, and the energy reward provided by the flowers. The temperature is important because to maintain flight a bee must have a thoracic temperature of greater than 30°C. Air temperatures are commonly below this and so energy is required to elevate body temperature by the shivering process already described. Foraging at 5°C requires an energy expenditure that is two or three times that at 26°C. Nevertheless if the energy returns in the form of nectar are sufficient it may still be economic to forage down to near 0°C. In one example (Heinrich, 1974) a 0·12 g bumble bee kept its thorax at 37°C early on a frosty morning with a 2°C air temperature, and foraged on flowers which provided approximately 6·3 J of heat equivalent. In conducting this activity the bee expended approximately 3·3 J min^{-1} (approximately 0·056 W). Thus as long as it visited a new flower about once a minute it was making an energetic profit. In fact if it waited until midday when ambient temperatures were higher, and energy costs were lower, it may not have been as beneficial because the flowers had been visited by other insects and their energy reward was only about 1·3 J. Equally a lightweight insect like a bee can make use of flowers with hundreds of tiny florets (each of which only provide a minute amount of nectar) because it can visit many florets in a very short period of time by walking, which only consumes about 0·3 J min^{-1} (approximately 0·006 W, or about 100 times less energy than flying). Such flowers with small florets would be unattractive to larger, heavier insects which would have to hover to gain the nectar. The energy expenditure would be just too great. If air temperatures are low the bee may not elevate its thorax temperature during its more prolonged stay on such a flower, only doing so prior to flight to the next. In these circumstances the bee is not ready for flight and, if knocked off the flower, will fall to the ground.

3 Climates of homeotherms

(a) GENERAL FEATURES

Before considering the special bioclimatic features of different homeotherms (birds, and mammals including humans) we will outline more fully their pattern of thermoregulation as expressed in Figure 6.8. In this generalized diagram the relationships between environmental temperature and both deep-body temperature (T_b) and metabolic heat production (Q_M) are identical to those in Figure 6.2. To these have been added the characteristic partitioning of the animal's energy balance (equation 6.1). The individual energy balance terms have been simplified to one non-evaporative heat loss term (i.e. lumping together heat losses from the animal by radiation (Q^*), convection (Q_H) and conduction (Q_G)), and one evaporative heat loss term (Q_E). Changes of heat storage by the animal's body (ΔQ_S) have been neglected. In compliance with equation 6.1 the algebraic sum of the three energy terms is zero at all temperatures. The relationships depicted are strictly only valid for an animal that is resting but free to move about in a controlled environment. Therefore they do not hold for activity in a natural setting, but they serve to illustrate broad thermoregulatory concepts.

The key characteristic of homeotherms is their ability to maintain a relatively constant deep-body temperature in the face of a wide range of external environmental temperatures (called the *zone of thermoregulation*, BE in Figure 6.8). Within this zone they are able to make physiological adjustments sufficient to at least temporarily maintain T_b within acceptable limits. It should be noted however that although T_b does not vary significantly, the temperature of peripheral tissues (e.g. skin) and appendages (e.g. arms, legs, tail, ears, etc.) may change quite considerably.

Within the thermoregulatory range the deep-body temperature is held constant by controlling heat gains and losses, and the mechanisms to achieve this are triggered by neural stimuli similar in action to a thermostat. If heat losses begin to exceed heat gains the body temperature starts to fall thereby triggering the low-set-point 'thermostat' which activates the machinery for reducing heat loss. What is required is an increase in insulation and this can be attained through control of blood vessels, changes in insulating pelage (ruffling hair, fur, feathers), or changes in posture. If this still fails to stem the heat drain sufficiently it is necessary to increase internal heat production by metabolism until T_b returns to normal. The environmental temperature necessary to activate increased metabolic heat production is called the *critical temperature*, T_{crit} (C in Figure 6.8). This

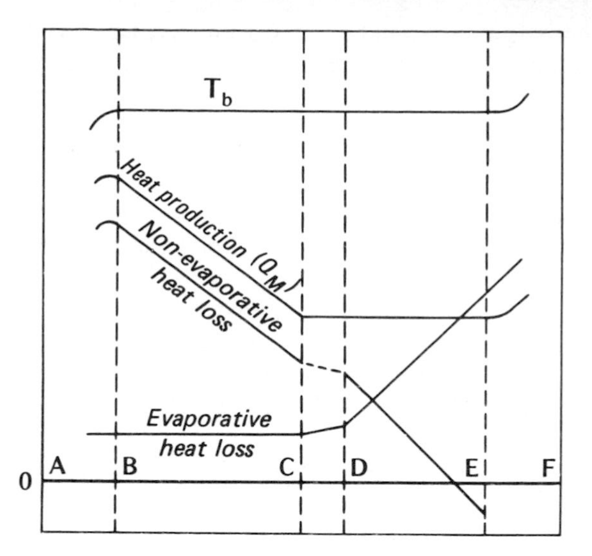

Figure 6.8 Generalized representation of the relationship between deep-body temperature and the energy balance components of a homeothermic animal over a range of environmental temperatures (after Mount, 1974).

Terminology:
- AB – zone of hypothermia
- BE – zone of thermoregulation
- CD – zone of least thermoregulatory effort
- CE – zone of minimal metabolism
- EF – zone of hyperthermia
- B – temperature of summit metabolism and incipient hypothermia
- C – critical temperature (T_{crit})
- D – temperature of marked increase in evaporative loss
- E – temperature of incipient hyperthermia

increase of Q_M cannot continue indefinitely however, and beyond a summit point the animal is unable to match heat losses by increasing heat production. This is the lower limit of thermoregulation (B in Figure 6.8) below which temperature the animal is defenceless and lapses into the zone of *hypothermia* where prolonged exposure is lethal.

If, on the other hand, heat gains edge ahead of heat losses and the body temperature starts to rise, this sets off the high-set-point 'thermostat' and activates the machinery for increasing heat loss through changes in blood flow or posture. Beyond a certain environmental temperature (D in Figure 6.8), these processes are insufficient and must be augmented by evaporative cooling (sweating and/or panting). At an even higher temperature (E in

Figure 6.8) the body has no means to prevent an increase of T_b, and the animal enters the zone of *hyperthermia*. If heat stress remains unalleviated death will ensue.

(b) BIRDS AND MAMMALS

The general structure of Figure 6.8 applies equally well to the thermoregulatory characteristics of both birds and mammals, and it is convenient here to consider them together. We will however deal with humans in a separate section (p. 188).

Even excluding humans the range of animal species and sizes to be covered under the heading is enormous. Therefore rather than attempt to examine their climates according to animal sub-divisions we will concentrate upon the methods by which thermoregulation is achieved, and illustrate these by the use of examples from many different species. A listing of these methods is given in Table 6.2, which also provides the framework for our discussion.

(i) *Responses to a cold environment*

In addition to the central 'thermostat' homeotherms are equipped with a

TABLE 6.2 Methods of thermoregulation in birds and mammals.†

Responses to cold environment	Responses to hot environment
Physiologic	
Insulation	Insulation
internal (core to skin); vaso-constriction, reduced heat rate, regional hypothermia	*internal*; vaso-dilation, changes in heart rate
external (skin to air); increased insulation of coat (piloerection, coat growth, lowered T_{crit})	*external*; changes in coat insulation, heat loss from uninsulated areas
	Storage – increased body heat storage
Metabolism – increased heat production (shivering)	Evaporation – increased water loss, sweating, panting
Hibernation, torpor	Estivation
Behavioural	
Posture – decreased surface area	Posture – increased surface area, orientation
Shelter – nesting, burrowing	Shelter – burrowing, movement to shade, nocturnal habit
Migration	Ingestion – intake of cool fluids
	Migration

† Including both immediate and long-term responses.

TABLE 6.3 Thermal resistance of the peripheral tissue and coats of animals; after Monteith, 1973.

Tissue	Resistance ($s\ m^{-1}$)	
	Vaso-constricted	Vaso-dilated
Steer	170	50
Man	120	30
Calf	110	50
Pig (3 months)	100	60
Down sheep	90	30
Coats	**Resistance ($s\ m^{-1}$) per mm depth**	**Percent of still air**
Air	47	100
Red fox	33	70
Lynx	31	65
Skunk	30	64
Husky dog	29	62
Merino sheep	28	60
Down sheep	19	40
Ayrshire cattle		
– flat coat	12	26
– erect coat	8	

series of peripherally-located 'thermostats' in the form of free nerve endings near the skin's surface which alert the central nervous system to changes in the external environment. If they sense a drop in temperature they start a chain of reflexes which lead to an increased insulation of the core by the body's shell. Animal insulation consists of three components: first, a layer of tissue fat and skin (internal insulation); second, a layer of air trapped within a coat of hair, fleece or feathers (external insulation); and third, the laminar boundary layer of air which surrounds all objects. The nervous impulses are capable of increasing the insulation provided by these three layers, and hence of creating an increased resistance to heat flow out through the shell.

The resistance of the first layer of insulation is increased by a reflex constriction of the fine blood vessels in the skin (Table 6.3). The withdrawal of blood from the periphery towards the core reduces the possibility of heat loss from the skin and conserves heat in the animal's interior. Correspondingly there is a slight rise of temperature in the core which reinforces the local skin reflexes so that the coat stands up more erectly. This process of *piloerection* effectively thickens the coat and leads to an increase in the thermal resistance of the second layer because a greater

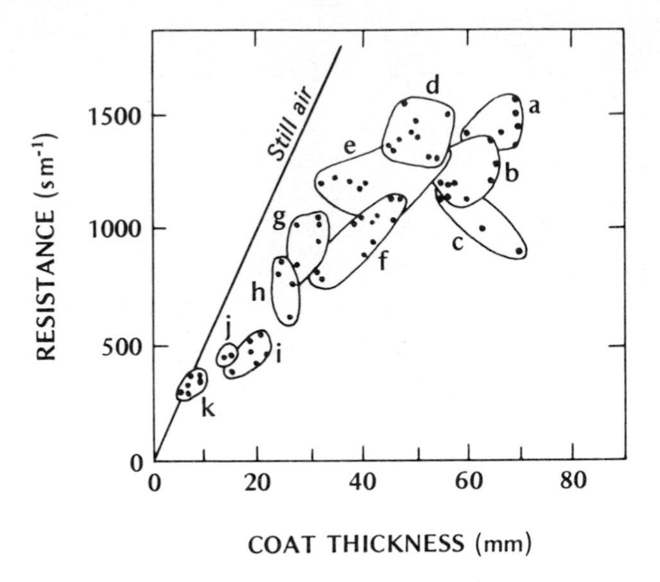

Figure 6.9 Thermal resistance of the coats of wild animals as a function of their thickness (after Scholander *et al.*, 1950).

(a – dall sheep; b – wolf, grizzly bear; c – polar bear; d – white fox; e – reindeer, caribou; f – fox, dog, beaver; g – rabbit; h – marten; i – lemming; j – squirrel; k – shrew)

depth of poorly conducting air is trapped. However, the increase is not linearly related to the thickness of the coat because free convection and radiation losses may increase as the spaces between the hairs, etc. open up. The third layer is affected by changes in the coat because the laminar boundary layer is usually thicker over rough surfaces which exert more drag on air movement. The thickness (and therefore the resistance) of this layer is however most strongly related to the wind speed (p. 103).

Permanent residents of cold habitats have thicker coats than related species from warmer areas, and the winter coats of animals subject to seasonal changes of temperature are thicker than in the summer. As Figure 6.9 shows, the thermal resistance increases with coat thickness for a wide range of wild animals. In these laboratory experiments the insulation efficiency was shown to be approximately 60% of that of still air (see also Table 6.3). In the natural environment the efficiency would be less, and a function of the wind speed. Body fat is also a very effective insulator and is usually found in greater amounts in animals from cold habitats.

It is not possible to keep all parts of the body equally insulated. In particular it is difficult to prevent heat loss from appendages such as arms, legs, flippers, ears and tail because of their relatively small mass:surface area ratios compared with the body trunk. One answer is to allow these extremities to cool well below the deep-body temperature so as to reduce the surface-to-environment temperature gradient and thereby curtail non-evaporative heat losses. Although this practice of *regional hypothermia* is helpful in terms of conserving body heat, it is potentially dangerous because the tissues will die if allowed to freeze. The problem is solved in many birds and mammals by the development of counter-current heat exchange systems (p. 168) in their appendages. An appropriate arrangement of blood vessels allows the warmer arterial blood to transfer much of its heat to the returning venous blood before it reaches the environmental heat sink, but still enables the blood to deliver oxygen to the peripheral tissues. In conjunction with seasonal adjustment of the fluid constituents of the appendages this system permits arctic birds and mammals (e.g. gulls, foxes and caribou) to stand on ice and snow surfaces that are well below freezing. Similarly arctic birds and seals (whose flippers are served with counter-current vessels) can swim in icy-cold water without distress.

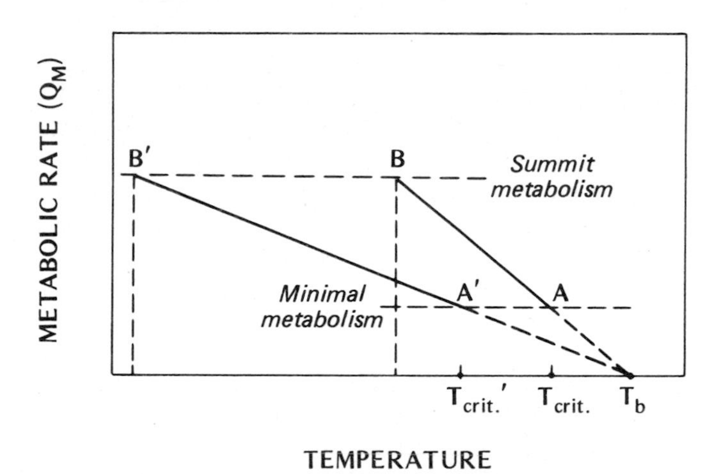

Figure 6.10 Generalized relationship between an animal's metabolic heat production and the environmental temperature. The different slopes *AB* and *A′B′* relate to the effect of different degrees of insulation (see text).

The fundamental importance of insulation to thermoregulation is demonstrated in Figure 6.10 which shows the shape of the metabolic heat production curve (Q_M in Figures 6.2b and 6.8) at two different degrees of animal insultation. As we noted earlier, below a critical environmental temperature (T_{crit}) a homeotherm must boost its metabolic heat output to keep pace with increasing non-evaporative heat losses, and thereby maintain T_b at a constant value. In Figure 6.10 this critical temperature is at point A. If temperatures continue to drop the metabolic output eventually reaches a summit value beyond which the animal is unable to match increasing losses (point B). This is the lower limit of thermoregulation below which it has no further physiologic defence. However if the animal's insulation were substantially increased the slope AB would become A′B′, pivoting on the same deep-body temperature, T_b. This has a number of related advantages. First, it lowers the value of T_{crit} to T'_{crit} (i.e. point A to A′) thus expanding the zone of minimal metabolism, and delaying the need to increase Q_M until it is much colder. As extreme examples the critical temperature of the Alaskan red fox is −10°C (Irving *et al.*, 1955), and values as low as −40°C have been reported for the arctic fox (Scholander *et al.*, 1950). The fur of these animals is an excellent insulator and they tolerate regional hypothermia in their appendages. Second, because of the lower slope of A′B′ the rate of heat production necessary to counteract a unit drop of environmental temperature is decreased. Third, the lower slope considerably lowers the temperature of summit metabolism and incipient hypothermia (point B′). It should be pointed out, however, that although in theory the lower limit is set by the animal's summit metabolism, in practice such high rates only apply to sudden rather than prolonged exposure because it is impossible to maintain them.

The outward signs of the need to increase Q_M are shivering, and increased activity. These are indications of reflexes attempting to increase the rate of muscular heat production (Q_{Mm}). Birds and a number of mammals shiver conspicuously. In addition many animals are able to increase Q_M without increased muscular activity. This feature is related to changes in body chemistry which develop after a period of exposure (of the order of days) to cold conditions.

Hibernation is another method of combating cold which is undertaken by small mammals (e.g. squirrels, mice, bats). Hibernation involves the temporary (usually cold season) abandonment of homeothermy – allowing T_b to drop thereby decreasing the metabolic costs for a given degree of insulation. It is a highly developed means of getting through the coldest part of the winter. It is especially necessary for small animals because they burn far more fuel per unit body weight than larger animals (p. 166), and the energy expenditure involved in foraging for sufficient food in a cold environment when food is scarce is simply prohibitive. Therefore the animal accumulates a large store of available food (either as an external hoard or as body fat), and then enters the energy-efficient state of dormancy. This involves a drop in body temperature at the rate of 2–5°C h^{-1} until it settles at about 1–4°C above the environmental temperature. Metabolism drops to about 1–5% of the normal rate, which is then approximately equivalent to that of a poikilothermic animal of the same body weight (see Figure 6.3). The rates of respiration and heart beat also slow down markedly. To prevent freezing hibernating mammals commonly burrow to below the depth of the annual frost line. The times of entry to, and arousal from, this state are probably related to certain threshold levels of environmental temperature. Other small animals (e.g. small rodents, bats, swifts, hummingbirds) undergo a daily cycle of activity and *torpor*. They maintain normal body temperatures when active but drop them during the cooler period when they are torpid.

Many of the behavioural means by which birds and mammals defend against excessive heat loss are the same as those utilized by poikilotherms. They can mostly be classed as attempts at avoidance. The long-distance migration of birds, bats, marine mammals and large-hoofed mammals clearly lies in this category. The building of nests and burrows can also be viewed as a behavioural means of extending body insulation. This is especially important in protecting the eggs and young of birds and mammals many of which are essentially poikilothermic for a period after birth. Their thermal balance is therefore largely dependent upon the heat balance of the parent, and the insulating properties of the nest, burrow, lair, etc. In birds, at the time of brooding the parent commonly develops a 'brood patch' on the thorax which is characterized by a lack of feathers (minimal insulation), and increased vascular development (enhanced heat flow by the blood). The patch is pressed against the eggs or chicks and provides an efficient pathway for heat to reach them from the parent via conduction. The feathered top of the parent and the light construction of the nest ensures maximum all round insulation. The attentiveness of the parent then depends upon such factors as the ambient air temperature, the effectiveness of the nest, and the ability to obtain food. The Q_M production of the embryo and young increases with size. Therefore attentiveness can be relaxed with time and this is necessary to permit foraging for the growing food demand. Finally, we should note that animals commonly assume postures which will minimize heat losses to the environment. The most common reaction to cold is to curl up thereby reducing the surface area from which heat loss can occur.

As a summarizing example consider the visible responses of a bird to extreme cold. In this situation a bird can be seen to fluff up its feathers in order to incorporate more air and hence increase its insulation properties; commonly it will shiver thereby increasing the muscular component of metabolic heat production; it may reduce its rate of breathing and tuck its beak under the feathers of its shoulder to prevent heat losses from its respiratory tract; and it may stand on one leg thus minimizing heat loss from its uninsulated extremities. The physical form that results from these actions is close to that of an insulated sphere, and that is the shape which possesses the minimum surface area per unit volume, and is ideally suited to energy conservation. A similar picture emerges in the case of the husky dog in a blizzard outlined earlier (p. 164). In fact the quality of the insulating shell formed by a curled up husky is so good that the snow underneath doesn't melt.

(ii) *Responses to a hot environment*

The deep-body temperature of most mammals is in the range from 35 to 40°C, and that of most birds is between 40 and 43°C. These values lie towards the upper end of the range of typical temperatures encountered in the atmosphere. It follows that the physiological mechanisms of mammals and birds have become reasonably well adapted to heat conservation (as we saw in the previous section), but should ambient temperatures rise above T_b their ability to dissipate heat to prevent over-heating is less well developed. This is particularly important when it is realized that the temperature of incipient hyperthermia lies less than 5°C above T_b for many species.

One of the initial short-term reactions to uncomfortably warm temperatures is to reduce the overall degree of insulation. Nerve stimulation on the periphery causes dilation of blood vessels and a corresponding drop in the thermal resistance of the skin (Table 6.3). These reflexes trigger the central 'thermostat' which further enhances outward heat flow by increasing the rate of blood circulation. The pilomotor muscles are relaxed and the depth of the coat becomes thinner thereby decreasing the external insulation.

On a seasonal basis most animals reduce coat thickness and the amount of dermal fat in the summer so as to facilitate heat loss. For example in horses and cattle the rougher winter coat is replaced by a thinner and sleeker-looking summer coat. In fact in addition to a lower resistance their summer coat has a higher albedo which helps to decrease the solar heat load. In the case of the camel the dermal fat (which is also a necessary food store) is restricted to the area of the hump. This permits the rest of the body surface to act more effectively in dumping heat by radiation and convection during the night. Some large mammals on the other hand exhibit *increased* insulation in hot conditions. The coats of camels and Merino sheep for example are exceptionally thick. This is helpful in insulating against heat flow from the exterior of the coat to the body. The very low conductivity of the hair (fleece) restricts the solar heating to the exterior of the coat where surface temperatures of 70°C have been measured on a camel, and 85°C on a Merino sheep. At the same time their deep-body temperatures remained at about 40°C.

As noted earlier the body covering of animals is necessarily incomplete. Whereas in cold habitats this presents a problem of heat loss, in hot conditions this provides a useful channel for heat dissipation. For example gophers, rats and beavers are able to utilize their large, uninsulated tails for thermoregulation. A gopher is capable of channelling up to 30% of its total heat loss through its tail even though it only represents 3·5% of its total body area (McNab, 1966). Elephants, rabbits and hares engorge their large ears with blood, and lose a large portion of their heat output via long-wave radiation to cooler surfaces, and/or by convection if the surrounding air is cooler than the ear surfaces. Similarly large losses occur from the legs of birds, and the less well insulated areas on the undersides of many mammals and birds. This may be aided in large animals by the fact that their shadow is cooler than fully irradiated surfaces, and therefore provides a 'built-in' sink for long-wave radiative cooling.

Despite insulative controls if an animal is faced with a situation where the environmental air is warmer than its body (i.e. the temperature gradient indicates a net heat flux from the environment to the animal) then it must accept one or both of two possibilities: it must allow heat storage in the body ($+\Delta Q_S$), resulting in an increase of T_b; it must utilize body water to sustain evaporative cooling (Q_E).

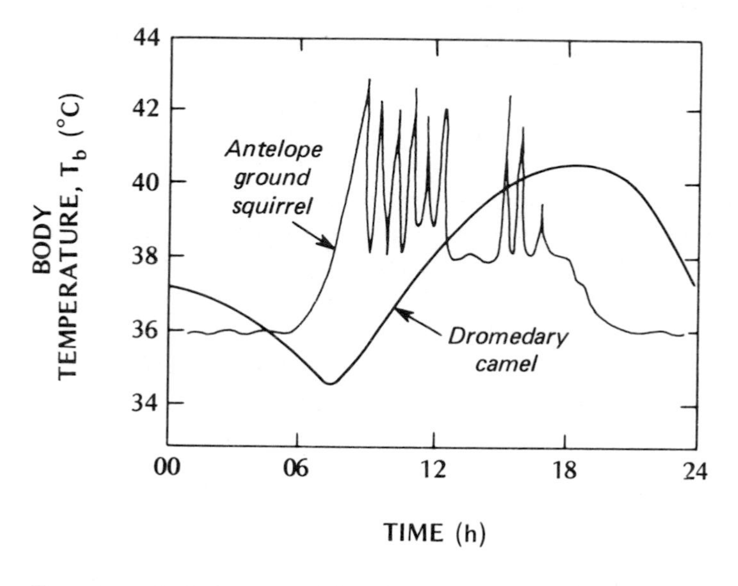

Figure 6.11 Daily patterns of deep body temperature in a large mammal (dromedary camel), and a small mammal (antelope ground squirrel) subjected to heat stress (after Bartholomew, 1964).

Most homeotherms cannot tolerate T_b fluctuations of more than a few degrees (except for short periods during intense activity), but some large mammals have sufficient mass that they can accept short period net energy storage without great distress. The dromedary camel for example allows its body temperature to rise as much as 6·5°C during the daytime (Figure 6.11). Since an adult camel weighs approximately 450 kg this represents a considerable amount of heat storage. The stored daytime heat load is dissipated by radiation and convection to the cooler surrounding environment at night. This facility has two major advantages: first, it reduces the need for the camel to use its precious water store for daytime thermoregulation; second, the increased body temperature reduces the daytime temperature gradient between the environment and the animal resulting in a decreased air-to-body heat flux.

Small mammals do not possess sufficient thermal inertia to behave in the same manner as the camel, yet the antelope ground squirrel (weight less than 0·1 kg) also utilizes body heat storage in combination with behavioural activity as a means of thermoregulation in the desert. By allowing its body temperature to rise to as high as 43°C it is able to create a temperature gradient from itself to the environment so that it can dissipate metabolic heat via conduction, convection and radiation when environmental temperatures are as high as 42°C. Bartholomew (1964) reports that the antelope ground squirrel is often active in the heat of the day, and this causes its body temperature to rise sharply. Then it disappears underground and unloads its heat in a cool burrow, following which it emerges again and resumes its activity. This results in the wild fluctuations of body temperature seen in Figure 6.11 whose periodicity is measured in minutes rather than the daily period of the camel.

Evaporative cooling (Q_E) becomes proportionately more important to thermoregulation as temperatures increase (Figure 6.8). It assumes this role because as the body and environmental temperatures converge the gradient necessary to support non-evaporative heat losses is progressively eroded. Indeed at very high temperatures the gradient is reversed, and sensible heat flow becomes an additional load upon the animal, and then Q_E is the *only* physiologic means of shedding body heat. However even this process breaks down if the animal's water store becomes depleted, or if the environmental vapour content is high enough to destroy the body-to-air humidity gradient.

Sweating is a process peculiar to mammals. It is well developed in the horse, donkey, camel, monkey and marmot, but less well in cattle, sheep and the pig. It is most effective if evaporation is restricted to the surface of the skin beneath the animal's coat. This enables the cooling to dissipate the metabolic heat production rather than being wasted in cooling the environmental air before it has reached the body. The coat insulation is therefore relied upon to buffer the skin from excessive heat input from the exterior.

Those mammals that either do not sweat, or only sweat weakly, usually accomplish evaporative cooling through their respiratory system by panting. Examples include the cat, dog, rabbit, guinea pig, pig, sheep and cattle. The main site of cooling is in the nasal sinuses, but can also occur in the mouth and lungs. The process is augmented by an increase in the respiratory rate in the form of rapid, shallow breaths. Dogs and sheep are capable of elevating their breathing from a normal rate of about 20 per minute to greater than 300 per minute during panting. This greatly increases the volume of air passing over moist respiratory surfaces. In the sheep panting can produce maximum cooling at the rate of about

60 W m^{-2} which represents 80% of the resting metabolic rate (Alexander, 1974). Birds are unable to sweat. To overcome this most pant, but others employ the process of *gular fluttering* whereby the bird opens its beak, and allows its throat to flutter rapidly, without affecting the rate of breathing. The main advantage compared with panting is that it does not require as much work (heat production).

Yet another way to achieve evaporative cooling is to wet the exterior of the coat. This can be achieved by licking, or by taking a water or mud 'bath'.

TABLE 6.4 Water balance of a kangaroo rat over a period of four weeks; after Schmidt-Nielsen, 1970.

Water gains (mg)			Water losses (mg)		
Oxidation water†	(*B*)	54·0	Evaporation	(*E*)	43·9
Absorbed water‡ (from the air)	(*B*)	6·0	Faeces	(*U*)	2·6
			Urine	(*U*)	13·5
	Input	60·0		Output	60·0

† Gained via the consumption and metabolism of 0·1 kg of barley.
‡ Atmospheric moisture absorbed by the barley grain.

Animals living in hot desert habitats continually run the risk of dehydration and cannot afford to expend water in thermoregulation. We have seen that body heat storage can be employed (e.g. camel and ground squirrel) in conjunction with insulative and behavioural adaptations. The kangaroo rat also demonstrates the possibility of maintaining a water balance without actually drinking water. Table 6.4 shows the water balance of this small desert mammal over about a four-week period. The total intake consists of the water it can extract from food both by oxidation in its metabolism by the body (water is always a by-product of the combustion of organic materials), and because dry grain absorbs water vapour from the air. The kangaroo rat is not equipped with sweat glands and minimizes the need for evaporative cooling by remaining in its burrow by day and foraging only at night. Even so it cannot avoid respiratory losses which form the main water output. Further water savings are effected by the ability to excrete very concentrated urine and almost solid faeces. These features allow the kangaroo rat to maintain a water balance without resort to drinking-water supplies.

A final physiologic means of combating excessive heat is *estivation*. This is a form of dormancy involving metabolic relaxation similar to hibernation from cold.

The behavioural responses to heat by birds and mammals are mainly aimed at simply avoiding stressful situations. These include changes of posture (stretching out to maximize the surface area exposed to heat exchange; lying on cool surfaces; orienting the body to minimize short-wave radiation receipt; flapping wings or ears to facilitate convective heat loss); seeking out shelter (burrowing to cooler depths, movement to shade behind vegetation, cliffs, caves, etc. and adoption of a nocturnal habit); and seasonal migration away from the stressful environment. Because of the disadvantageous mass:surface area relationship of small animals they face particularly severe problems which most are unable to solve physiologically. Their only means of survival then lies in ingenious patterns of behavioural response such as was outlined for a lizard (p. 171). This enables them to uncouple their thermoregulation from the normal homeothermic 'safety-valve' of evaporative cooling.

(c) HUMANS

The model of mammalian thermoregulation represented by Figure 6.8, and the list of physiologic and behavioural responses given in Table 6.2, need little modification to fit the case of humans. Perhaps the main physiologic differences relate to our erect posture, lack of a significant insulating coat, and the extent to which we employ sweating as a means of cooling. In terms of distinctive behavioural practices the provision of elaborate means of insulation (clothing) and shelter (housing) are the most notable. Aspects of all of these features will be mentioned here except for building climates which are covered in Chapter 7.

The deep-body temperature of humans is about 37°C, and a resting, clothed individual feels 'comfortable' at an environmental temperature of about 20–25°C. At this point the body is able to maintain a balanced heat budget with the least thermoregulatory effort. If the subject is indoors and the air movement is weak (say less than $0{\cdot}1\ \mathrm{m\ s^{-1}}$) the metabolic heat production is mainly dissipated to the environment by net radiation (60%), but also by convection of sensible heat (15%) and evaporation from the lungs and through the skin (25%). Based on the basal metabolic rate of $60\ \mathrm{W\ m^{-2}}$ for an adult this means that $Q^* \simeq 36\ \mathrm{W\ m^{-2}}$, $Q_H \simeq 9\ \mathrm{W\ m^{-2}}$ and $Q_E \simeq 15\ \mathrm{W\ m^{-2}}$. But if winds are higher the losses via Q_H and Q_E will be proportionately greater and those from Q^* reduced.

As temperatures fall below this comfortable range the body activates its thermoregulatory machinery to maintain balance. One of the first responses is the constriction of peripheral blood vessels thereby increasing the thermal resistance of the skin (Table 6.3) and reducing heat loss from the blood. As a result the skin appears whiter in colour. This is commonly accompanied by the appearance of 'goose bumps' on areas of exposed skin. The bumps are associated with the erection of skin hairs in an attempt to increase insulation in the same way as animals and birds fluff up their fur or

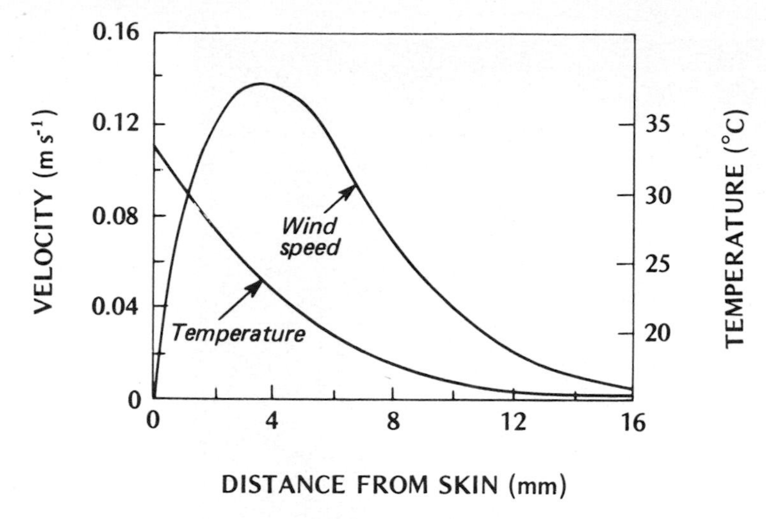

Figure 6.12 Profiles of air temperature and velocity in the laminar boundary layer adjacent to the skin of a human (after Lewis *et al.*, 1969). Note – skin temperature 33°C; ambient air temperature 15°C; ambient air calm.

feathers. However because of the sparse hair cover it is unlikely to be very effective although it may help to increase the thickness of the laminar boundary layer.

The boundary layer around the human body has been investigated by Lewis *et al.* (1969). They observed the nature of the layer around a subject, wearing only briefs, in a room with an air temperature of 15°C and no air movement. The boundary layer was found to grow in thickness with height, and to be in constant upward motion. Figure 6.12 illustrates the structure of the laminar layer. The temperature difference across the first 10 mm is 17°C (i.e. 1700°C m^{-1}!) which dramatically shows the thermal resistance provided by air. The velocity profile shows an increase with distance away from the retarding surface (cf. Figure 2.9), and a peak at about 4 mm where the upward flow is concentrated. It returned to zero at greater distances because the room air was calm. From the feet to a height of 1 m the layer was laminar (p. 33) but above 1·5 m (mid-chest) it became turbulent with a laminar layer immediately adjacent to the skin. A human heat 'plume' was seen to extend to a height of about 0·5 m above the head before merging with the ambient air. Over the face the boundary layer is disrupted by respiration of air from the nose and/or mouth (Figure 6.13). Some of the boundary layer is inhaled with each respiratory cycle, and since this layer contains up to four times more micro-organisms than ambient air, this flow provides a means of transporting bacteria to the body's interior. It may for example provide a connection between skin diseases and respiratory

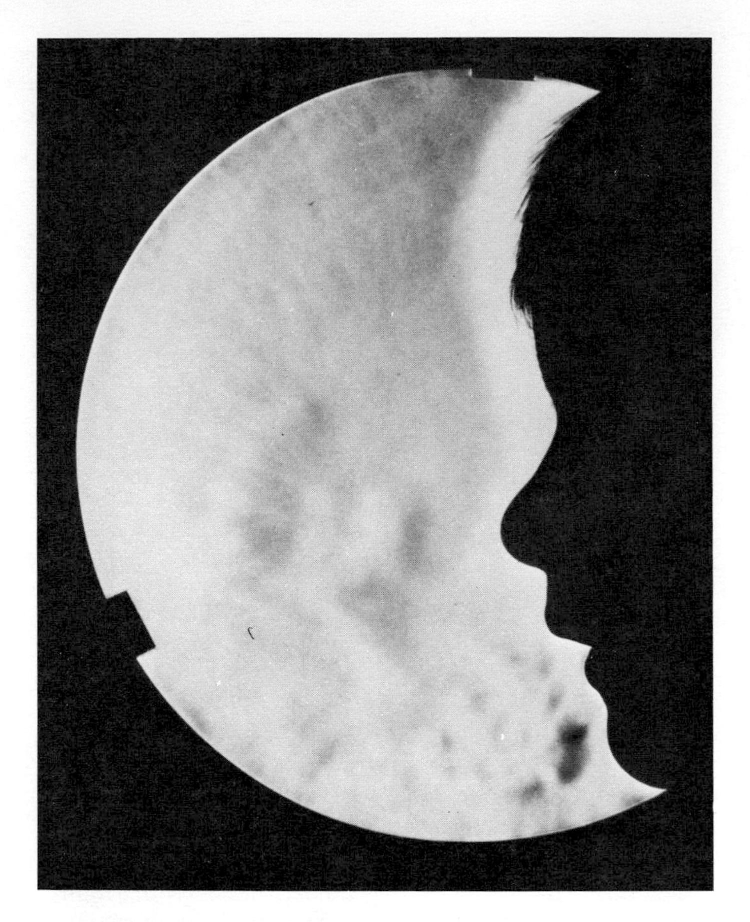

Figure 6.13 Schlieren photograph of a child's face showing regions of relatively warm air (light) and cool air (dark). The light layer immediately adjacent to the face is the laminar boundary layer (after Gates, 1972).

ailments (e.g. skin eczema in children is often followed by asthma). Similarly, since the velocity of the layer is related to the skin-to-ambient-air temperature difference it may explain why the number of respiratory infections tends to increase a few days after a drop in air temperature (Lewis *et al.*, 1969). Although the above results relate to a semi-unclad subject the same investigators report that the boundary layer is also present on the outside of clothing.

At a *skin* temperature of about 19°C most clothed humans start to shiver signalling the need to increase metabolic heat production through muscular activity in order to balance the heat losses to the environment. There also tend to be behavioural responses such as putting on more clothes, curling up, or extra activity, although these are triggered by more than just the climatic environment. As cooling progresses the blood flow to the appendages is reduced and they are allowed to cool. The arrangement of arterial and venous blood vessels in the arms and legs is conducive to counter-current heat exchange (p. 168) thus conserving heat in the core but maintaining sufficient blood flow to keep tissues alive. However, if skin temperatures fall more than a few degrees below freezing-point the area may suffer 'frostbite' including the actual formation of ice crystals in the cells. Because of their peripheral location, and small mass: surface ratios, the fingers, toes, nose and ears are most susceptible. If heat loss continues the deep-body temperature must eventually drop. When T_b drops below 35°C thermal control is lost, and below 26°C death is almost inevitable.

For humans the coldest atmospheric conditions are associated with both low air temperatures and strong winds. The rate of heat loss by radiation and especially convection then becomes extremely high. The term *windchill* is used to express this cooling effect and the *windchill equivalent temperature* (obtained by an empirical formula, Table 6.5) gives some impression of how cold a given wind/temperature combination would feel on exposed

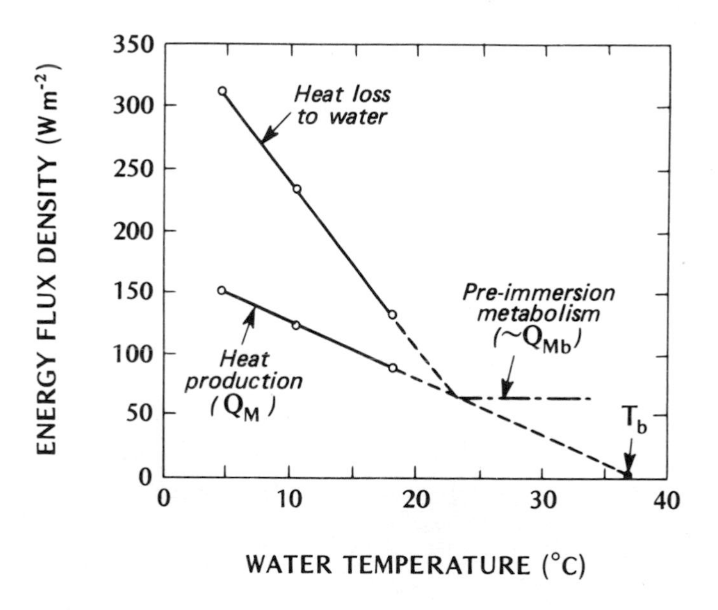

Figure 6.14 Comparison of calculated rates of human heat loss, and measured rates of heat production, at different water temperatures. Subjects wore light clothing and remained still with their head above water. The water current was approximately 0·1 m s^{-1}. The pre-immersion rest metabolism is slightly greater than the basal rate (Table 6.1). Data from Hayward *et al.* (1975) converted to flux density values using Du Bois area (footnote, Table 6.1).

flesh. Table 6.5 shows that at an actual air temperature of −10°C with a 15 m s^{-1} wind it would feel as cold as if it were −25°C with a 2·23 m s^{-1} wind.

TABLE 6.5 Windchill equivalent temperature† (°C); after Steadman, 1971.

Actual temperature (°C)	Windchill equivalent temperature			
	Actual wind speed (m s^{-1})			
	Calm	5	10	15
0	1	−2	−7	−11
−5	−4	−9	−13	−16
−10	−9	−13	−19	−25
−15	−13	−19	−26	−33
−20	−18	−26	−34	−42
−30	−28	−37	−50	–
−40	−37	−50	–	–

† Giving the temperature at which a wind of 2·23 m s^{-1} would give equivalent cooling.

An even greater challenge is presented by immersion in cold water because it provides a very efficient heat sink. Figure 6.14 shows that a human immersee who remains still in the water loses heat by conduction at about twice the rate of metabolic heat production. Thus the colder the water, the greater the energy imbalance, and the shorter the period of survival. It might therefore be thought that swimming would be beneficial by increasing the rate of heat production. But on the contrary Hayward *et al.* (1975) show that although swimming does increase Q_M by about 250%

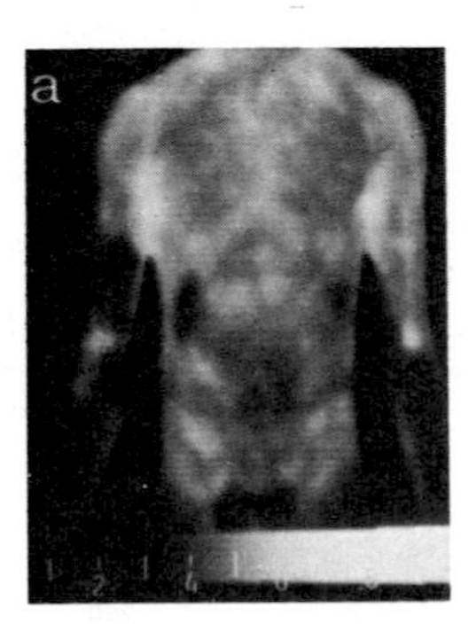

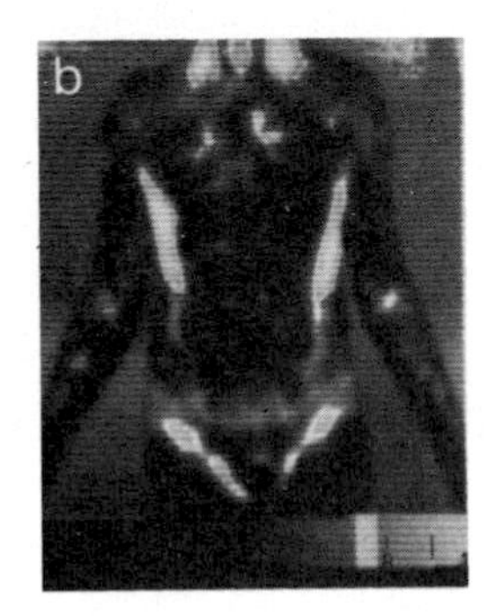

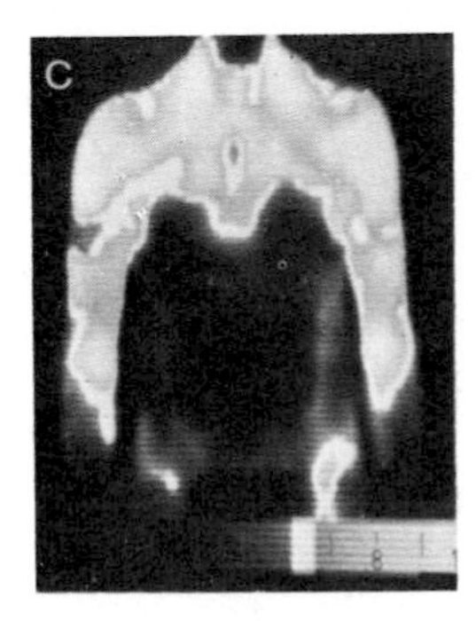

Figure 6.15 Thermograms of the front view of the trunk of male adults (a) before immersion in water at 7·5°C, (b) after holding still in the water for 15 minutes, and (c) after swimming for the same period. The lighter areas are warmest and indicate areas of high heat loss. A bright isotherm outlines the warmest regions (after Hayward *et al.*, 1973).

this is more than offset by greater body heat losses. In their experiments the net result of swimming was to produce a 35% greater rate of body cooling (i.e. the *difference* in slope between the two lines in Figure 6.14 was increased). The areas of greatest heat loss were revealed by the use of thermograms of the body taken prior to immersion, and both after holding still and after swimming in 7·5°C sea water for 15 minutes (Figure 6.15). Before immersion (a) the body shows a fairly uniform temperature distribution, but after holding still in the water for 15 minutes (b), there is a concentration of warm areas in the lower neck, central chest, lower abdomen and especially in the lateral thorax and groin areas. These regions are characterized by having little muscle and fat between them and the heat core of the body. After swimming (c), the warm areas are very much more extensive over the upper chest and arms. It follows that unless an accidental immersee is sure to reach safety by swimming for a short distance it is better to prolong survival by remaining still. In 12°C water the maximum distance that can be swum without protection (such as a wet suit or body grease) is about 1·3 km. The ideal posture to prolong survival is a huddle, with the arms close to the sides of the thorax and the legs drawn up to decrease heat loss from the groin.

On the other hand if environmental temperatures rise above the comfortable range the body seeks to lose sufficient heat to maintain balance. But this time it cannot regulate the metabolic output because it is already at the minimum level, and as temperatures rise this heat production becomes an increasing liability. Non-evaporative losses provide little relief because as external temperatures rise the body-to-air thermal gradient diminishes. At about 35°C convective heat losses become zero, and radiative losses become small unless there is an available cool spot in the environment to act as a long-wave radiative sink. Above 35°C sensible heat convection becomes a net heat *source* for the body thereby adding to the problem.

The body has two main physiologic means of combating an increasing heat load. Initially it can decrease internal insulation by dilating blood vessels in the skin (Table 6.3). The increased blood flow to the periphery causes the skin temperature to rise thereby increasing the skin-to-air temperature gradient, and aiding non-evaporative losses, but this is limited to skin temperatures below 35°C. This process is responsible for the progressive reddening of skin colour in hot conditions.

The second and most powerful response is to increase evaporative cooling by sweating. Below about 25°C evaporation cooling is due to losses through the lungs, and by diffusion through the skin, but above about 28°C true sweating occurs wherein water is exuded on to the skin, and then evaporated. At environmental temperatures above 35°C all body heat (both metabolic and non-evaporative heat gains) are expended as Q_E. The rate of evaporation cooling depends on the body-to-air vapour pressure

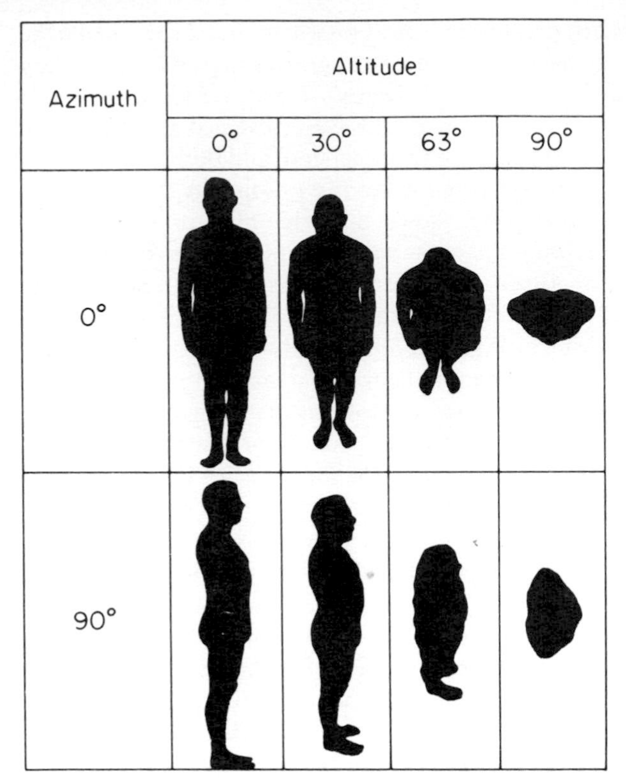

Figure 6.16 Silhouettes of an erect male corresponding to the areas illuminated by direct-beam short-wave radiation at different values of solar azimuth and altitude (after Underwood and Ward, 1966).

gradient and the rate of air movement. Therefore in hot, dry, windy environments Q_E is easily able to pump latent heat away from the body. But in a hot, humid atmosphere with little ventilation even this channel of heat loss is stifled. The situation may be further aggravated by the fact that sweat contains salt which depresses the saturation vapour pressure in comparison with that of pure water. This further limits sweating unless salt accumulation is removed by sweat drips.

One feature of humans which is of help in minimizing heat gain is our erect posture. Figure 6.16 shows the area of a human that is presented to the Sun's beam at different angles of solar altitude and azimuth. Note that at the times when solar radiation is most intense the body presents its smallest area. In contrast, four-legged animals expose their maximum area at solar noon. The advantage of an erect posture is therefore most clear in the tropics where the Sun is high in the sky for most of the day, and near the zenith at noon. On the other hand it might seem to be a paradox that dark-skinned peoples populate the tropics because the albedo of black skin is about 0·18, whereas for white skin it is 0·35. The crucial point however is not the albedo but the depth to which the radiation penetrates. In black skin short-wave radiation penetrates to a maximum depth of 0·4 mm (i.e. not piercing the epidermis), but in white skin it reaches 2 mm (well into the dermis). Thus although black skin absorbs better, the heat is contained near the surface of the skin where it can be more easily lost, whereas in white skin the heat is taken into the blood and contributes more to the general body heat storage problem.

Sweating gives relief from the heat load but places stress upon the body's water balance. Table 6.6 shows the daily water balance of a man doing light

TABLE 6.6 Water balance of an adult man doing light work over a period of one day (based on data of Slager, 1962 and Schmidt-Nielsen, 1970).

Water gains (kg)		Water losses (kg)	
Oxidation water† (*B*)	0·35	Evaporation – sweat (*E*)	0·50
		– lungs (*E*)	0·30
Water in food (*B*)	0·30	Faeces (*U*)	0·10
Drinking water (*B*)	1·50	Urine (*U*)	1·25
Input	2·15	Output	2·15

† Gained via the consumption and metabolism of 0·55 kg of food.

work in a moderate environment. Increased activity or larger external heat loads would result in much greater sweat output. The maximum rate of sweating in a normal man is about 1 kg h^{-1}, giving a maximum evaporative loss of about 375 W m^{-2} for an adult (Monteith, 1973). Even higher rates have been measured for short periods in extreme heat but they cannot be sustained because of the danger of dehydration. An average man weighs about 80 kg and when he loses more than 2% of his body weight by sweating he becomes extremely thirsty, at 4% his throat becomes dry and he feels apathetic and impatient, at 8% speech becomes difficult, beyond 12% he cannot care for himself or swallow and at 18–20% dehydration is lethal. Therefore to maintain a reasonable water balance humans need easy access to drinking water, but to replenish water loss adequately it is also necessary to regulate the rate of uptake.

The primary thermoregulatory role of clothing is insulation (i.e. giving a greater peripheral thermal resistance). In cold climates it prevents heat loss from the body and in hot climates it prevents excessive heat gains. As with an animal's coat the degree of insulation depends upon the structure of the clothing, including not only its thickness but also the amount of air it encloses. The insulating quality is also dependent upon the amount of

TABLE 6.7 Clothing design characteristics for use in extreme climates (modified after Mather, 1974).

Cold, dry (frozen)	Cold, wet (unfrozen)
Layered clothing, small air spaces	Layered clothing, small air spaces
Absorptive layer next to skin	External vapour barrier
Cover extremities	Absorptive layer next to skin
Head covering	Head covering
Respiratory preheating	Loose fit to prevent overheating
Loose fit to prevent overheating	
Face mask to prevent frostbite	
Light colour with darker layer beneath to absorb radiation *within* clothing	
Hot, dry	**Hot, wet**
Close weave	Open weave
Light colour for reflection	As thin as possible
Moderate thickness	Minimum coverage
Body and head covered	Good fit
Loose fit	Minimum underclothes
Underclothes desirable	

moisture contained, and the wind speed. If the clothing becomes wet, either as a result of rain or snow, or by perspiration, the thermal resistance drops sharply.

Table 6.7 gives some basic design characteristics for clothing to be worn in extreme climates. In cold, dry conditions the aim is to control heat losses. Clothing should be of many loose-fitting layers to maximize air content, but of a close-weave construction to reduce wind penetration. An absorptive layer next to the skin prevents perspiration from wetting the main insulating layers, and the head must be covered since this can be a site of major heat loss. A face mask helps to prevent frostbite and if it partially covers the nose and/or mouth it helps to reduce cooling in the lungs by mixing the inspired air with the warmer exhaled air. In cold, wet climates it is essential to maintain an exterior barrier to moisture.

In hot, dry climates clothing remains important but mainly to provide body shade, and to reflect solar radiation. The fabric should be closely-woven to prevent radiation penetrating to the skin, yet thick enough to provide some insulation against conductive gains. All clothes must be loose thereby allowing sufficient circulation for sweating. In hot, wet conditions clothing should be very light in weight and with a minimum of layering.

PART 3

Man-modified atmospheric environments

This part of the book deals with the consequences of human interference in otherwise natural climatic systems. In some cases the intervention is planned to 'improve' the atmospheric environment for specific human uses (e.g. frost protection). This is classed as intentional modification *of climate and is covered in Chapter 7. It will be shown that the modification results from changes in the solar and hydrologic cascades and/or the local wind flow.*

In many other cases the intervention is not planned. The atmospheric modification occurs as an unintentional side-effect of human activity. This inadvertent modification *of climate is brought about either through alteration of the surface cover (e.g. by farming, forestry, urbanization, etc.), or by direct atmospheric contamination by pollutants. The often subtle effects of surface change are dealt with in Chapter 8, and the problem of air pollution in Chapter 9.*

In practice the dividing line between intentional and inadvertent climate modification is obscure. For example a house is planned to provide a controlled interior climate, but its presence on the landscape leads to unplanned alterations to the exterior thermal and wind environments.

CHAPTER 7

Intentionally modified climates

1 Surface control

(a) ALBEDO CONTROL

The surface albedo (α) is a fundamental surface property and one which can be relatively easily altered by simple surface treatment. The value of α directly determines the absorptivity of an opaque surface. Thus for a given solar input ($K\downarrow$) it regulates the surface short-wave absorption (K^*), and this in turn dominates the daytime net radiation budget (Q^*). This sets the limit to the surface energy balance, and the inter-related water balance, and thereby controls the thermal and moisture climate of the surface and the adjacent air and soil layers. Manipulation of α therefore invokes a considerable climatic chain reaction.

Table 7.1 gives an example of the changes brought about by the application of a surface dressing of magnesium carbonate (white powder) to a bare soil in Israel. The treatment increased α from 0·30 to 0·60, thereby doubling the short-wave reflection ($K\uparrow$) and approximately halving K^*. Despite the decreased net long-wave loss (L^*) from the cooler surface of the whitened soil, the net all-wave radiation was cut from 6·2 to 1·1 MJ m^{-2}. The reduced availability of energy and the lower surface temperature combined to reduce the evapotranspiration rate (Q_E) by about 20%. The existence of this 'cool' plot in an otherwise hot environment produced a micro-'oasis effect' (p. 142) with sensible heat from the atmosphere (Q_H) acting as a heat source for the surface. On a daily basis the soil heat flux

TABLE 7.1 Radiation and energy balance components for bare and whitened soil. Mean daily values (MJ m^{-2} day^{-1}) for July and August at Gilat, Israel (based on data from Stanhill, 1965).

Component	Bare soil	Whitened soil
$K\downarrow$	27·2	27·2
$K\uparrow$	8·2	16·3
K^*	19·0	10·9
L^*	−12·8	−9·8
Q^*	6·2	1.1
Q_H	1·9	−2·5
Q_E	4·1	3·4
Q_G	0·2	0·2
T_0 (°C)	33	28
Derived terms		
α†	0·30	0·60
β†	0·46	−0·74
Q_E/Q^*†	0·66	3·09
E (mm)	1·80	1·50

† Non-dimensional

(Q_G) was not different between the two plots, but over shorter periods the whitening produced a marked reduction and the maximum surface cooling reached 10°C.

The results of a similar experiment over short grass are given in Figure 7.1. One plot was whitened with talc powder, one was blackened with carbon-black, and a third remained untreated as a control. On a fine summer day the maximum near-surface (10 mm depth) temperature in the black plot was more than 6°C warmer, and in the white plot more than 8°C cooler, than in the control. At night when the direct influence of albedo was absent the differences were small, although the coolness of the white plot was retained.

Clearly albedo-induced changes in the thermal climate of the soil can be substantial, and as a by-product soil moisture can be affected because of evapotranspiration changes (Table 7.1). In the case of the plots in Figure 7.1, after three weeks treatment the black plot had 50% less soil moisture in the uppermost 10 mm, and the white one had conserved 50% more, when compared with the control.

Lowering α by surface blackening is especially effective in altering the absorptivity of ice and snow surfaces which otherwise reflect the majority of $K\downarrow$. This fact has been utilized in projects to hasten snow melt from fields and catchment basins, and to melt sea ice and icebergs to keep shipping lanes open.

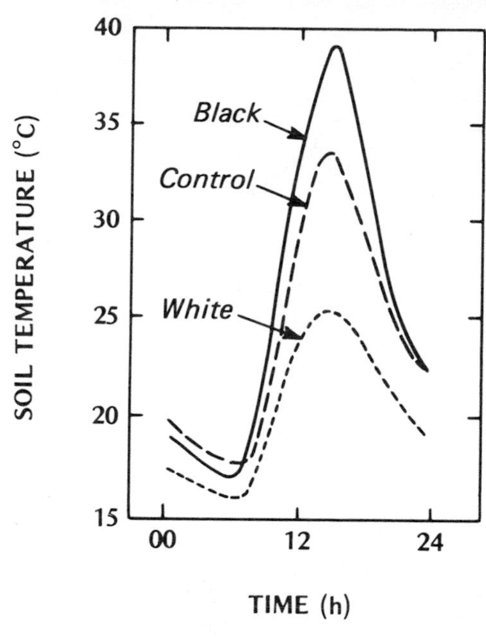

Figure 7.1 Effect of albedo change on near surface (10 mm) soil temperatures. Data from white-, black-coloured and control (short grass over fine sandy loam) plots on 16 July 1964 at Hamilton, Ontario (43°N) (modified after Oke and Hannell, 1966).

The effects of different albedos can also be usefully employed in the choice of building materials (e.g. walling and roofing, paints, window glass), clothing, aircraft and car exteriors.

(b) GEOMETRY CONTROL

Better use of available short-wave radiation can often be accomplished by manipulating the geometry of receiving surfaces to take advantage of the cosine law of illumination (equation 5.2). One obvious example is the practice of making ridge and furrow geometry in agricultural fields (i.e. by considering micro-topography). At locations outside the tropics horizontal surfaces never experience direct-beam solar radiation (S) in the zenith, and hence never receive the maximum radiation intensity. By making ridges the sunlit slopes are capable of receiving S at or near their local zenith (i.e. Θ is small: Figure 7.2a). This is especially useful in the spring when heating is critical for soil drying and seed germination, but when the Sun's elevation is low. The row orientation (local aspect) is also important. In N–S rows both facets of the ridge are irradiated during the course of a day, whereas in E–W rows in the northern hemisphere the south-facing slope is particularly favoured (see Figure 5.9).

Ridge and furrow geometry also provides a radiative 'trap' for both S and the outgoing long-wave, $L\uparrow$ (Figure 7.2b, c). Convoluting the surface tends to trap S because after initial reflection from a facet of the system there is at least some chance of the reflected (diffused) radiation encountering another surface before exiting to the atmosphere. This

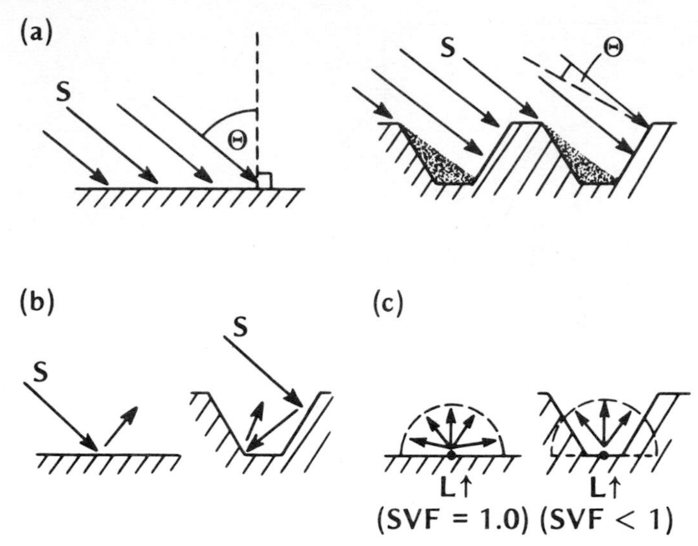

Figure 7.2 Role of surface geometry in radiation exchange. Comparison of horizontal and convoluted (e.g. by ridge and furrow) surfaces in terms of (a) receipt of direct-beam short-wave radiation (S), (b) reflection of S, and (c) emission of long-wave radiation ($L\uparrow$).

increases the chance of absorption and hence decreases the albedo of the field compared with the level case. On the other hand the emission of $L\uparrow$ from *within* the furrow is relatively restricted due to the reduced sky view factor (SVF, p. 112). For all locations in the furrow the SVF is less than unity because the overlying hemisphere they subtend includes at least some view of other parts of the furrow. Hence for these positions the cold sky sink is 'contaminated' by warm furrow surfaces. Trapping of S by day tends to increase the maximum soil temperature; and trapping of $L\uparrow$ tends to reduce surface cooling. The overall response is to increase soil temperature in a ridge and furrow field system.

The influence of radiation geometry has been exploited in a wide range of other practical applications, for example crop-row spacing and orientation can be managed so as to maximize or minimize the penetration of light and the trapping of short- and long-wave radiation. Light-coloured stone walls can be used to reflect extra short-wave onto plants, and to reduce their SVF so as to decrease $L\uparrow$ losses. The high heat capacity of stone (Table 7.4) also makes the walls a good heat source at night. Solar energy collectors for cooling or heating obviously require careful orientation, or even a continuous means of tracking the Sun in order to optimize receipts.

(c) MULCHING CONTROL

Mulching is the practice of placing a moisture or heat barrier over the top of

TABLE 7.2 Components of the radiation and energy balances of a bare soil plot, and three similar plots covered by mulches. Data (W m^{-2}) are for 11 June 1959 at 1144 h at Hamden, Connecticut (41°N) (modified after Waggoner *et al.*, 1960).

	Bare soil	Black plastic	Paper	Hay
Radiation budget				
K^*	819	993	631	840
L^*	−177	−282	−199	−233
Q^*	642	711	432	607
Energy balance				
Q_H	363	635	349	488
Q_E	195	0	42	84
Q_G	84	77	42	35
Derived term				
α†	0·24	0·08	0·42	0·22

† Non-dimensional, estimated

the soil. The purpose is usually to conserve soil moisture by reducing evaporation (E), but it may also be used to enhance soil warming or to prevent excessive cooling, depending on the mulch used. The traditional mulch consists of a well-aerated, and therefore poorly conducting, surface cover. This can be achieved by tilling the upper soil layer so as to introduce more air, or by covering the surface with insulating materials such as hay, straw, leaf litter, moss, wood chips, sawdust or gravel. More recently artificial mulches have become common including foam, plastic films (opaque, translucent or coloured), paper, and aluminium foil.

Table 7.2 and Figure 7.3 show the energy balance and thermal changes produced by three different mulches (black plastic, paper and 60 mm of hay) over a fine sandy loam in Connecticut. The bare soil (control plot) energy balance is typical of a fairly dry site, with approximately 57% of Q^* lost via Q_H, 30% as Q_E, and 13% as Q_G. The soil thermal conductivity was relatively low (approximately 0·42 W m^{-1} K^{-1}). This probably accounts for the rather low value of Q_G, and the fact that the soil surface temperature was as high as 38°C.

As a result of its lower albedo (Table 7.2) the black plastic absorbed radiation more efficiently than the bare soil and its surface temperature rose to about 50°C. However this heat could not easily be passed down into the soil because of the insulation provided by the air trapped beneath the sheet. Neither could it be dissipated as latent heat because the plastic was dry and impervious to water (Q_E set at zero). The only remaining channel for heat loss was upwards as sensible heat (Q_H). The soil surface

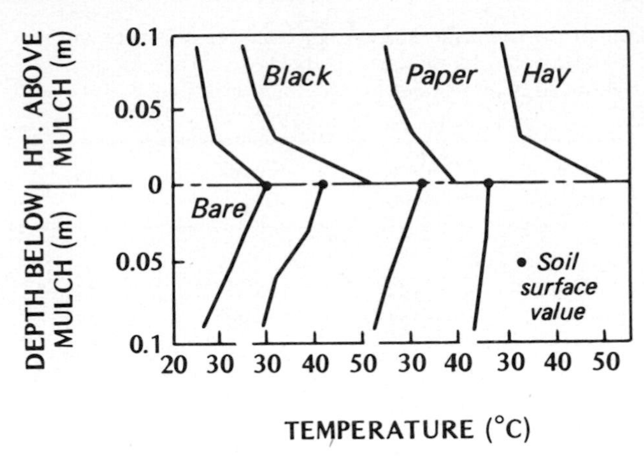

Figure 7.3 Profiles of air and soil temperature associated with a bare soil and three different mulches near midday at Hamden, Connecticut on 11 June 1959 (modified after Waggoner *et al.*, 1960). (Note – vertical scale does not include depth of mulch.)

temperature remained very slightly warmer than the bare soil, especially at night (not shown) when the plastic would have prevented much of the L^* loss from the soil surface. The mulch therefore conserved soil moisture, kept soil temperatures equable, and acted as a source of sensible heat for the atmosphere.

The paper mulch exhibited a very much lower radiation budget (Q^*) than any of the other surfaces primarily because of its relatively high surface albedo. Despite this the surface temperature of the mulch (approximately 40°C) was equivalent to that of the bare soil because the heat was not effectively conducted downwards. Therefore the soil heat flux was halved, the soil surface temperature was about 6°C lower, and the evaporation (and therefore water loss) was considerably reduced in comparison with the control plot.

Hay is characterized by a very low conductivity. Thus although the hay mulch albedo was similar to that of the bare soil its surface temperature was almost as warm as that of the black plastic. The heat transmission through the hay was so poor that the soil heat flux was the weakest, and the soil surface temperature the lowest of any of the plots. However the hay did not suppress water loss to the air as effectively as the other covers.

At night thermal differences between the plots were much smaller. In general the top of the mulches were cooler than the bare soil, but the soil surfaces under the mulches were warmer. This points to a possible disadvantage if plants rather than the soil are being protected. If the plants extend above the mulch their tips will experience a more extreme climate being hotter (even scalded) by day, and colder by night.

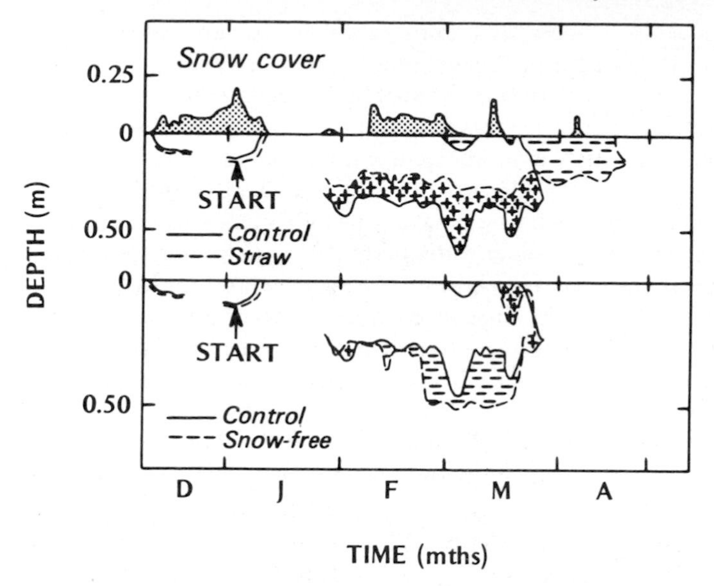

Figure 7.4 The effect of snow and a straw mulch upon the depth of frost penetration (0°C isotherm) at Hamilton, Ontario. The depth of frost in the control plot (short grass plus undisturbed snow cover) is compared against that in a straw-covered plot in the upper graph, and a plot kept free of snow in the lower graph. The depth of snow on the control plot is given at the top (modified after Oke and Hannell, 1966).

The preceding refers mainly to summer conditions where the aim of a mulch is to conserve soil water. In the winter mulches can be used to conserve soil heat and thereby prevent or delay frost penetration. Figure 7.4 shows the depth of the frost-line (0°C isotherm) during the winter in a fine sandy loam soil in Southern Ontario. One plot was left as a control, a second was covered with a 0·1 to 0·15 m straw mulch, and a third had its surface kept free of snow. The effect of the straw mulch was to reduce the depth of frost penetration by about 40% in comparison with the control, and it also kept the climate more stable. However it can be seen that if the mulch is not removed in the spring it becomes disadvantageous because it prevents the penetration of the warming wave. Therefore ideally the mulch should be applied soon after the time of the autumn soil temperature 'turnover' (p. 41) so as to retard the loss of the summer heat storage, and be removed early in the spring.

In the other plot (Figure 7.4) the removal of the insulating snow cover allowed the frost-line to penetrate approximately 20% deeper than in the control. The line was on average about 80 mm deeper in the snow-free plot, and the average snow depth was 70 mm. This is in agreement with the 'rule-of-thumb' that snow reduces frost penetration by an amount approximately equal to its own depth (Legget and Crawford, 1952).

The most recent form of moisture 'mulching' is the use of surface films applied by spraying. In the case of wet soil, and open water, the practice is to spray the surface with a monomolecular layer that is impermeable to water. If the film is complete this virtually eliminates evaporation. Then at least some of the energy which would have gone into latent heat is available as sensible heat to warm the soil or water body. This is helpful in the case of paddy fields where the water temperature is very important to rice growth. The conservation of water in large reservoirs is another application, but there is a problem if winds are greater than 2 m s^{-1} because the film is moved and broken up. Films have also been used to diminish transpiration losses from plants. In some cases the attempt is to seal off the stomata, but since they are mainly located on the underside of the leaf this makes field application difficult. Other anti-transpirants operate by chemically inducing the stomata to close, or decrease their aperture thus increasing stomatal resistance (p. 103). All such materials must of course still allow carbon dioxide exchange if the plant is to remain healthy.

(d) MOISTURE CONTROL

Just as mulching employs the excellent insulating properties of air, irrigation and flooding can be seen as exploitation of the special thermal properties of water, particularly its high heat capacity and large latent heat (pp. 26–7).

Irrigation is usually undertaken in order to re-stock the soil moisture store, and thereby to reduce the potential for plant moisture stress. It can also be used to provide a more stable soil climate. The addition of water to most soils results in an increase in the thermal diffusivity (κ_s, Figure 2.5). This promotes the diffusion of heat in the soil and offsets extremes of both daytime heating and nocturnal cooling (p. 39). The increased availability of water usually enhances evaporation, and the associated uptake of latent heat provides an additional daytime cooling effect.

Complete flooding of some crops is practised. Here the aim is to establish the very conservative climate associated with water bodies (Chapter 3). This is discussed further in the next section.

2 Frost protection

The principles of frost protection can best be viewed in the framework of the nocturnal energy balance (Figure 1.12b). The essential aim is to maintain the temperature (energy status) of a soil-plant-air volume above some critical temperature value because below this damage to sensitive plants may occur. This can be achieved in three ways. First, energy loss from the system can be retarded; second, existing energy can be redistributed within the system; third, new sources of energy can be added to the system by artificial means.

Frost is said to occur when the surface temperature (of the ground or plant, etc.) falls below 0°C. The conditions in which protection measures may be helpful are those where the surface temperature may dip below 0°C for some hours before recovering again. Such frosts are either due to *in situ* radiation cooling with clear skies and light winds, or to the advective introduction of cold air to a site accompanied by stronger winds. Accordingly they are called *radiation frost*, and *advective frost*, respectively. Protective measures are most effective against the former because it depends on local site properties and processes which are more easily amenable to control.

Before considering remedial protective methods it should be pointed out that the incidence of frost can be minimized by considering frost hazard potential at the time of site selection. The conditions for radiation frost are also ideal for katabatic airflow (p. 154). It is therefore advisable to avoid sites where cold air can stagnate and accumulate. These include low-lying areas such as valleys, basins and other terrain depressions, and behind obstructions to downhill flow such as walls, hedges, large buildings and road or railway embankments. If such barriers exist the problem can be lessened if gaps or diversionary channels are provided for the cold air to break through (e.g. gateways in walls). In large valleys use should be made of the 'thermal belt' (p. 155) rather than the valley bottom. Unfortunately the soil quality is often best on the valley floor and the value of this has to be weighed against the potential frost risk.

(a) RADIATION CONTROL

The surface net long-wave radiation loss (L^*) is the driving force behind nocturnal cooling, and frost is most prevalent on cloudless nights because the atmospheric 'window' (p. 17) is open to the transmission of $L\uparrow$. Therefore one method of protection is to try and 'close the window' by placing a radiative screen above the surface. As with cloud this barrier will absorb much of $L\uparrow$ from the surface and re-radiate some portion back so that $L\downarrow$ at the surface is greater than with a clear sky, and L^* heat losses are correspondingly reduced. Artificial clouds of mist or fog (from water sprays) and smoke (from smudge pots or burning car tyres) have been used to provide this radiation control. The former is clearly preferable on air pollution grounds. In other cases cheese-cloth, wooden slats, or glass covers are used to restrict the sky view factor for long-wave radiation losses, without unduly hindering daytime solar input.

(b) SOIL HEAT CONTROL

There are two basic approaches to controlling soil heat for frost protection. The first is to apply a mulch to the surface. This moves the active cooling surface (and therefore the site of maximum frost risk) to the top of the mulch. Thus the soil heat reservoir remains 'untapped' and any heat that does move up through the soil is trapped by the mulch and retained near the soil surface. Ideally the mulching materials should be applied in the evening of a frost-risk night. This allows maximum daytime soil heat storage to occur. (Cloudless, light wind nights conducive to frost are usually preceded by fine sunny days.) The mulches outlined previously for moisture and heat retention are also suitable for frost protection.

The second approach is to increase the thermal conductivity of the upper soil layers so as to maximize the upward transmission of soil heat. This may be done by adding moisture through irrigation (Figure 2.5), or by rolling the soil so as to exclude soil air. This technique is acceptable in the autumn when the soil heat store is well stocked, but it clearly cannot be repeated frequently.

Another related approach is to completely flood the soil and/or crop. Certain plants are able to withstand this treatment (e.g. cranberries). The energy balance is then similar to that of a paddy field (Figure 3.13a) which leads to a more stable thermal environment.

(c) LATENT HEAT CONTROL

The actual temperature at which plants are harmed is usually below 0°C, so that there is a small range of sub-freezing temperatures within which no lasting damage occurs. The continuous spraying approach to frost protection takes advantage of this margin of safety, and utilizes the fact that when water freezes it releases latent heat of fusion, L_f (p. 27). When the plant parts have cooled slightly below 0°C they are submitted to a fine water spray. As long as the water freezes it will release L_f which helps to retard cooling. The addition of water raises the heat capacity of the plants and this also slows the cooling rate. Provided that the spraying and freezing are continuous the temperature will stay close to 0°C and the crop will not be harmed (assuming the plant structure is sufficiently robust to support the weight of the ice without breaking).

The method requires careful control. If too little water is added the temperature could drop because of evaporation/sublimation into the surrounding air. If too much is added the amount of L_f liberated may not be enough to warm the enlarged ice + plant mass sufficiently to offset the radiative and convective cooling from its exterior. If the spray is prematurely terminated heat will be drawn from the plant and damage may result.

(d) SENSIBLE HEAT CONTROL

During radiation frost nights the lower atmosphere is characterized by a radiation inversion (i.e. there is a considerable store of warm air aloft). The base of the inversion is at the ground over bare soil (Figure 2.2), and near

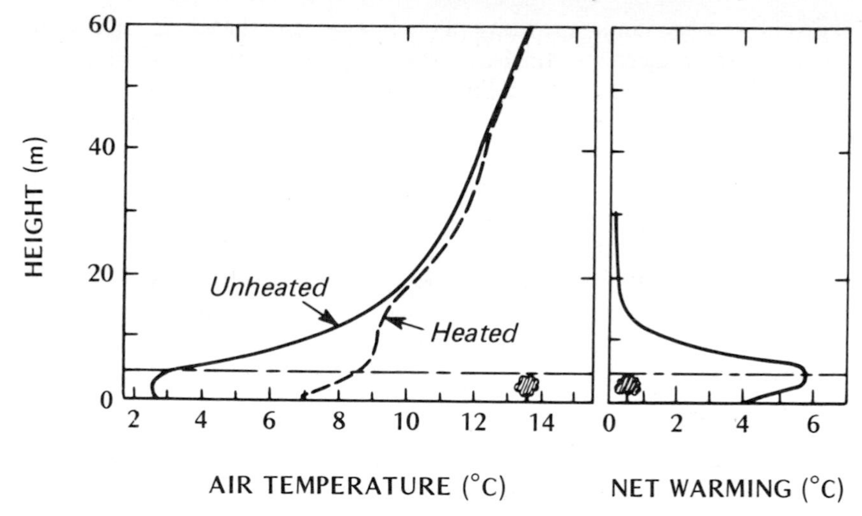

Figure 7.5 Modification of the temperature profile in the centre of a citrus orchard produced by fuel-burning heaters (after Kepner, 1951).

the top of the canopy over vegetation (Figures 4.15, and 7.5). Vertical mixing in combination with this distribution gives rise to the input of sensible heat (Q_H) to the surface. By increasing turbulent mixing in the surface layer this downward flux can be enhanced, and the average temperature of the lower layers is raised. The amount of the temperature increase depends on the depth and intensity of the inversion. Suitable mixing can be provided by large motorized fans, fixed to revolving mounts on the top of towers, or by the downwash from a hovering helicopter.

(e) DIRECT HEATING

The final approach is to supplement the natural energy balance with heat released by combustion. This can be achieved in the soil by installing electrical heating cables. A similar practice is employed in roads and airport runways to prevent icing.

The heating can be supplied to a vegetation-air volume such as a crop or an orchard by fuel-burning heaters. These are capable of off-setting the plant or tree cooling through both radiation and convection. The radiation heating arises out of the long-wave emission from both the heater itself, and the warm plume of gases rising above it. The impact is however restricted to the foliage and fruit able to 'see' the heater and its plume. Further benefit can result from the protective cloud of pollutants which commonly accompanies this practice as was outlined under radiation control (p. 207).

The warm gases released from the heaters also warm the vegetation convectively. Since the air is likely to be stable above the crop or orchard, the injected heat will remain in the lowest layers and through mixing will raise the average air temperature. This strengthens the air-to-vegetation temperature gradient and further stimulates the downward sensible heat flux (Q_H) to the radiatively cooling vegetation. In the example shown in Figure 7.5 heaters in a citrus orchard were able to raise the air temperature in the tree zone by 4–6°C in comparison with a similar, but unheated, adjacent orchard. The heaters also promote convective mixing within a deeper layer above the orchard. This stirring provides a beneficial redistribution of heat as described in (d), and in Figure 7.5 it appears as though the layer up to at least 15 m was involved. Both the radiative and the convective warming processes are aided by the deployment of many small heaters rather than a few large ones.

3 Fog clearance

A fog is defined as any surface cloud which envelopes the observer such that horizontal visibility is restricted to a distance of 1 km or less. Fogs can form under a variety of conditions and we have already outlined the physical conditions under which the radiation (p. 58), and advection (both cold-water and warm-water, p. 142) types develop. To these we could add *upslope fog*, formed when a moist air mass is cooled below its dew-point by being forced to rise up a hillside; and *warm-rain fog* which occurs when rain falls through a much colder layer near the surface and the consequent evaporation of the droplets saturates the layer.

Fogs remain a major problem especially to air, sea and land transportation, and this has led to attempts to clear them by surface and atmospheric control measures. The fog types listed above are classified according to their genesis, but from the point of view of clearance it is better to recognize a more simple grouping according to temperature, viz: *warm fog* (above 0°C); *super-cooled fog* (0 to −30°C with at least some liquid droplets); and *ice fog* (below −30°C with only ice crystals).

(a) CONTROL OF WARM FOG

The vast majority of fogs are of this type, and there are three basic approaches to their dispersal:

Mechanical mixing – this method is based on the fact that above the fog there is usually drier, cleaner and warmer air. If this air is forced downwards and mixed with the fog the layer may drop below saturation and the droplets will evaporate. This can be achieved by the downwash from a helicopter in the same manner as frost protection. The approach is simple, and relatively inexpensive, but is only effective in clearing isolated patches in shallow fog.

Hygroscopic particle seeding – this approach seeks to deplete the water balance of the fog layer. Hygroscopic particles (substances such as sodium chloride and urea having a strong affinity for water) are spread over and upwind of the area of desired clearance by an air-borne 'seeder'. The particles absorb water by condensation, grow in size, and fall out in about 5 min. The removal of water from the layer 'dries' the air sufficiently for many of the remaining droplets to evaporate. The maximum effect on visibility occurs about 10 min after seeding. The cleared area advects across the desired area with the wind, and later re-fills. The size of the particles is critical: if they are too large they fall out quickly and little or no condensation occurs; if they are too small they remain in suspension and actually cause a further deterioration in visibility.

Direct heating – if sufficient heat is added to the fog layer the vapour 'holding' capacity of the air is increased (Figure 2.15), and the droplets evaporate. Jet engines installed along the side of airport runways have been found to be effective, but are costly to install.

(b) CONTROL OF SUPER-COOLED FOG

Super-cooled fogs are the easiest to disperse. The means of clearance are designed around the fact that the saturation vapour pressure over an ice surface is slightly less than that over a water surface at the same temperature (see inset Figure 2.15). The difference is small (~20 Pa) but significant. It means that if ice crystals and water droplets co-exist in the same cloud, there is a vapour pressure gradient directed from the droplet to the crystal. As a result the droplets shrink due to evaporation, and the crystals grow by vapour deposition. The technology of super-cooled fog dispersal involves the 'seeding' of materials that will act in the same manner as ice crystals. The most common substances used are dry ice (solid carbon dioxide) and liquid propane. The former is released from an aircraft above the fog, whereas the latter is injected from ground-level. Upon release liquid propane vaporizes, expands and therefore cools, to form freezing nuclei. In either case the nuclei grow at the expense of the fog water droplets and are precipitated to the ground as snowflakes.

(c) CONTROL OF ICE FOG

Ice fog is almost totally attributable to human activities (i.e. it is an example of inadvertent climate modification, Chapter 8) but at present there are no economical means developed for its clearance.

4 Shelter effects

The use of barriers to provide shelter from the wind is an old and well-developed procedure. The barrier may be a line of trees (a shelter belt) or any other arrangement of trees, bushes, hedges, soil embankments, stone walls or fences. The principal aim is to reduce the horizontal wind speed near the ground in areas open to damaging or otherwise undesirably strong winds. Such areas include coastlines and open prairie landscapes where frictional retardation is weak because of small terrain roughness, or locations open to topographic or other local wind systems possessing undesirable features (e.g. strong and/or cold katabatic winds). The objects of protection could be sensitive agricultural crops, domestic animals, houses and farm buildings, transportation routes, or the conservation of such properties as the snow cover, soil moisture and the top soil. The climatic effects of shelter are not restricted to the simple reduction of wind speed. Therefore in the following section we will not only investigate the typical pattern of wind changes caused by barriers, but also the related modification of the energy and water balances, and their thermal, moisture and biological implications.

(a) WIND AND TURBULENCE EFFECTS

We have considered some of the typical responses of airflow to an isolated obstacle in relation to Figure 5.13; here we will extend that discussion to the case of flow over a barrier placed in its path. Figure 7.6a shows the mean *streamlines* (lines that are parallel to the direction of flow at all points and therefore indicate the flow at a given time) as airflow encounters a solid barrier placed normal to its original direction. Figure 7.6b is a general classification of the flow zones which result. Even before the air reaches the obstacle it begins to react because of the pressure build-up ahead. The bulk of the flow is forcibly displaced up and over the barrier. Immediately above the barrier the streamlines are forced to converge as the same mass of air

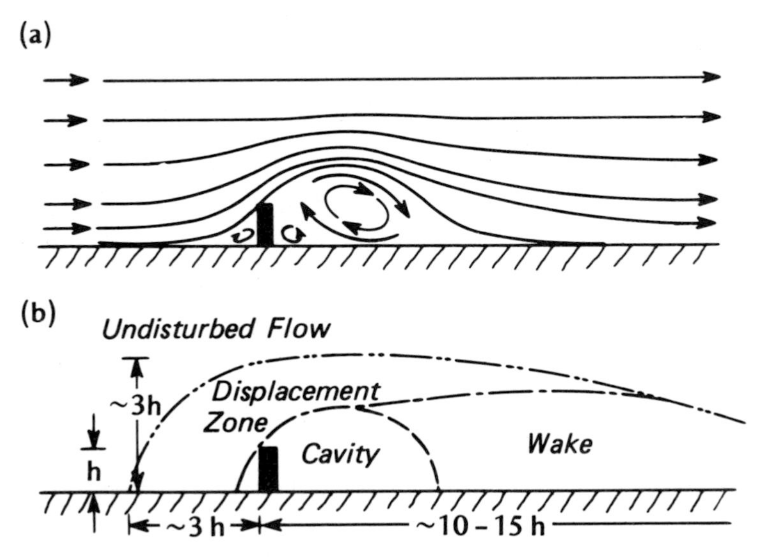

Figure 7.6 (a) Streamlines and (b) generalized flow zones associated with the typical pattern of airflow induced by a solid barrier placed normal to the flow. Dimensions expressed as multiples of the barrier height, *h*.

attempts to 'squeeze' over, causing an acceleration or jet, but once over it is able to expand again and decelerates accordingly. This is the flow found in the *displacement zone*. After crossing the barrier the room available for expansion suddenly increases but the fluid cannot immediately react to fill it. The flow therefore separates from the barrier's surface and its organization breaks down into a much more turbulent condition in the low pressure or *wake zone* which extends downwind from the barrier. Immediately behind the barrier the pressure is low and thus tends to 'suck' air into a semi-stationary lee eddy or vortex. This part of the wake is known as the *cavity zone*. The large lee eddy structure is dissipated into the smaller turbulent eddies of the wake zone, before finally settling down and reassuming conditions similar to those of the upwind flow. Within the wake the separation of the flow lessens the force of the wind on the ground and the near surface wind speed is decreased – this is the sheltering effect.

In order to be able to compare the effects of different-sized barriers it is common to represent horizontal and vertical dimensions in terms of the barrier height (h). In these units the barrier is seen to affect flow to at least 3 h above the surface, and to the same distance in front of the barrier (Figure 7.6b). The distance of influence downwind of the barrier depends upon the density of the barrier, defined as the percentage ratio of the open area of the barrier as viewed normal to its axis, to its total vertical area (i.e. a totally impermeable barrier has the maximum density of 100%). The distance of downwind influence is usually judged in terms of the percentage reduction of horizontal wind speed compared to the upwind (open) value at the same height. The effect of barrier density upon the distance of downwind shelter is illustrated in Figure 7.7 from measurements at a height of ~0·25 h in the vicinity of shelterbelts. If the barrier is very dense the reduction in wind speed is seen to be considerable immediately to the lee because there is little or no penetration, but the wind regains its former value relatively quickly because the strength of the cavity is intense and the faster moving air above is quickly drawn in. If we assume that a 10% reduction is the least value of significance then a dense barrier can be seen to extend its influence to about 10–15 h downwind. As might be expected the low density barrier provides the least protection near its base because it allows considerable throughflow, but its recovery is slower than the high density one because the air passing through provides a 'cushion' in the cavity zone and the flow assumes a more even aerodynamic shape. The point of 90% recovery occurs about 15–20 h from the obstacle. Thus it emerges that the medium density barrier provides the best overall shelter by combining maximum retardation compatible with aerodynamic 'cushioning' in the lee. The effects of such a barrier extend to about 20–25 h based on the 90% recovery criterion, but reduced wind speeds may even be observed as far as 40 h downwind.

The above relationships apply to the ideal case of flow normal to a long

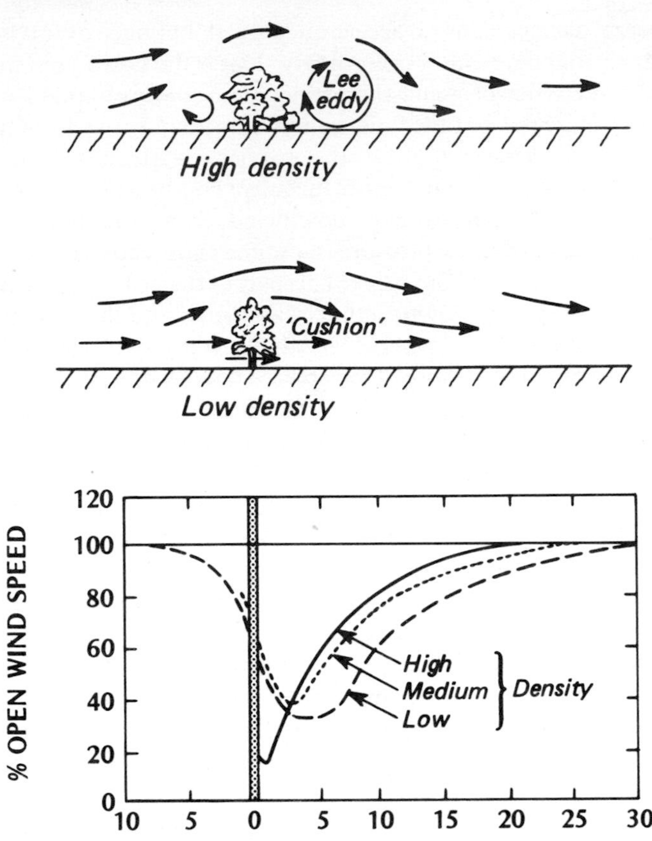

Figure 7.7 Wind speed reduction in the vicinity of shelterbelts with different densities (after Nägeli, 1946).

(in the across-wind direction) barrier. If the angle-of-attack of the flow is oblique then the area of shelter is proportionately reduced until with parallel flow the shelter is negligible, except due to friction on either side of the barrier itself. At the ends of a barrier the wind tends to curl around and form turbulent vortices which 'eat' into the sheltered area. Equally if there are significant gaps in the barrier the wind will jet through giving speeds *greater* than in the upwind flow. If the width of the barrier is significantly larger than h then throughflow is virtually eliminated. Therefore the pattern of wind speed in its lee conforms more to that of a high density barrier (Figure 7.7).

(b) ENERGY AND WATER BALANCES

The radiation budget in sheltered areas is unlikely to be significantly different to that in the open except immediately adjacent to the barrier

TABLE 7.3 Energy balance components in open and sheltered irrigated soybeans at Mead, Nebraska in July. Data are energy totals (MJ m^{-2}) for daylight hours (after Miller *et al.*, 1973).

	(Q^*-Q_G)	Q_H	Q_E	β†
Slight advection day				
Open	16·9	−2·4	19·4	−0·12
Shelter	16·9	−0·7	17·7	−0·04
Strong advection day				
Open	15·4	−9·5	24·9	−0·38
Shelter	15·4	−3·6	19·0	−0·19

† Dimensionless

where the radiation geometry is changed. If the barrier is oriented E–W areas to the north will be in shade especially when the Sun is low, but areas to the south may receive extra short-wave radiation due to reflection. If the barrier is oriented N–S the effects are likely to be small. Near the barrier the sky view factor (SVF) is reduced and this may have an effect in reducing net long-wave radiation losses.

Unfortunately the energy balance of sheltered areas has received relatively little attention probably because the inhomogeneity of the wind field renders many flux measurement approaches invalid. From the scanty evidence available it appears as if the sensible and latent heat fluxes (Q_H and Q_E) are reduced by the dampening of turbulence in the sheltered zone, and since net radiation (Q^*) is relatively unaffected this implies a slight increase in the soil heat flux (Q_G). Table 7.3 shows some estimates of the reduction of Q_H and Q_E in the sheltered area provided by a slat-fence in an irrigated field of soybeans in Nebraska. This irrigated field, in an otherwise semi-arid area, exhibits evidence of advection due to the 'oasis-effect' (p. 142), because Q_E is greater than Q^* and Q_H is negative. Despite this complication it is clear that turbulent heat transfer is diminished in the sheltered zone.

If we assume no major differences in run-off between open and sheltered areas we may analyse differences in the water balances of the two areas in terms of the relation:

$$\Delta S = p - E \tag{7.1}$$

The deposition of precipitation, p (both rain and snow) is inversely related to horizontal wind speed. Therefore in the lee of a barrier deposition is enhanced relative to that in the open. However, snow subsequently undergoes considerable spatial redistribution by drifting before entering the liquid water balance as meltwater. Snow fences and shelterbelts are often erected in order to retain the snow cover and there is little doubt that this is achieved. Therefore it seems that shelter can at least locally lead to an increase in precipitation receipt.

We have just noted that evapotranspiration (E) in sheltered areas is at least marginally less than in the open. The differences are probably greatest for bare soil and least with mature crops because sheltered crops tend to show lower stomatal resistances to vapour flow than their unprotected counterparts (Brown and Rosenberg, 1970). On the other hand dewfall is considerably enhanced in sheltered areas due to a number of factors including reduced wind speed, greater humidity, and colder nocturnal temperatures (see next section). Dew tends to form earlier and to evaporate later, and can represent as much as a 200% gain compared with deposition in the open. Only in the area next to the barrier is dewfall diminished, and this is due to the reduced nocturnal radiative cooling brought about by the restricted SVF. Summarizing the effects of shelter upon E it appears that daytime water losses are reduced, and nocturnal water gains are increased.

Considering equation 7.1 it follows that soil moisture storage (ΔS) is enhanced by the provision of shelter because it tends to increase water input (precipitation and dewfall), and decrease water output (evapotranspiration). Further, since these water balance changes are all related to wind speed reduction we may anticipate that the increase of soil moisture will decay with distance downwind from the barrier (i.e. in conformity with the curves in Figure 7.7).

(c) CLIMATIC EFFECTS AND APPLICATIONS

As we have just seen, decreased turbulent diffusion in the flow behind a barrier leads to a reduction in the fluxes of heat and water vapour, and we may anticipate the same to be true for carbon dioxide. The decreased transport and mixing of these entities means that the microclimatic profiles of temperature, water vapour and CO_2 will be steeper than in the open, and that their mean concentrations will be different. The diminished turbulent activity decreases the interaction between the layers next to the surface and those above. By day the sensible heat output is therefore used to heat a relatively shallow layer and gives higher air temperatures than in the open. At night the surface radiative heat losses are not as efficiently replenished from the atmosphere and air temperatures are lower. Except during dewfall the surface is a moisture source, therefore reduced transport results in higher humidities both by day and by night. For CO_2 the surface is a sink by day (assimilation) and a source by night (respiration), therefore the effect of shelter is to produce a relative CO_2 deficiency by day and an enhancement by night.

Shelter is constructed for many practical purposes, some of which have already been touched upon. Shelterbelts, hedges, fences and even grass are used to reduce soil erosion and arrest the movement of sand dunes by reducing the speed of near-surface winds which transport the soil or sand

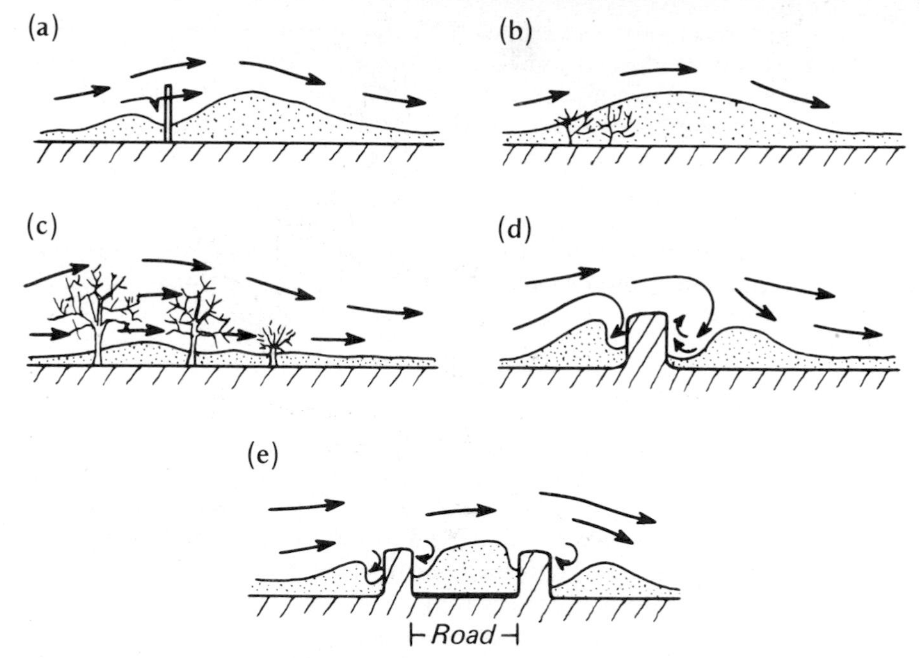

Figure 7.8 Snow accumulation in the vicinity of (a) a snow fence, (b) an open low plant cover, (c) a wide low density shelter-belt, (d) a wall and (e) two solid barriers (e.g. two hedges bounding a roadway).

particles. Similarly barriers are used to retain the snow cover on open terrain because of its value as an insulator and water source. They are also used to control snow drifting which may lead to blockage of transportation routes (roads, runways and railway tracks), or restrict access to buildings.

Some examples of the relationship between snow drifting and shelter are given in Figure 7.8. In (a) a snow-fence or narrow, medium density shelterbelt is seen to give maximum snow accumulation in the lee of the barrier at the position of greatest wind reduction. Therefore to provide protection for a transportation route such a barrier should be placed about 10 h upwind. In (b) even a low, open plant barrier can be seen to provide a sufficient braking-effect to stabilize a drift. In (c) the open shelterbelt is so wide that its primary role is as a simple trap for drifting snow, the lee effects are less significant. In (d) a solid barrier, like a wall or earth bank, provides an upward snow dam, and a downwind drift, but strong eddying next to the wall leads to an erosion by scouring. In (e) two closely-spaced solid barriers provide overlapping drift accumulation, so that the road in between becomes inundated.

Shelter is of course often established to guard against physical damage to buildings and crops, but the reduction in turbulent heat transfer can also be beneficial in conserving metabolic heat losses from exposed animals such as cattle and sheep, and can reduce the consumption of energy for space heating in isolated farms by up to 45%. Not least significant, the combined effects of shelter can give increases in crop yield. Most work suggests increases are especially found between 5 and 15 h downwind, but near the barrier the effects of shade, and competition for moisture from shelterbelt trees, may actually decrease yields for about 0·5 h.

Not all the effects of shelterbelts are beneficial. For example the lower nocturnal temperatures impart a greater frost risk to the sheltered area, and the combination of lower winds and higher humidities can lead to increased fungal disease.

5 Glasshouse climate

The thermal benefits of glasshouse construction have been recognized for a long time, but their physical explanation still remains to be completely elucidated. In general terms a glasshouse provides two forms of control: first, it is a radiative filter; and second, it reduces turbulent heat losses because it gives almost complete wind shelter. The radiative filter was always held to be the dominant heating mechanism, but more recent work suggests the shelter may be more important.

The classic explanation of glasshouse warmth related to the spectral absorption properties of glass which readily transmits short-wave, but absorbs most incident long-wave radiation. It was reasoned that incoming short-wave ($K\downarrow$) is free to enter the glasshouse and be absorbed by the soil and plant surfaces (allowing for normal reflection), but that the long-wave radiation emitted by these warmed surfaces ($L\uparrow$) is not free to leave because it is largely absorbed by the glass, and then re-radiated back inside the system. The glass therefore acts as a radiative 'trap' by spectral filtering, and this has been called the '*greenhouse-effect*'. It now seems that this analysis is too simple and that a better framework for viewing glasshouse radiative exchange is one similar to that for a forest stand system (Figure 4.17). That is, the glass acts somewhat like an elevated canopy by attenuating short-wave transmission from above and below, and by providing both a barrier and an elevated source for long-wave radiation.

The short-wave properties of glass vary with the angle of incidence of the radiation. At local zenith angles between 0 and 40° the typical albedo (α) is 0·07 to 0·08, with an absorptivity (a) of approximately 0·05, leaving a transmissivity (t) of about 0·87 (i.e. the glass attenuates about 13% of the direct-beam input). At greater zenith angles the albedo increases significantly so that at 60° $\alpha = 0{\cdot}16$, $a = 0{\cdot}06$, and $t = 0{\cdot}78$, and at 80° $\alpha = 0{\cdot}52$, $a = 0{\cdot}13$, and $t = 0{\cdot}35$. These values relate to the characteristics of clean glass and direct-beam radiation. In practice the glass becomes soiled, and with the inclusion of diffuse-beam radiation the attenuation of $K\downarrow$ can be from 15% in clean rural areas, to as high as 50% in heavily polluted districts. Thus it should be noted that although glass allows light to penetrate, it is far from transparent, and hence it reduces the short-wave radiant input to the vegetation and soil inside the glasshouse.

In the long-wave region of the spectrum glass is indeed a good absorber. Thus $L\downarrow$ from the atmosphere and $L\uparrow$ from the enclosed surface are prevented from passing directly through. The net long-wave budget (L^*) of the vegetation and soil inside the glasshouse is therefore dependent upon the temperature difference between them and the inside surface of the glass. This difference is likely to be considerably less than that between similar vegetation and soil surfaces outside and the sky. Therefore the effect of a glasshouse is to reduce the net long-wave radiation budget.

In comparing the net all-wave budget (Q^*) inside a glasshouse with that in the open during the day the drop in K^* has to be weighed against the drop in L^*. Such comparison depends upon many factors, but it is quite possible for Q^* to be *less* inside the glasshouse. Under these circumstances the 'greenhouse-effect' is clearly non-existent. At night with only long-wave exchange it is common for the glasshouse to reduce L^* losses to less than 10% of those in the open.

The daytime warmth of glasshouses is now mostly attributed to the sheltering role of the structure. The almost total exclusion of the external wind virtually eliminates forced convection. This allows strong soil-to-air, or plant-to-air temperature gradients to develop, and turbulent transfer of heat and water vapour is dominated by free convection. Being effectively trapped inside a limited volume, the heat and water vapour accumulate, thereby increasing both the air temperature and humidity. Similarly the higher surface temperatures promote a stronger soil heat flux. Figure 7.9

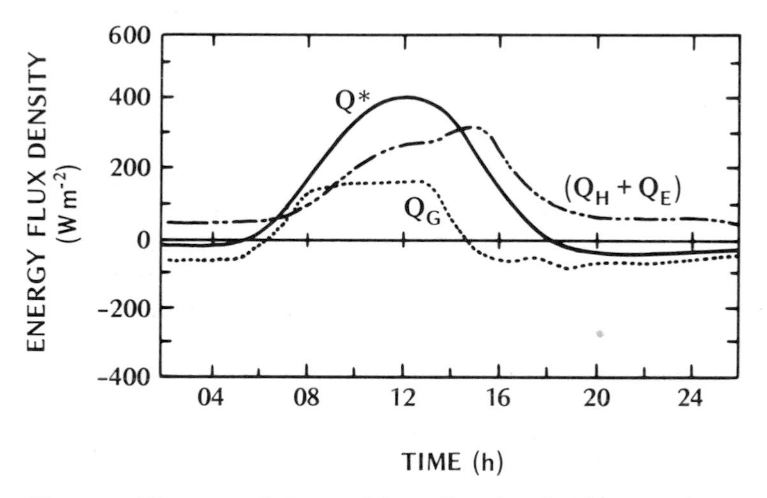

Figure 7.9 The energy balance of the soil surface inside a greenhouse in Japan on 30–31 July 1959 (after Yabuki and Imazu, 1961).

shows the diurnal energy balance at the soil surface in a glasshouse in Japan. Notice that Q_G is equal to or greater than the combined turbulent losses ($Q_H + Q_E$) in the morning, and that the soil is absorbing at rates of almost 200 W m^{-2} for a number of hours. Later in the afternoon when the soil heat reservoir is well stocked, and surface temperatures are at a maximum, turbulent transfer ($Q_H + Q_E$) becomes more important. At night Q_G serves as the major source of heat and is sufficient to cover both the radiative deficit and to continue heating the glasshouse air by convection. Naturally the air volume cools by radiation and conduction out through the glass, but the soil heat release is sufficient to act as a buffer against the rate of cooling experienced in the open.

In practice the climate in a glasshouse is not always ideal. For example in summer the daytime heating may be too strong. To alleviate this ventilation fans and openings may be installed so as to reduce interior/exterior differences. Another approach is to cut down the solar gain by painting the glass, or placing straw mats on the roof. If humidities are low enough it may be possible to increase evaporative cooling by spraying water on the soil and plants. Conversely, in the winter and by night the glasshouse may become too cold. This can be mitigated by adding straw matting to the roof to diminish radiative losses, or by providing direct heating such as ducted warm air or electrical soil heating cables. If light levels are insufficient artificial illumination can be provided, and if the carbon dioxide concentration becomes depleted (especially by day when the glasshouse air is partially sealed-off from the atmospheric source) it can be restored or even enriched directly from compressed gas cylinders or through fossil fuel combustion.

6 Internal building climate

Buildings are mainly constructed to provide a safe and controlled atmospheric environment for humans and domestic animals, and are the most elaborate form of behavioural thermoregulation (p. 163). Part of the need is to gain shelter from undesirable weather elements such as high winds and precipitation and this protection is fairly simple to construct. The more sensitive requirement is to provide a low stress thermal climate. From Chapter 6 we know that ideally this means the provision of conditions wherein the energy balance and constant deep-body temperature of a homeotherm can be maintained with a minimum of thermoregulatory effort (zone CD in Figure 6.8) despite the fact that exterior temperatures are well above or below this range.

Here we will consider the general principles of energy exchange and the thermal climate of buildings, and then some examples of building practices designed to provide amenable interior conditions in otherwise stressful climatic environments.

A full appreciation of the factors bearing on the thermal climate of homeotherms inside a building requires a consideration of the three interactive relationships shown in Figure 7.10. These are the interaction between (a) the external climate and the building; (b) the building shell and the internal living space; and (c) the living space and the occupant.

(a) ENERGY BALANCE OF A BUILDING

The energy balance of a complete building and its contained air volume (Figure 7.10a) is given by:

$$Q^* + Q_F = Q_H + Q_E + Q_G + \Delta Q_S \qquad (7.2)$$

where, Q^* – net all-wave radiation of the building exterior, Q_F – total

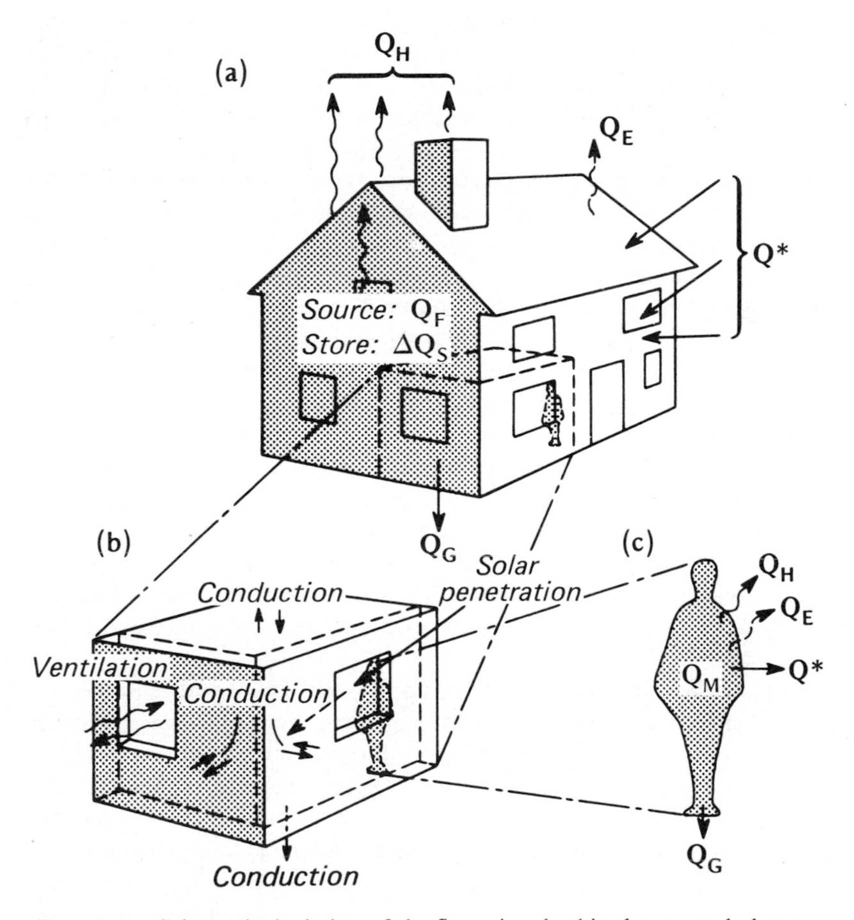

Figure 7.10 Schematic depiction of the fluxes involved in the energy balance of (a) a complete building volume, (b) a room in a building and (c) a person in a room.

internal anthropogenic heat release, Q_H, Q_E – sensible and latent heat exchange with the external air, Q_G – heat conduction between the building and the underlying ground, ΔQ_S – net change of energy storage by the building materials and the enclosed air volume. As with the balance of an animal (equation 6.1) equation 7.2 refers to the complete system, and the fluxes are spatial averages referenced to the total external area of the building-air volume. The analogy with the energy balance of an animal is in fact rather apt as we shall see.

The input of direct-beam short-wave radiation around a house is very uneven because of its three-dimensional geometry. An appreciation of the complexity of the input is gained by re-considering the remarks in Chapter 5 regarding radiation loading on horizontal and inclined surfaces (e.g. Figures 5.8 and 5.9). Clearly in the northern hemisphere east-facing walls receive an early peak in S soon after sunrise because their receiving surfaces are illuminated with the Sun almost in their local zenith (i.e. Θ is small). In the middle of the day the south-facing wall is most favoured, and in the afternoon the west-facing wall. North-facing walls only receive S near the time of the summer solstice. Figure 7.11 illustrates the role of aspect on solar radiation receipt, utilizing data from a house in Pretoria, South Africa (25°S) on a calm, cloudless winter day. (Note that in this example the north-facing wall receives the greatest input because it is a southern hemisphere location.) The south wall receives only diffuse-beam input because it is in shade all day. This 'background' level of diffuse beam input is also evident in the traces for the east and west walls when they are in shade. The receipt of S by roofs depends upon their inclination and aspect in the same manner as sloping topography (Chapter 5). In the tropics the roof becomes relatively more important because of the high elevation of the Sun. The absorption of $K\downarrow$ by a building depends upon the albedo of the materials and the area of windows allowing penetration to the interior. In strong radiation environments the use of paints and materials with a high albedo greatly help to reduce heat loading. A building is usually warmer than its surroundings and hence its net long-wave budget is always negative. Because of its high SVF the roof is the most important site of radiative heat loss at night.

The net radiation budget of a building is hard to assess because of the uneven distribution of the short- and long-wave exchanges over its surface in space and time. By night the budget is almost invariably negative, but by day the solar loading can be sufficient to give a substantial net radiative gain for the building. The magnitude of the loading depends upon the radiation at the location and the building's geometry and materials.

The other potential energy source for a building is anthropogenic heat (Q_F). This is heat released inside the structure either intentionally as space heating (fires, heaters, furnace-heated air, etc.), or as a by-product of other activities (cooking, lighting, electrical appliances, etc.) including the metabolic releases of the human and/or animal occupants. Some space-heating in cold climates is thermostatically controlled to ensure internal temperatures do not fall below a set value. This system provided the analogue used in describing the control of metabolic heat release (Q_M) in homeotherms in order to maintain a constant body temperature. In both cases the system has an internal energy source which can be regulated to provide comfortable conditions. If an almost constant internal building temperature is required throughout the day then Q_F will vary in response to changes in the energy balance of the building-air volume. Over a complete heating season the total energy required to maintain internal comfort is closely related to the external air temperature. Conversely in hot weather the amount of energy extracted by air-conditioning systems is inversely related to the external air temperature.

Convective sensible heat losses (Q_H) from the exterior of a building depend upon the wind speed and the building-to-air temperature gradient. The wind controls both the thickness of the laminar boundary layer, and the degree of turbulent motion around the building. With light winds the laminar layer is about 10 mm in thickness and combined with weak turbulence the heat loss for a given building/air temperature difference is relatively small. Using the electrical analogy (p. 102) the thermal resistance is therefore large. On the other hand, in windy conditions the laminar layer may contract to less than 0·5 mm and its insulating role is greatly diminished. Similarly the more turbulent state of the rest of the atmosphere (partly induced by the building, Figure 8.1) creates a very efficient means of heat loss. Averaged over its exterior area a building is usually warmer than the surrounding air (Figure 7.11) and therefore the direction of heat transfer is almost always outwards. (This may not be true of the roof at night especially if its materials have a high emissivity, and if the roof is well insulated from the ceiling below.) The heat loss occurs as a result of both heat conduction through the walls and windows, and direct air seepage through chimneys, windows and doors. Lack of air-tightness is a major problem in cold climates, and loss via this pathway is greatest with gusty winds which produce interior/exterior pressure pulsations. On the other hand in hot climates such ventilation losses may be welcome because they help to dissipate uncomfortably warm interior heat loads.

Evaporation cooling of a building (Q_E) is not usually as effective as Q_H. It can however be significant if the building has been wetted (by rain or irrigation), has its roof flooded, or if there is a substantial vegetation covering such as clinging vines or creepers. A continually wetted fence with vines covering it can be a very effective air cooler when located upwind of a house in dry, windy areas. The system does of course depend upon the availability of an abundant water supply.

Sub-surface conduction losses (Figure 7.10a) depend upon the degree of building/ground contact as well as the thermal properties of the building

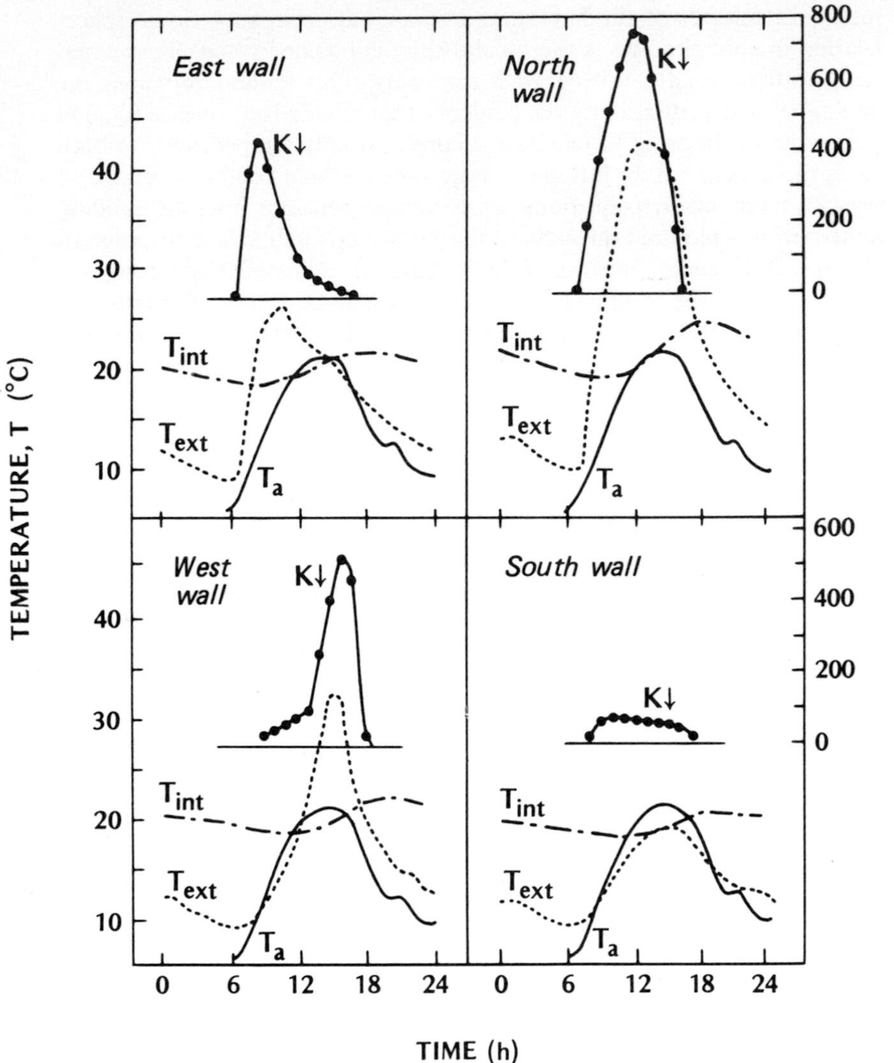

Figure 7.11 Diurnal variation of the incoming short-wave radiation ($K\downarrow$), the interior and exterior wall temperatures (T_{int}, T_{ext}) and the air temperature (T_a) for walls of different exposure at Pretoria, S. Africa (25°S) (modified from a diagram constructed by Landsberg, 1954).

and the ground, and the temperature gradient between them. This mode of heat loss from the building can be beneficial or wasteful depending on whether the building's heat load is excessive or insufficient. In areas of permafrost (permanently frozen ground) buildings are elevated on stilts to minimize heat conduction to the ground. This is not to conserve building heat but to prevent melting of the ground ice which would cause subsidence of the foundations.

(b) THE ENERGY BALANCE AND CLIMATE OF A ROOM

The energy balance of a room (Figure 7.10b) depends upon the nature of the energy balance on the exterior; the extent of any anthropogenic heating on the interior; and the facility with which the building shell allows interior/exterior interaction. During periods of strong solar heat loading on a building the exterior is warmer than the interior and a room will receive an inward-directed heat flux (e.g. Figure 7.11). On the other hand, at night, and with weak exterior irradiance, the gradient and the flux are directed outwards. This is always the case in cold climates where Q_F maintains an outward energy drain. If the construction of the building shell allows easy exchange between the interior and the exterior then the building provides very inefficient climate control. On the other hand if exchange is prevented then large inside/outside thermal differences can be maintained with a minimum of effort.

Interior/exterior heat exchange is likely to occur via three main pathways (Figure 7.10b). First, solar radiation may enter the room through openings and glass windows. The importance of this heat gain depends upon the size and orientation of the openings, and the nature of the incident solar radiation in respect both of its intensity, and of its directional character (e.g. the proportions of S and D). Second, heat may leave or enter as a result of ventilation through windows, doors, cracks and other openings. Third, heat may be conducted through the building fabric (walls, windows, ceiling, floor). This flow depends upon the thermal properties of the building materials, and the strength of the interior/exterior temperature gradient. The most important thermal properties are the thickness and thermal conductivity of the materials, and the thickness of the laminar boundary layer adhering to the interior and exterior surfaces. From the electrical analogy the heat flow is therefore seen to be directly proportional to the temperature difference, and inversely proportional to the total building resistance (composed of the wall and boundary layer resistances).

In thermally uncomfortable climates these pathways for heat flow must be controlled if a building design is to be successful. In cold environments the interior must be heated, and the primary aim is to prevent this energy being lost to the exterior. To achieve this it is essential to cut losses via the ventilation and conduction pathways. Losses by direct seepage are minimized by sealing all openings tightly. Examples include the use of weather stripping around doors and windows, and the double and even triple glazing of windows. Conduction losses are reduced by insulation. The idea is to encapsulate the room (or whole building) with a protective 'blanket' of poorly conducting materials. The ideal substance for this purpose is still air, as is evident from Table 2.1. This protection can be

TABLE 7.4 Thermal properties of materials used in building and urban construction.

Material (dry state)	Remarks	ρ Density (kg m^{-3} × 10^3)	c Specific heat (J kg^{-1} K^{-1} × 10^3)	C Heat capacity (J m^{-3} K^{-1} × 10^6)	k Thermal conductivity (W m^{-1} K^{-1})	κ Thermal diffusivity (m^2 s^{-1} × 10^{-6})
Concrete	Aerated	0·32	0·88	0·28	0·08	0·29
	Dense	2·40	0·88	2·11	1·51	0·72
Stone	Av.	2·68	0·84	2·25	2·19	4·93
Brick	Av.	1·83	0·75	1·37	0·83	0·61
Wood	Light	0·32	1·42	0·45	0·09	0·20
	Dense	0·81	1·88	1·52	0·19	0·13
Steel		7·85	0·50	3·93	53·3	13·6
Glass		2·48	0·67	1·66	0·74	0·44
Plaster	Gypsum	1·28	1·09	1·40	0·46	0·33
Gypsum board	Av.	1·42	1·05	1·49	0·27	0·18
Insulation	Polystyrene	0·02	0·88	0·02	0·03	1·50
	Cork	0·16	1·80	0·29	0·05	0·17

Source: van Straaten, 1967.

incorporated into a building by ensuring that sufficient air space exists between interior and exterior surfaces as in the use of cavity-walling, double glazing and the provision of adequate attic and basement space. The effectiveness of these features is greatly enhanced if artificial insulation (Table 7.4) is installed in these spaces. These cellular materials keep the air motionless, thus preventing convection. It is also essential that the insulation remains dry if its beneficial properties are to be retained.

In hot environments the primary aim is to keep interior conditions cool and this means preventing the external heat load from getting inside too rapidly. The building design in hot and dry areas shows the attempt to control heating via all three pathways. Solar radiation input is cut down as much as possible by keeping all openings very small, and by the use of shade from window blinds, verandahs, overhanging eaves, trees, or adjacent buildings. Mutual shading is provided by placing buildings very close to each other so that the intervening streets are narrow and in shade most of the time (e.g. Marakesh). It is also important to shade surrounding areas to keep them cool. In modern buildings the window glass is designed to reflect or absorb, rather than transmit, short-wave radiation. Conduction gains are offset by the use of thick walls made of high heat capacity materials such as earth, brick, or stone (Table 7.4). In this way the exterior heat load is absorbed by the wall and its transmission delayed so that it reaches the interior well after the period of maximum external temperatures. This lag effect is illustrated in Figure 7.11 where it will be noted that it takes 5–7 h for the peak of the exterior temperature wave to reach the interior. The use of wall storage (ΔQ_S) is a useful delaying practice but the heat will still arrive on the interior in the evening. It is then necessary to increase interior/exterior ventilation to remove the excess. Conduction losses to the cool soil are also maximized by ensuring good contact with the ground. By day the ventilation pathway is restricted by the small openings and shutters, and any anthropogenic releases (e.g. by cooking) are vented to the exterior. At night ventilation is maximized so as to remove the wall heat, and to replace it with the now cooler exterior air.

In hot and humid areas the preceding solution is unworkable because whereas in a dry environment only slight air movement is required for evaporative cooling to be effective, with high humidity vigorous motion is required. The characteristic hot/humid design therefore stresses shade and openness. Shade is even more important than in the hot/dry case because very humid areas are often also cloudy so that the diffuse-beam input is relatively large. This necessitates providing shade from the complete sky and not just the solar disc. Openings are oriented to make best use of prevailing winds and local breezes, and the whole building is placed on stilts in order to take advantage of the natural increase of wind speed with height (Figure 2.9). Similarly much of the surrounding vegetation is removed to reduce the possibility of air stagnation. Interior/exterior exchange is artificially promoted by fans.

The preceding examples of traditional building practices in different climatic regions incorporate simple but sound micro-climatic principles. More recently there has been a tendency to override such considerations so that buildings in widely varying climatic contexts have very similar features. This is made possible by offsetting thermal imbalances through the use of heating or air-conditioning systems. However, this is inefficient in its use of expensive energy resources and it is to be hoped that this 'brute force' approach will again yield to one which is sensitive to the local climatic environment.

Within an individual room the temperature distribution depends upon the arrangement of heat sources and sinks. For example in cold conditions a heated room usually has cooler areas near the windows, floor and poorly-insulated or especially exposed walls on the building periphery. These cool surfaces then act as long-wave radiation sinks for warmer surfaces. Surface coverings such as carpets, curtains and tapestries can help to alleviate this form of cooling. It is not uncommon to find differences of at least 5°C within a heated room and this is sufficient to create air movement. Air tends to rise near heat sources and to spread out as a warm layer near the ceiling. Correspondingly air sinks near cool walls and windows and accumulates as a cool air pool across the floor. An open fire is a strong convergence node and the influx of cool air across the floor often creates uncomfortable 'draughts'.

(c) THERMO-REGULATION PROVIDED BY BUILDINGS

The reaction of homeotherms to thermally stressful environments was covered in Chapter 6; here we will amplify the nature of the relief provided by buildings. In a cold climate a homeotherm is in danger of expending too much metabolic energy (Q_M) to maintain a constant core temperature due to energy losses via Q^*, Q_H, Q_E and Q_G. Provision of heated shelter slows this drain by surrounding the homeotherm with surfaces at approximately the same temperature (thereby decreasing Q^* and Q_G losses), and by giving shelter from the wind and cold air (thereby decreasing Q_H and Q_E losses). It is important to realize the importance of these processes and not to confine the idea of thermo-neutrality in terms of air temperature alone. For example it is quite possible for a person to feel uncomfortably cool in a room whose air temperature is 20°C if there is a cold surface acting as a radiative sink (e.g. uncovered window whose interior temperature may be at least 20°C below the person's surface temperature).

In a hot, dry climate a homeotherm has a problem in remaining cool due to the extra heat load on the body from Q^*, and the inability to cool effectively by Q_H because of the lack of a body-to-air temperature gradient. The primary role of a building is then to provide shade, and to keep air temperatures from rising too high. Even slight air motion is sufficient to allow body heat to be shed via Q_E, as long as the body has sufficient water. Heat may also be dissipated by long-wave radiation to cool interior surfaces.

The problem of body overheating is more acute in a warm, humid climate. In addition to the problems of hot, dry areas the Q_E channel is limited because even slight evaporation causes the body to become enveloped in saturated air thus destroying the necessary body-to-air humidity gradient for further evaporation. Comfort can then only be achieved by forcefully removing this envelope by air movement (e.g. with fans).

CHAPTER 8

Inadvertent climate modification

The climatic side-effects of human activities are many and varied. They are the result of interference in the operation of natural systems. Tampering with natural energy and water cascades often results in rather complex ramifications throughout the system including feedback effects. In many cases the full web of cause and effect linkages is so large that the climatic impact of altering a part of a system is largely unknown. Hopefully our knowledge of the inter-relationships will increase so that we may develop models which accurately mimic the operation of natural systems. Only then will it be possible to predict the climatic effects of pursuing alternative land-use or management strategies, and hence to avoid undesirable inadvertent modification.

1 Non-urban modification

The removal of vegetation markedly alters the surface properties of an area, and hence will modify the energy and mass cascades. The removal may be temporary as in harvesting crops or trees, it may be permanent as in land clearance for agriculture, or it may be accidental as a result of fire, disease or over-grazing. If the area involved is extensive the altered heat and water balances may give rise to local, mesoscale or even larger scale changes in climate and hydrology. For example the removal of vegetation often leads to adjustments in the local water balance because the interception role of the canopy is lost, evapotranspiration may be reduced, the snow cover distribution and duration is changed and run-off may be increased. Removal can also be expected to upset the radiation budget by presenting a new surface geometry, and albedo. Equally the energy balance partitioning is likely to be modified by a new set of thermal, moisture and aerodynamic characteristics.

Large-scale irrigation can modify the climate and hydrology, especially in otherwise arid or semi-arid areas. In such locations the development of an 'oasis-effect' (p. 142) leads to cooler summer temperatures, an increase humidity and of course changes in the local water balance. Similar but more pronounced effects are likely to result from the creation of artificial lakes due to dam construction or other major earthworks such as those associated with road and railway construction. The climatic effects of lakes are the same as those for extensive irrigation, but the large mass of water also imparts its conservative thermal influence as seen near other large water bodies (Chapter 3). In summer the surrounding shore areas are cooler, and in winter warmer, than before the flooding. The thermal lag of the lake also causes the seasons to be delayed (i.e. the onset of the autumn cooling and the spring warming occur later). If the lake becomes ice-covered the ameliorating influence of the water is lost. Indeed the surface climate may become more hostile than before (e.g. wind speeds may increase because of the reduced roughness of ice compared with most surfaces). In very cold areas (air temperature less than −30°C) the release of relatively warm deep lake water from behind a dam may lead to the development of ice-fogs downstream. Ice-fog forms because water evaporates from the relatively warm water into the very cold air where it condenses and freezes (or sublimates) almost immediately. Evaporation occurs because the vapour pressure at the warm water surface is much greater than in the very cold, dry air above. For example, if the surface temperature of the water is only 0°C its surface vapour pressure will be the saturation value ($e^*_{(T)}$) which is approximately 600 Pa, whereas the maximum vapour pressure of the air can only be about 50 Pa at an air temperature of −30°C (Figure 2.15). Therefore the addition of only a small amount of vapour is sufficient to cause saturation and a fog of ice crystals. Similar problems occur in the vicinity of cooling ponds where water warmed by industrial processes is left to cool.

The effects of vegetation clearance, irrigation and flooding are some of the more obvious examples of activities leading to climate modification. There is however a wide range of activities which alter climates either in a subtle fashion or only affect small areas. Examples of the former include changes in agricultural cropping or grazing patterns and practices, and the change in ocean properties due to pollution such as oil slicks. In the latter category are the surface changes brought about by the extension of transportation routes (e.g. roads, railways, pipelines and the opening up of ocean lanes by ice-breaking) and the changes caused by open-pit mining, land reclamation, refuse dumping etc. Individually such modifications are small, but their integration over space and time represents a process of continuous change which is being conducted without intelligent control.

2 Modification by buildings

The placement of a building on the landscape gives rise to radiative, thermal, moisture and aerodynamic modification of the surrounding environment. The most important radiative effects are a decrease in the solar radiation receipt by areas in shadow, a local increase in solar receipt by reflection from sunlit walls, and the reduction of net long-wave cooling from surfaces near the building due both to a reduction in $L\uparrow$ (caused by the reduced SVF) and to an increase in $L\downarrow$ from the usually warm building. Also in the immediate vicinity of a building, soil and air temperatures are often warmer than in the open due to heat losses from the building (p. 223), and as a result of the wind shelter provided (p. 211).

The water balance around a building is upset because of spatial variability in precipitation receipt (caused by differences in interception and wind shelter), soil drainage, and evaporation. Most important however are the airflow changes produced by the building acting as an obstacle to the wind. The remainder of this section will be devoted to a discussion of these aerodynamic effects.

(a) AIRFLOW AROUND BUILDINGS

The pattern of airflow around an isolated flat-roofed building placed normal to the wind (i.e. with its windward side at a right-angle to the direction of flow) is given in Figure 8.1a–c. This figure is based on wind-tunnel studies using a cube to represent a building. Wind measurements around full-scale structures generally show that such modelling gives a good approximation to the real situation as long as the upwind flow conditions (wind and turbulence profiles, perturbations due to other buildings and topography) are also simulated. Notice that the flow zones (A – undisturbed, B – displacement, C – cavity and D – wake) conform to those of Figure 7.6 for flow over a solid barrier.

Upon encountering the impermeable building the air is either deflected over the top, or down the front (Figure 8.1a) or around the sides (Figure 8.1c). The air 'pushing' against the building gives relatively high pressures over much of the surface of the windward wall. Maximum pressure occurs near the upper middle part of the wall where the wind is actually brought to a standstill, and pressure decreases outwards from this *stagnation point* (Figure 8.3b). Near the outside edges of the windward face the accelerating flow actually produces areas where the pressure is below that of the undisturbed atmosphere (i.e. suction). If the building has sharp corners the flow accelerating over the top and around the sides becomes separated from the surface. Therefore the sides, roof and leeward wall experience suction. Since air moves from high to low pressure, these areas are characterized by reverse flows (i.e. in the opposite direction to the main stream). This is reponsible for the lee eddy circulation in zone C (Figure 8.1a) which extends up into the strong suction zone above the roof. In plan view (Figure 8.1c) the cavity zone is characterized by a double eddy circulation at ground-level which incorporates the side wall suction areas into a horseshoe-shaped pattern.

The wind velocity profiles associated with this flow pattern are shown in Figure 8.1b. In the undisturbed upwind flow (profile 1) the standard logarithmic shape (cf. Figure 2.9) is evident. Immediately over the building (profile 2) the profile is sharply distorted. In the displacement layer above the wake boundary there is a pronounced jet of high velocity air as the streamlines converge. Below this the velocity decreases very sharply, and in the lowest layers the roof return flow is seen. Leeward of the building (profile 3), the jet is less pronounced as the streamlines begin to diverge, and the cavity zone lee eddy gives a return flow near the ground. Averaged over the depth of the cavity velocities are obviously less than in the corresponding

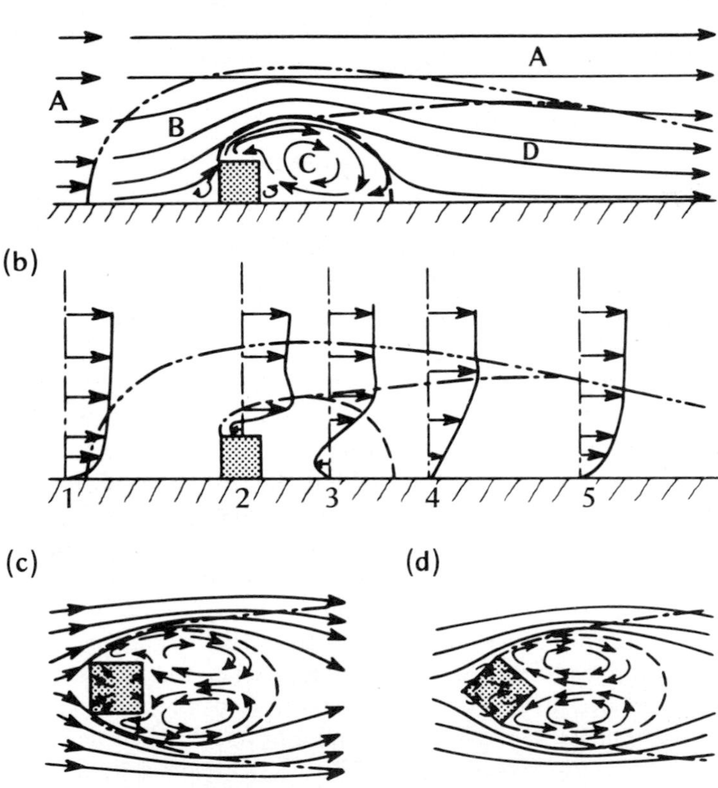

upwind layer, but it should be pointed out that these data hide the fact that it is more turbulent. At greater distances downwind (profiles 4 and 5) the shelter is progressively lost, and the jet merges with the flow which is readjusting towards its undisturbed form. Full adjustment has not been attained at profile 5 because the velocity gradient near the surface is not as steep as at profile 1. This indicates that residual turbulence in the wake is continuing to facilitate momentum transport at a rate greater than normal for the terrain.

Other building shapes and orientations produce variations upon this basic pattern. If the same cubic form is oriented diagonally with respect to the wind (Figure 8.1d) there are two windward and two leeward walls oriented obliquely to the flow. This tends to reduce the strength of the suction zones especially on the roof. If the roof has a pitch the point of flow separation usually occurs at its crest, but the double-eddy pattern still results in a horseshoe shape in the downwind zone (Figures 8.1d and e). If the pitch is greater than 20° the windward face is under increased pressure

Figure 8.1 Flow patterns around a sharp-edged building. Side view of (a) streamlines and flow zones, and (b) velocity profiles and flow zones with the building oriented normal to the flow. Plan view of streamlines with the building oriented (c) normal, (d) diagonally to the flow (modified after Halitsky, 1963), and (e) Flow made visible by snowdrifts around a grain elevator near Boise City, Oklahoma.

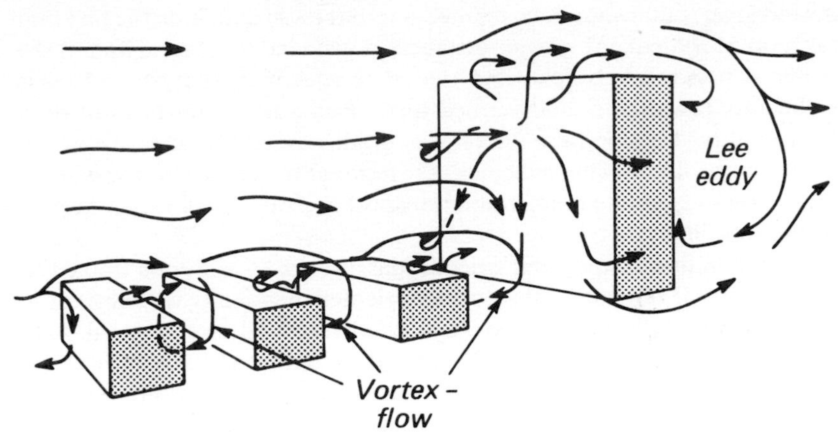

Figure 8.2 Flow patterns over and between buildings of approximately the same height (left), and in the vicinity of a relatively much taller building (right).

but the leeward face is under suction. With rounded buildings (e.g. a silo) the flow disturbance is less, but the basic pattern remains.

If instead of being isolated the building is part of an urban area with buildings of similar size upwind and downwind, then the flow may be as on the left-hand side of Figure 8.2. With winds normal to the buildings a *vortex-flow* develops between them. This circulation is an augmented form of the cavity zone lee eddy where the downward pull of the suction zone is re-inforced by deflection down the windward face of the next building downstream. The winds at ground level in the intervening streets are sheltered but sometimes more turbulent than in the open at the same height. The relative shelter is greatest with very light or very strong above-roof winds. In the former case the vortex is poorly developed and ground/above-roof interaction is weak, and in the latter the winds tend to skim over the roof tops. If the buildings are oriented at an angle to the wind the vortex takes on a 'cork-screw' motion with some along-street movement. If the flow is parallel to the buildings, shelter is destroyed and channelling of the wind may cause a jet-like effect so that speeds are greater than in the open.

The situation is different if a particularly tall building juts above the general roof-level, as in the right-hand side of Figure 8.2, and Figure 8.3. The oncoming wind impacts against the windward face of the tall building and produces a stagnation point in the centre at about three-quarters of the building height (Figures 8.3a and b). The air diverges from this point. Some passes over the top and gives a lee eddy in the cavity zone, whilst much of the rest streams down the windward face. This enhances the lee eddy of the upwind low building and produces a strong vortex near the surface. The rest is deflected around the building sides as *corner-streams* which wrap around the back to give the characteristic horseshoe-shape

(Figure 8.3c). This photograph of the flow simulation shows that the horseshoe wakes of the two buildings have merged, but that the tall building is dominant. The leeward pattern is in excellent correspondence with that of Figure 8.1c. If the building is raised above the ground on pillars, or if there is a walkway under it, then the descending windward stream will produce a jetting *through-flow*.

The problem with this arrangement is that the obstacle deflects the faster moving upper air (Figure 2.9) down to the ground. Therefore instead of shelter there is an increase in low-level winds especially in the vortex-flow, through-flow, and corner-stream areas (Figure 8.3b). The numbers in these areas show the winds likely to be encountered at pedestrian height as a ratio of those at the same height in the open. They indicate that the building can create conditions three times as windy as in the open, and therefore many times greater than in sheltered streets nearby.

(b) APPLICATIONS

Knowledge of the wind environment around buildings is useful in order to protect against wind damage and to economize on wind-related maintenance and running costs. It is also of importance to the safety and comfort of the occupants and nearby pedestrians, and to the dispersion of atmospheric pollutants. We will consider aspects of each of these applications.

The greatest amount of work on buildings and the wind has been related to the assessment of wind loads (force exerted by the wind). Structures are designed to withstand prescribed maximum loads which are calculated on the basis of maximum wind speeds (mean and gust) expected in the region, and the nature of the structure. The total load is assessed as the force of the wind averaged over the area presented to the wind. This is necessary to ensure that the structure (building, mast, tower, bridge) is strong enough to avoid being literally blown over. The load on a structure depends on whether it is 'clad' (e.g. a building enclosed by roof and walls) where flow is diverted around the whole structure, or is 'unclad' (e.g. a bare frame, or open tower) where the air may pass through. In the case of a clad structure the wind load on any component depends on the pressure difference between the two faces of the component (commonly this is the interior/exterior pressure difference). For example roofs are prone to being ripped off in strong winds because the outside experiences suction whereas the interior pressure may be positive. In combination they produce an effective lifting force. This is augmented by the lift exerted under the roof overhangs (eaves) on both the windward and leeward sides as a result of rising air currents (Figure 8.1a). Interior pressures depend upon interior/exterior ventilation. If windward openings allow inflow but leeward ones are closed, the interior pressure is increased. Conversely, interior pressures are decreased by closing windward openings and opening leeward ones. Opening both sets to permit cross-ventilation encourages equalization of interior/exterior pressures but may be unacceptable on heat loss and other practical grounds. The areas most prone to damage are those where flow separation occurs, and therefore suction is greatest (e.g. the corners of buildings and roofs). This can be sufficient to lift roofing materials, and pluck window glass and cladding panels off the face of a building.

The wind environment is important with regard to driving rain (forcible impaction of rain on buildings), which affects moisture uptake and weathering. Rain drops in flight, and water already deposited, to some extent follow the path of surface airstreams. Therefore, on the windward face where most rain is impacted, the rain transport is likely to be similar to that of the smoke in Figure 8.3a (i.e. radially from the stagnation point). Observed patterns of pollution soiling and chemical weathering around a building relate to these airstreams and to the path followed by water due to the force of gravity.

Wind may affect the access to buildings. For example the operation of doors is hampered, especially if they are located in a through-flow region (Figure 8.3b). In some cases suction rips the doors open during strong winds, whereas in other cases the external pressures are so great that the doors can only be opened with difficulty. Access may also be impeded by the accumulation of drifted snow. Therefore in snowy climates the location

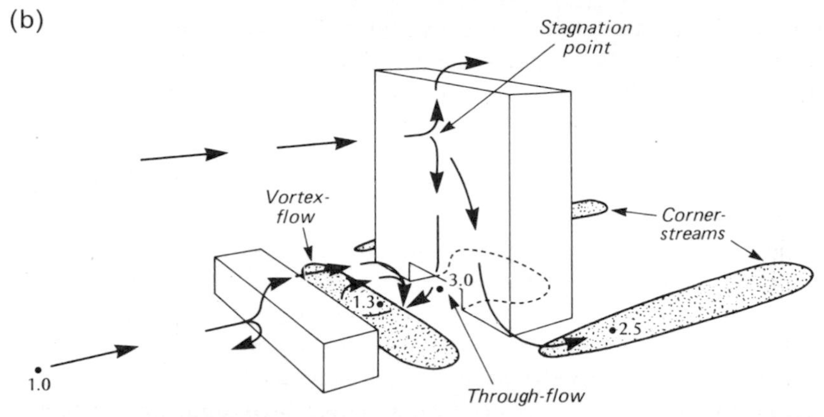

Figure 8.3 Flow around a tall building with lower buildings upwind. (a) Flow over the windward face visualized by the deflection of smoke jets emanating from a wind tunnel model. (b) Illustration of the three main regions of increased wind speed at pedestrian level (stippled), see the text for the meaning of the numbers. (c) Plan view of flow at ground level around a tall building (rectangle on the right) with a low building upwind. Horseshoe pattern is made visible by surface film of pigmented oil around wind tunnel models (photographs after Penwarden and Wise, 1975, Crown copyright by permission of the Director, Building Research Establishment).

of doors, sidewalks, loading bays and car parks should be sited with this possibility in mind. When designing in such areas the effects of the mutual interaction between buildings must also be incorporated. Snow accumulation on roofs can present a related structural loading problem. In calculating the snow load the density as well as the depth is important. Figure 8.1e illustrates how the snowdrifts around a single building can affect transportation routes.

The comfort and safety of pedestrians around the base of tall buildings is intimately connected with the wind environment. The increased winds and turbulence found in the stippled areas of Figure 8.3b can create a hostile environment. In cold climates thermal comfort is decreased because of the reduction in sunshine due to shade, and the increased loss of body heat. The increase in chill-factor (Chapter 6) may make such areas unbearable in winter. Conversely in hot climates the increased ventilation may be favourable since stagnation is to be avoided. The whirling eddies around buildings also tend to accumulate dust, leaves and litter, but most important they pose a potential threat to the safety of pedestrians, especially if they are old or infirm. Emerging from a sheltered area a person may well experience a sudden, four-fold increase in wind speed, near the base of tall buildings. Since the force of the wind increases with the square of its speed, this implies a sixteen-fold increase in the force upon a pedestrian. Unfortunately, people have been fatally injured as a result of being blown over in such locations.

Solutions to these problems are not simple. The best answer is not to build tall buildings in the first place. Failing this, some partial solutions are illustrated in Figure 8.4. They are designed to minimize the ground-level effect of the streaming down the windward face. One approach is to place the main building slab on a podium of one or two storeys in height (Figure 8.4a). High winds are then confined to the roof of the podium. This can be further helped if the tower is raised off the podium to provide an elevated through-flow (Figure 8.4b). In the same way a canopy and vent space provide some ground-level shelter (Figure 8.4c).

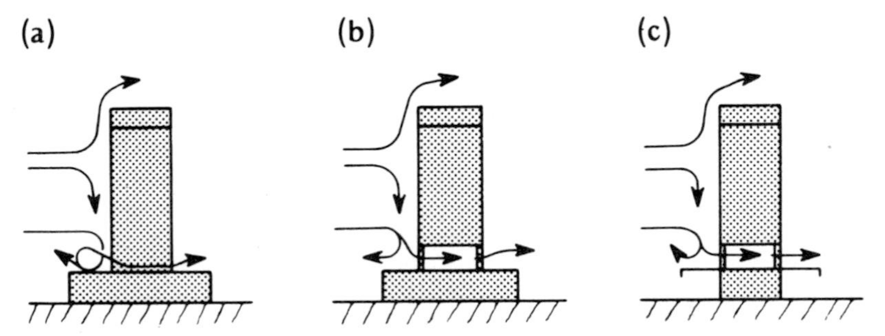

Figure 8.4 Tall building designs helpful in alleviating undesirably increased wind speeds at pedestrian level (modified after Hanlon, 1972).

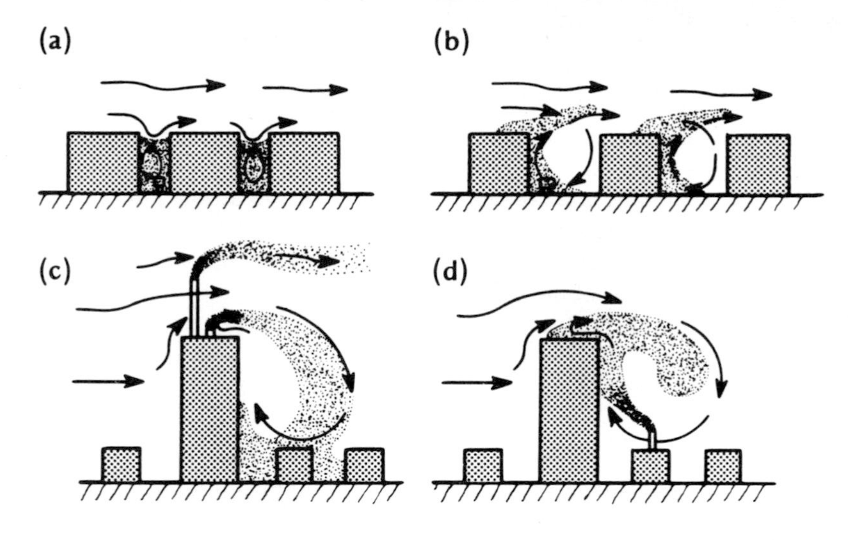

Figure 8.5 The influence of building air flow on pollution dispersion.

Researchers at the Building Research Establishment in Great Britain have been involved in extensive research on model and full-scale flows around buildings. They find that wind becomes an annoyance at about 5 m s^{-1} by disturbing hair and causing clothing to flap. At 10 m s^{-1} it is definitely disagreeable and dust and litter are picked up, and by 20 m s^{-1} it is likely to be dangerous. In their studies they found that the design speed of 5 m s^{-1} was exceeded less than 5% of the time in low-rise areas such as the left-hand side of Figure 8.2, but around tall buildings it was exceeded 20% of the time. They generally find complaints are received about conditions around a building if it is more than 25 m (approximately 6 storeys) in height, or if it is more than twice the height of the surrounding buildings.

Wind and turbulence are vital to the dispersion of air pollutants. In areas characterized by low buildings the exchange between street-level where car pollutants are emitted, and above roof-level depends upon the width of the streets relative to the height of the building. If the streets are narrow air exchange is restricted (Figure 8.5a) compared with that in a more open arrangement where the vortex circulation aids street-level flushing (Figure 8.5b). Severe problems can arise in the 'downwash' (p. 159) behind a tall building. This can be due to a source placed in the suction zone above the roof of the tall building (Figure 8.5c), or located near the surface in the eddy of the cavity zone (Figure 8.5d). The former situation can be alleviated by constructing a taller stack so that the effluent is carried downwind in the displacement zone flow, but there is no simple remedy for the latter, short of eliminating the source.

3 Modification by urban areas

The process of urbanization produces radical changes in the nature of the surface and atmospheric properties of a region. It involves the transformation of the radiative, thermal, moisture and aerodynamic characteristics and thereby dislocates the natural solar and hydrologic cascades. For example the seemingly inevitable increase of air pollution affects the transfer of radiation, and supplies extra nuclei around which cloud droplets may form; the dense urban construction materials make the system a better heat store, and waterproof the surface; the block-like geometry creates the possibility of radiation trapping and air stagnation, and gives a very rough surface; and the heat and water released as 'waste' products of human activities supplements the natural sources of heat and water in the urban system.

Considering these major changes it is hardly surprising that urban areas exhibit the clearest signs of inadvertent climate modification. Settlements are continually expanding to accommodate the influx of migrants from rural areas and the natural increase of population, and by the year 2000 it is estimated that 60% of the world's people will live in towns with 5,000 or more inhabitants. This makes study of urban climates doubly important; first to ensure a pleasant and healthy environment for urban dwellers, and second to see that the effects of urbanization do not have harmful repercussions on larger scale (even planetary) climates.

Estimation of the size of any 'urban effect' often proves difficult. Ideally one would wish to have an extensive set of pre-urban measurements of the climate of a region against which present observations could be compared. Only in rare cases is this possible. Instead it is common to compare the climatic data from the centre of an urban area with those from rural (or non-urban) stations in the surrounding area. Such urban/rural comparisons are at best only an approximation of the urban modification. In

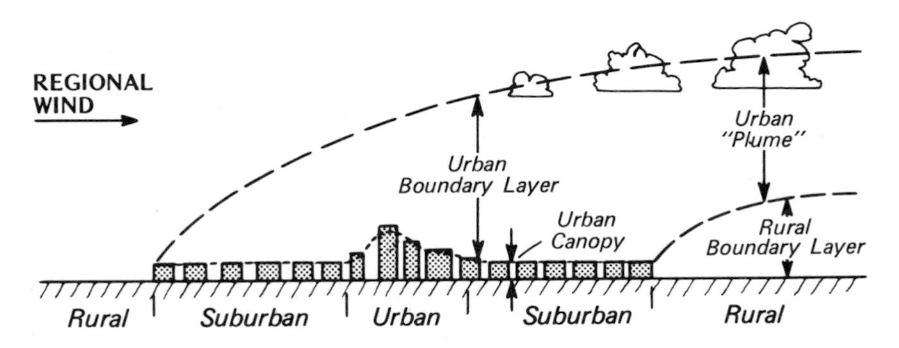

Figure 8.6 Schematic representation of the urban atmosphere illustrating a two-layer classification of urban modification (after Oke, 1976a).

selecting station pairs it is particularly important to try to eliminate extraneous effects due to topography, water bodies and the downwind effects of the urban area itself.

As air flows from the countryside to the city it encounters a new and very different set of boundary conditions. Thus in accord with Figure 5.1 an internal boundary layer develops downwind from the leading-edge of the city (Figure 8.6). The *urban boundary layer*, which is based at about roof-level, is a local to meso-scale phenomenon whose characteristics are governed by the nature of the general urban 'surface'. Beneath roof-level is the *urban canopy layer*, which is produced by micro-scale processes operating in the streets ('canyons') between the buildings. Its climate is an amalgam of microclimates each of which is dominated by the characteristics of its immediate surroundings. In the following discussion we will first consider the cycling of energy and water through an active surface visualized to be a plane at about roof-level (i.e. the interface between the two layers defined above), then consider the climate in each layer.

(a) ENERGY AND WATER BALANCE OF A BUILDING-AIR VOLUME

The energy balance of a building-air volume such as that illustrated in Figure 8.7a is given by a relation similar to that for a single building (equation 7.2):

$$Q^* + Q_F = Q_H + Q_E + \Delta Q_S + \Delta Q_A \qquad (8.1)$$

and the water balance (Figure 8.7b) by:

$$p + F + I = E + \Delta r + \Delta S + \Delta A \qquad (8.2)$$

where, F – water released to the atmosphere by combustion, I – urban water supply piped in from rivers or reservoirs, and ΔA – net moisture advection to/from the city air volume. These balances apply to volumes which extend to sufficient depths that vertical heat (Q_G) and water (f) exchange is negligible. The terms ΔQ_S and ΔS refer to heat and water storage changes in the ground, the buildings and the air contained within the volume, and Q_F and F refer to heat and water sources in the city that are associated with combustion. Note that Q_F, F and I are energy and mass flows that are directly controlled by human decisions and respond to activity rhythms only indirectly related to the solar cycle. The advective terms are due to the net horizontal transfer of sensible and latent heat (ΔQ_A), and of water droplets and water vapour (ΔA) through the sides of the building-air volume.

(i) *Anthropogenic heat and water sources*

Table 8.1 gives an idea of the size of Q_F for a number of cities, in a range of climates. These data are based on estimates of energy use within a city from all sources (electricity, gas, coal, solar conversion, gasoline, wood etc.) for

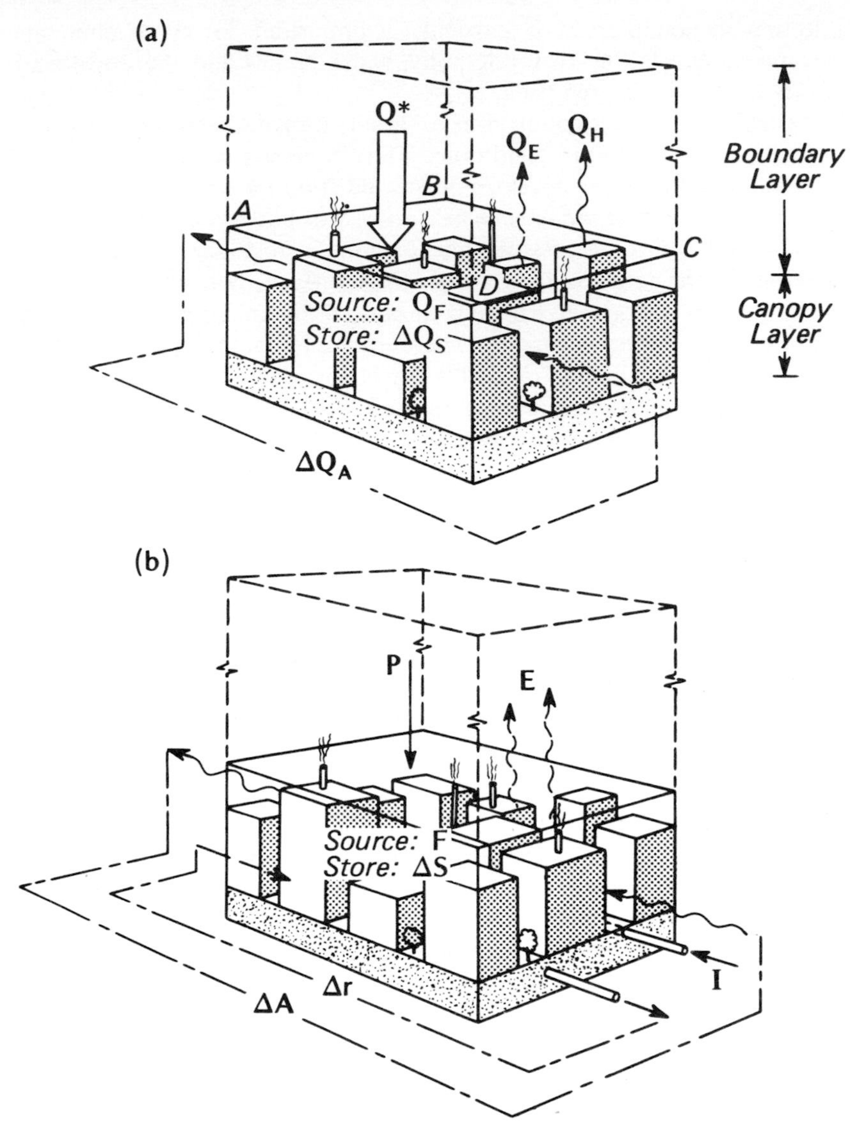

Figure 8.7 Schematic depiction of the fluxes involved in (a) the energy and (b) the water balance of an urban building-air volume.

the purposes of space heating, manufacturing, transportation, lighting etc. The average anthropogenic heat flux density (Q_F) depends upon the average energy use by individuals, and the city's population density. The per capita energy use depends on many factors including the affluence and nature of the economy, and the need for winter space heating. Clearly in a number of cities Q_F is a significant energy source, approaching or

TABLE 8.1 Average anthropogenic heat release (Q_F) from selected urban areas

Urban area	Year	Period	Population (× 10⁶)	Population density (persons km⁻²)	Per capita energy use (MJ × 10³)	Q_F (W m⁻²)	Q^* (W m⁻²)
Manhattan (40°N)	1967	Year	1·7	28,810	128	117	93
		Summer				40	
		Winter				198	
Montréal (45°N)	1961	Year	1·1	14,102	221	99	52
		Summer				57	92
		Winter				153	13
Budapest (47°N)	1970	Year	1·3	11,500	118	43	46
		Summer				32	100
		Winter				51	−8
Sheffield (53°N)	1952	Year	0·5	10,420	58	19	56
West Berlin (52°N)	1967	Year	2·3	9,830	67	21	57
Vancouver (49°N)	1970	Year	0·6	5,360	112	19	57
		Summer				15	107
		Winter				23	6
Hong Kong (22°N)	1971	Year	3·9	3,730	34	4	~110
Singapore (1°N)	1972	Year	2·1	3,700	25	3	~110
Los Angeles (34°N)	1965–70	Year	7·0	2,000	331	21	108
Fairbanks (64°N)	1965–70	Year	0·03	810	740	19	18

Sources: Oke, 1974; Kalma and Byrne, 1975.

surpassing the net radiation, especially in the winter. In both Manhattan and Fairbanks, Alaska, Q_F is greater than Q^*, but for very different reasons. The per capita use in Manhattan is quite moderate considering it is an economically developed region with a cool winter climate (e.g. compare with Los Angeles), so that here it is the high population density which is important. In Fairbanks, on the other hand, the very high per capita energy use is spread over a very dispersed settlement.

The average annual data conceal important time and space variations. A city with a mild winter climate exhibits a much smaller seasonal variation; for example compare Vancouver (mild) with Montreal (cold) in Table 8.1. There is relatively little information concerning daily variations but in mild climates it seems that Q_F is largest in the daytime with peak periods in the morning and evening. Spatially Q_F varies greatly within a particular city, with the highest values usually found in the city core.

Considerable amounts of water vapour are released when fossil fuels such as natural gas, gasoline, fuel oil and coal are burnt. The use of water to absorb 'waste' heat from power plants and other industrial processes also greatly enhances vaporization from cooling towers, cooling ponds, rivers and lakes. In combination these provide a preferential source of vapour for the urban atmosphere (*F*). The importation of water to the city (*I*) is

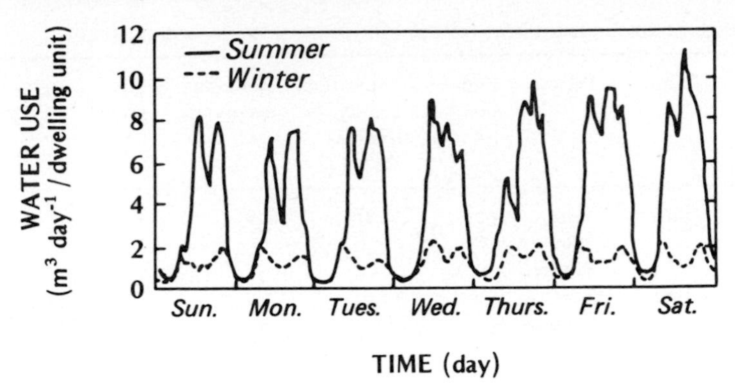

Figure 8.8 Winter and summer patterns of daily water use by the small community of Creekside Acres, California (after Linaweaver, 1965).

necessary to meet demands from residential, industrial and other users. This mass input to the city system (Figure 8.7b) can be fairly easily monitored, and Figure 8.8 is an illustration of the seasonal and diurnal variations of I in a small (mainly residential) community in California. The strong seasonal difference is due to the summer use of water for lawn and garden sprinkling, swimming pools, car washing, etc. Peak-use is concentrated during the day, with peaks in the morning and evening. Ultimately this water is lost from the system via evapotranspiration or run-off.

(ii) *Water balance*

Let us compare the water balance of an urban building-air volume such as that in Figure 8.7b, with that of a corresponding soil (Figure 1.14b) or soil-plant-air volume (Figure 4.1b), in the surrounding countryside. To simplify matters consider both to exist in an extensive area of similar composition, so that we may neglect ΔA for both.

The water input to the urban system is greater because its precipitation (p) is augmented by F and I, for which there are no rural counterparts (if we ignore irrigation). On the other hand, it seems likely that urban evapotranspiration (E) and sub-surface storage (ΔS) are less than in the rural situation. Evapotranspiration is expected to be reduced because of the removal of vegetation and its replacement by relatively impervious materials (although some building materials are quite efficient water stores). The few measurements available tend to support this view. Although the convoluted surface of the city presents a large interception area it seems that the poor infiltration properties of urban materials outweigh this benefit and thus water storage is smaller than in the rural case.

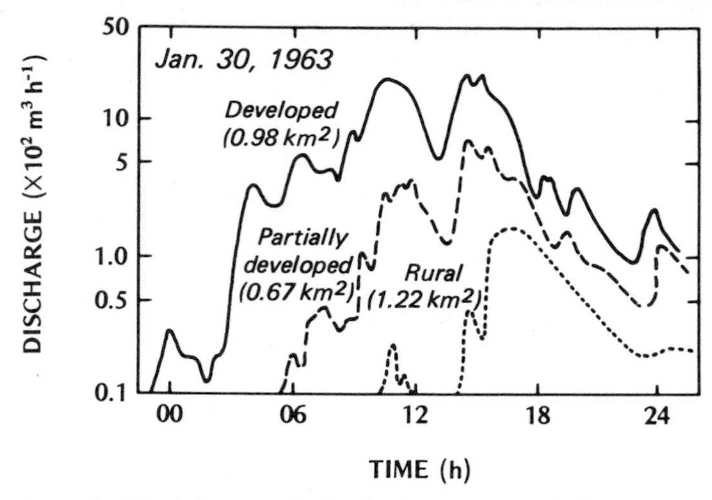

Figure 8.9 The influence of urbanization on storm drainage illustrated by data from three drainage basins (one rural, one partially and one fully developed) near Palo Alto, California (after Moore and Morgan, 1969).

It follows from these considerations, and equation 8.2 that the urban run-off (Δr), is greater than in rural areas. Part of this is simply due to the disposal of a portion of I as waste water (via sanitary sewers). The remaining increases are due to the surface waterproofing and artificial run-off routing (e.g. storm sewers) that accompanies urbanization. These effects are illustrated in Figure 8.9 which shows the storm discharge from three small basins; one rural, one partially developed and one urbanized. Before any development took place the discharge curves for the three basins were similar. They all showed a peak discharge at approximately the same time after a storm, and the magnitude of the discharge increased with basin size (drainage area). After development the curves show two major changes. First, the urbanized basins respond much faster to the water input from a storm, so that the peak discharge occurs much earlier. Second, the amount of run-off increases with the degree of urbanization, and not with basin size. In fact the largest basin gives the smallest discharge both in terms of the peak value, and when integrated over time. These characteristics of urban run-off mean that storm sewers must be designed to accommodate very large volumes of water in short periods of time. Other hydrologic studies show that urbanization leads to greater sediment loads in rivers and decreased water quality.

(iii) *Energy balance*

Now let's compare the energy balance of an urban building-air volume (Figure 8.7a) with that of a nearby rural soil-plant-air volume (Figure

4.1a), in the same qualitative way that we gauged the effects of urbanization on the water balance. We will consider the sites to be free of advection, so that ΔQ_A can be ignored. This is acceptable for a central urban site surrounded by fairly uniform building density, whereas it would not be reasonable for an area near the rural/urban boundary, or where land-use is variable. Having set the net energy flow through the sides and base of the volume to zero, we are only concerned with vertical fluxes averaged over the plane ABCD (Figure 8.7a), and any internal storage changes.

The short-wave input ($K\downarrow = S + D$) to urban areas is considerably altered in its passage through a polluted atmosphere. The attenuation of the incoming radiation depends upon the nature and amount of the pollutants, but in a large city the annual receipt of $K\downarrow$ is usually 10–20% less than that of surrounding rural areas. The size of this reduction is linked to the seasonal variation of pollution concentration (Figure 9.13a). On especially polluted days, and at times of low solar elevation, the reduction in $K\downarrow$ may be in excess of 30%.

In addition to the overall diminution of short-wave radiation, its spectral and directional composition are also changed. Pollutants tend to filter out the shorter wavelengths preferentially. In the ultra-violet portion of the spectrum it is common to lose 40%, and on occasion as much as 90%, due to scattering and absorption. This is likely to be important to plants (reduced photosynthesis) and humans (less tanning, skin cancer and vitamin D production). The greater scattering and reflection by pollutants also increases the proportion of $K\downarrow$ that arrives as diffuse sky light (D). This is helpful in providing better interior lighting of buildings, but it also is responsible for decreases in urban visibility and colour perception. These effects are most obvious when viewing an object towards the Sun. Then the air between the object and the observer is illuminated by sunlight scattered into the line of sight, which tends to obscure details of distant objects, and causes them to appear lighter in colour the more distant they are. The colour of the cloudless sky also depends on the size of the atmospheric constituents. The diameter of particles of a clean atmosphere is smaller than most of the wavelengths of light, therefore they only scatter the shorter wavelengths of violet and blue light, so that the sky is blue. The particles of polluted urban air are larger and scatter and reflect all of the visible wavelengths more equally. Thus the urban sky tends to appear a much paler blue or white because the whole of the visible spectrum is affected.

The reflection of short-wave radiation from a building-air volume depends both on the albedo of the individual reflecting surfaces, and on their geometrical arrangement. The albedo of some typical urban materials is given in Table 8.2, where it can be seen that in comparison with most rural surfaces (Table 1.1) they are rather low. Other things being equal the effect of the urban geometry (blocks separated by street canyons) is to

TABLE 8.2 Radiative properties of typical urban materials and areas

Surface	α Albedo	ε Emissivity
1. Roads		
Asphalt	0·05–0·20	0·95
2. Walls		
Concrete	0·10–0·35	0·71–0·90
Brick	0·20–0·40	0·90–0·92
Stone	0·20–0·35	0·85–0·95
Wood		0·90
3. Roofs		
Tar and gravel	0·08–0·18	0·92
Tile	0·10–0·35	0·90
Slate	0·10	0·90
Thatch	0·15–0·20	
Corrugated iron	0·10–0·16	0·13–0·28
4. Windows		
Clear glass		
zenith angle less than 40°	0·08	0·87–0·94
zenith angle 40 to 80°	0·09–0·52	0·87–0·92
5. Paints		
White, whitewash	0·50–0·90	0·85–0·95
Red, brown, green	0·20–0·35	0·85–0·95
Black	0·02–0·15	0·90–0·98
6. Urban areas†		
Range	0·10–0·27	0·85–0·95
Average	0·15	?

† Based on mid-latitude cities in snow-free conditions.
Sources: Threlkeld, 1962; Sellers, 1965; van Straaten, 1967; Oke, 1974.

decrease the albedo in comparison with the value for a horizontal surface. The decrease is due to radiation trapping within the canyons in a manner similar to that outlined in vegetation stands (p. 110) and ridge-and-furrow fields (p. 201). On the basis of available measurements it appears that these features combine to produce average urban albedos of about 0·15 which is lower than most rural landscape values except for forests, and areas with dark soils. These values apply to mid-latitude cities in the absence of snow. The effect of snow is to accentuate the better absorptivity of the urban area, because whereas most rural surfaces are coated with highly reflecting snow, the city has all of its vertical facets free of snow. Further, the urban snow cover is quickly removed by snowploughs and artificially-assisted melting (salt, heated streets and the urban heat island, p. 254), and the albedo of the remaining snow is reduced by pollution soiling. The albedo of low-latitude cities is less certain. The materials and paints used in these areas are often specifically chosen to increase reflection, and the geometric arrangement of buildings is designed to minimize penetration of sunshine into the streets. These factors should combine to give higher urban albedos than in the mid-latitude case, but urban/'rural' differences will also depend upon the albedo of the surrounding landscape and this can vary considerably (e.g. desert $\alpha \simeq 0{\cdot}35$, jungle $\alpha \simeq 0{\cdot}12$).

In summary if we restrict consideration to the mid-latitude case it is expected that $K\downarrow$ will be less in the urban area compared with the surrounding countryside, but that this deficit is partially offset by a lower urban albedo. Most evidence suggests that the average net short-wave gain (K^*) is slightly lower for the city.

As we shall see later the urban area is usually warmer than its environs at night, and as a result it emits more long-wave radiation to the atmosphere ($L\uparrow$). This is the case despite the probability that the urban emissivity is less, and the fact that the canyon geometry restricts the emission of long-wave to the atmosphere from within the canyons, because of their reduced sky view factor (p. 112). The return long-wave flux from the atmosphere ($L\downarrow$) is also greater in the urban area because of the overlying pollution layer. Pollutants help to close the atmospheric 'window' in the same manner as was outlined in connection with the use of smoke for frost protection (p. 207). The urban increase in $L\downarrow$ at night is not quite sufficient to balance the increase in $L\uparrow$ and hence the net long-wave loss (L^*) is slightly greater than in the rural case. Daytime urban/rural differences of long-wave radiation are not yet clear, so only tentative suggestions can be made. It seems possible that L^* is about the same for both areas, or that the rural loss is greater. This may be brought about by the relative increase in $L\downarrow$ to the urban surface caused by solar warming of the pollutants, whereas urban/rural temperature (and therefore $L\uparrow$) differences are small.

In summary we expect L^* differences to be small between urban and rural areas. By night L^* is likely to be greater in the city, but by day it is possible that the rural loss is larger.

If we combine these changes we find that the net all-wave radiation flux density (Q^*) is not greatly affected by urbanization. At night the urban radiative deficit is probably slightly greater. By day the solar absorption is decreased in the city, but the long-wave loss may also be reduced, so that the urban/rural Q^* difference is small. This simple analysis therefore suggests that although urbanization alters every component flux of the radiation budget, the net effect on urban/rural radiation differences is small. This apparently arises from a fortuitous arrangement of input/output changes which offset each other.

The energy balance results in Figure 8.10 form an approximation to the energy balance of a building-air volume. They were obtained from measurements conducted in a mid-latitude city, during a cloudless summer day. The city surface was moist (five days after the last rain) and had good tree cover. The terms Q^* and Q_H were measured directly; ΔQ_S is approximated by $\Delta Q_S = 0{\cdot}25Q^*$ (see p. 250); and Q_E is the solution of the equation:

$$Q_E = Q^* - Q_H - \Delta Q_S$$

It has been assumed that ΔQ_A is negligible, and that Q_F (see Vancouver values in Table 8.1) is incorporated in the measured radiative, convective and storage terms.

It is perhaps surprising that the gross partitioning of energy for the urban system (Figure 8.10, Table 8.3) is not greatly dissimilar to that of many rural sites. In general terms it appears as though ΔQ_S and Q_H are greater,

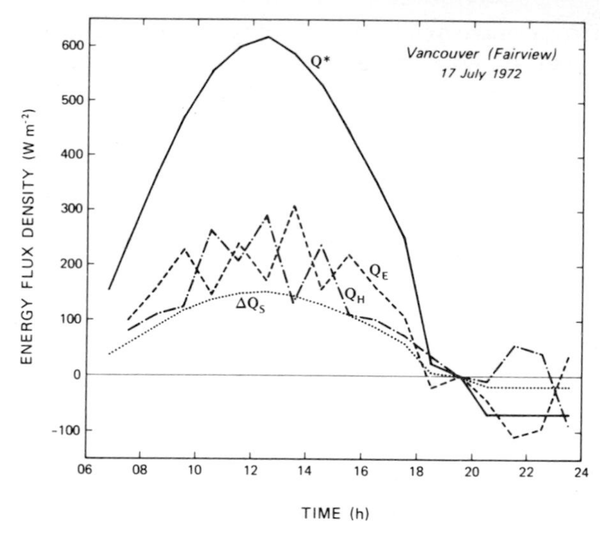

Figure 8.10 Variation of the energy balance components for an urban building-air volume in Vancouver, B.C. on 17 July 1972 (after Yap and Oke, 1974).

but that Q_E is somewhat reduced in the urban case. This would agree with our expectation that city materials are good energy but poor water storers.

In more detail the situation is less clear. Heat storage depends upon the ability of the materials to transmit and retain heat. Direct comparison of the appropriate thermal properties (k, C, and κ) for soils and urban building materials (Tables 2.1 and 7.4 respectively) does not reveal great differences, especially if the soils are moist. However, this ignores the presence of an insulating layer of vegetation or leaf litter over many rural soils which considerably reduces storage (e.g. Figures 1.11, 4.13 and 4.21), whereas in the urban case it neglects the role of surface geometry which tends to increase storage due to the greater surface area available.

TABLE 8.3 Energy balance components of an urban building-air volume in Vancouver, B.C. (49°N). Energy totals for the period 07 to 17 h on 17 July 1972 (data from Yap and Oke, 1974).

Component	Energy total ($MJ\ m^{-2}/10\ h$)	Percent of Q^*
Q^*	16·9	100
Q_H	5·7	34
Q_E	7·0	41
ΔQ_S	4·2	25

Experience suggests that on average ΔQ_S is 15–30% of Q^* for a building-air volume, but more like 5–15% for crops, forests and grass, and 25–30% for bare ground (Sellers, 1965). Urban/rural snow cover differences (p. 247) tend further to accentuate heat storage differences between the two areas.

Whereas ΔQ_S usually represents a reasonably fixed proportion of Q^*, the energy partitioned into the turbulent fluxes are much more variable especially as a result of surface water availability and air mass (humidity and temperature) characteristics. For example results from the site in Figure 8.10 showed that Q_H dropped to only 20% of Q^* following heavy rain, whereas after a dry period it rose to 750%. The partitioning is also likely to exhibit considerable spatial variability related to intra-urban land use differences. During dry periods Q_E must be almost absent from the centre of some cities where waterproofing is almost total, yet at the other extreme it is reasonable to hypothesize that Q_E is greater than Q^* and that an 'oasis-effect' may prevail in urban parks and lawns, especially if they are irrigated. It is also difficult to know the relative contributions made by the roofs and the canyons. It might be expected that turbulent exchange to and from the sheltered canyons is suppressed; however this tendency must be weighed against the fact that at certain times of the day the canyon air is warmer than that above roof-level thereby promoting instability.

The nocturnal energy balance has not been sufficiently researched to provide general characteristics with any certainty. Most work seems to suggest that the turbulent fluxes are small in magnitude, but that their direction is interesting. Rural Q_H fluxes at night are almost always from the atmosphere to the surface, but in cities it is not uncommon to observe the opposite because the surface remains warmer than the air. In the case of latent heat transfer (Q_E), urban areas appear to show greater nocturnal evapotranspiration and reduced dewfall. Both of these turbulent heat

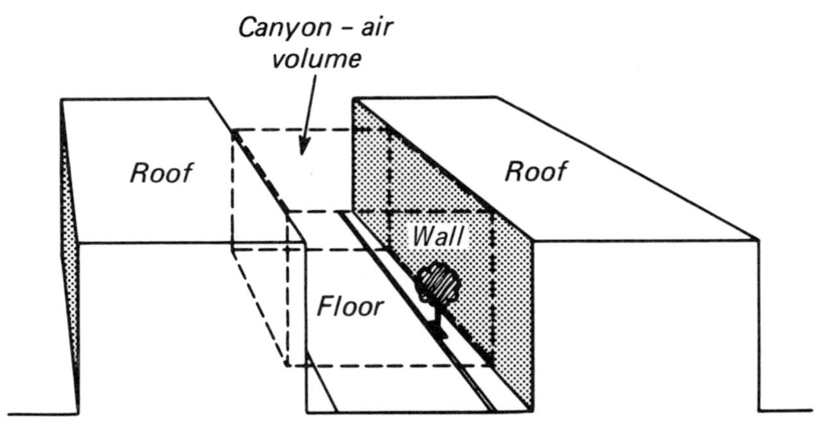

Figure 8.11 Schematic cross-section of the urban/atmosphere interface including an urban canyon and its contained canyon-air volume.

transfer changes require the city to be a nocturnal energy source and this must be furnished from volume storage and anthropogenic releases.

(b) MICROCLIMATE OF THE URBAN CANOPY LAYER

The urban canopy layer (Figures 8.6 and 8.7a) is characterized by considerable complexity, mainly deriving from the convoluted nature of the active surface. It is acceptable to neglect this when discussing the total building-air volume, but not if we are to understand climates within the canopy.

The problem can be simplified somewhat by considering active surface *units* whose basic form is repeated throughout the urban canopy. The principal such unit is the *urban canyon* consisting of the walls and ground (usually a street) between two adjacent buildings (Figure 8.11). The canyon can also be seen to contain a *canyon-air volume* which has three sides with active surfaces (walls and floor), and three open sides (one being an imaginery 'lid' near roof-level and the other two being 'ends' through which along-canyon flow may take place). This arrangement recognizes the three-dimensional nature of the urban canopy, and allows the inclusion of the interaction between buildings rather than treating them as isolated objects. In a city with a grid-like street pattern there are two canyon orientations offset by 90°, and each will possess a different microclimate as a result of differences in the angle of solar incidence and the angle-of-attack of the wind. Such differences are in addition to those created by the radiative, thermal and moisture characteristics of their construction materials, and the canyon geometry (e.g. the ratio of the canyon width and height). Although there is relatively little published work regarding climatic processes within the canopy environment, the following section briefly outlines the probable nature of the energy exchanges inside a canyon.

(i) *Energy balance of an urban canyon*

Considering the range of possible orientations, width-to-height ratios, construction materials, moisture availabilities etc., it is impossible to designate a truly representative canyon. Here we will consider the characteristic energy exchange in one reasonably typical canyon and later mention some of the deviations to be anticipated in other arrangements.

The canyon whose energy balance is illustrated in Figure 8.12 is oriented with its long axis in a N–S direction; it has a width to height ratio of approximately 1 : 1, concrete walls (painted white) with no windows, and the floor is gravel with sparse vegetation. The data are spatial averages for each canyon surface (east- and west-facing walls and the floor). The measurements were conducted over a three-day period with cloudless skies and light winds. Under these conditions advection (ΔQ_A) was found to be negligible, and Q_F was assumed to be included in the ΔQ_S values for the

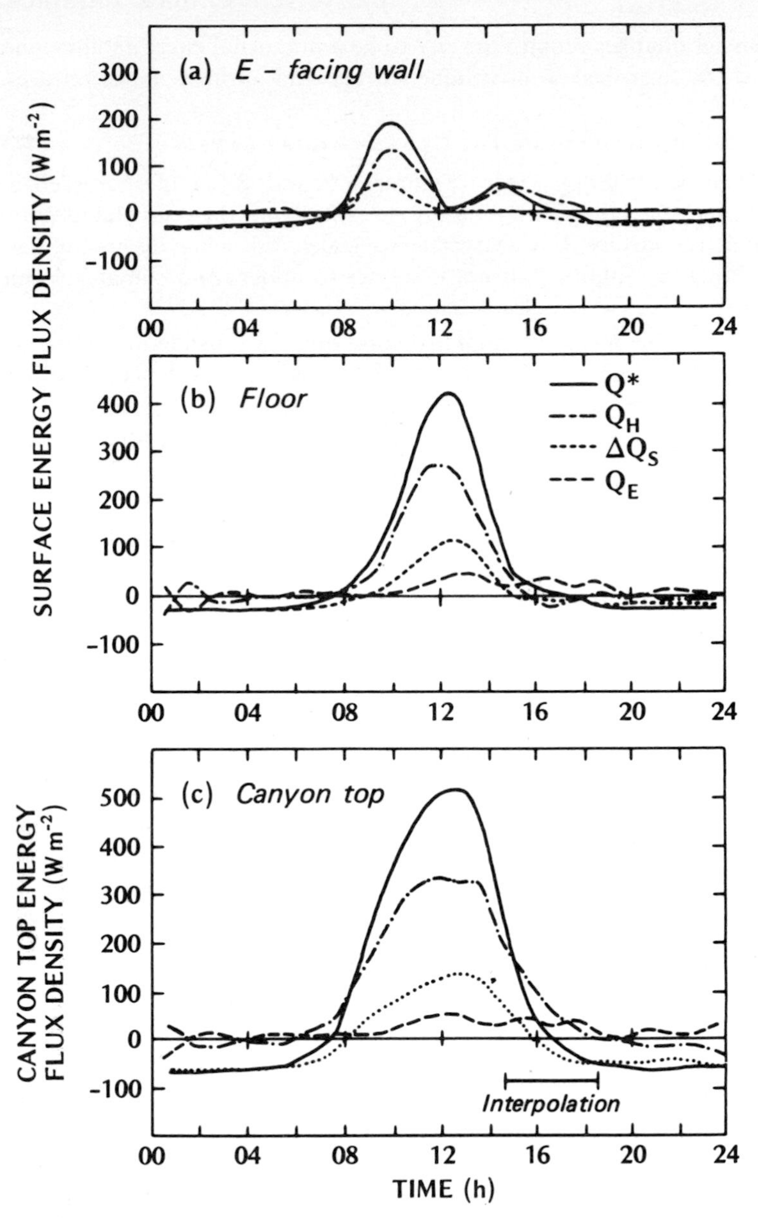

Figure 8.12 Diurnal variation of the energy balance components of a N–S oriented urban canyon including the surface balances of (a) an east-facing wall, (b) the canyon floor, and (c) the energy balance of the complete canyon system expressed as equivalent fluxes through the canyon top. Note: the situation for the west-facing wall (not shown) is almost a mirror-image of (a) during the day. Data from measurements in Vancouver, B.C. in the period 9–11 September 1973 with cloudless skies and light winds (modified after Nunez and Oke, 1977).

walls. Thus the energy balance of the dry walls was:

$$Q^* = Q_H - \Delta Q_S$$

and for the floor:

$$Q^* = Q_H - Q_E - \Delta Q_S$$

The results can best be viewed by following through a diurnal sequence. As the Sun rises above the local horizon of the canyon, the east-facing wall is the first surface to become irradiated, whilst the floor and west-facing wall remain in shade. At first only the uppermost part of the wall is sunlit, but this area receives the direct-beam input almost in its local zenith, and thus the receipt per unit area is maximized. For example at 0830 h the net radiation was approximately 360 W m^{-2} near the top of the east-facing wall, but since over the rest of the wall Q^* was close to zero the spatially-averaged wall value was only 65 W m^{-2} (Figure 8.12a). Thereafter Q^* increases until all of the wall is sunlit, whereupon it declines because the Sun's local angle of incidence becomes increasingly less favourable. After midday the east-facing wall is in shade and only receives diffuse-beam short-wave radiation, but it does experience an afternoon sub-peak which coincides with the time of maximum irradiation of (and therefore maximum reflection from) the opposite wall. The floor is sunlit only during the middle of the day, and its Q^* curve is symmetrical about solar noon (Figure 8.12b) and is greater in magnitude than for the walls because its albedo is considerably less. At night the net long-wave budget of all surfaces is rather small (20 to 30 W m^{-2}) in comparison with most other surfaces we have considered (with the exception of snow and ice). This is due to the reduced sky view factor for positions within the canyon. In comparison with an open horizontal surface (e.g. roof-top or rural) locations inside the canyon have a portion of the cold sky radiative sink replaced by canyon surfaces that are very much warmer.

By day 70–80% of the radiant energy surplus of all surfaces was dissipated to the air via turbulent transfer and the remaining 20–30% was stored in the canyon materials. At night the release of this heat (ΔQ_S) was sufficient to offset almost entirely the net radiative deficit, and turbulent exchange was minor.

Figure 8.12c shows the diurnal course of the energy balance of the complete canyon system by expressing the values of the three canyon surfaces as equivalent fluxes through the canyon top (assuming no storage in the air volume). Note that the diurnal curve for each term is smooth despite the fact that the canyon geometry produces radically different timing of maximum energy exchange for each component surface. Again it emerges that by day the radiative heat surplus is mainly convected out of the canyon and the remainder is stored in the fabric. But at night radiative losses are almost completely supplied by conduction from heat storage.

In other canyons we can expect to find different results. If the canyon orientation was E–W, only the south-facing wall and the floor would receive appreciable solar radiation in the northern hemisphere. This would result in asymmetric wall climates (cf. north and south walls in Figure 7.11). If the canyon width-to-height ratio were significantly greater or less than 1 : 1 there would be changes in the ability of solar radiation to penetrate, in the trapping of outgoing long-wave radiation, and in the amount of wind shelter. If the construction materials were different they would probably change the canyon albedo and the capacity for canyon heat storage. If water was more or less easily available the turbulent energy partitioning would be affected. In the case of Figure 8.12 the value of β at midday is approximately 6.4. This is likely to represent about the upper limit in urban areas; with lawns and trees more typical β values of 0·75–2·0 may be expected.

Different weather conditions will also change the canyon exchange conditions. For example, greater wind speeds would enhance the role of turbulence and advection and tend to reduce inter-canyon, canopy/urban boundary layer, and canopy/rural differences. Similarly increased cloud cover would reduce energy availability differences, and equalize long-wave radiation exchanges because the role of the sky sink in sky view factor differences would be reduced.

(ii) *Urban heat island*

The air in the urban canopy is usually warmer than that in the surrounding countryside. This *urban heat island* effect is probably both the clearest and the best documented example of inadvertent climate modification. The exact form and size of this phenomenon varies in time and space as a result of meteorological, locational and urban characteristics. To simplify matters we will restrict consideration initially to the heat island of a large

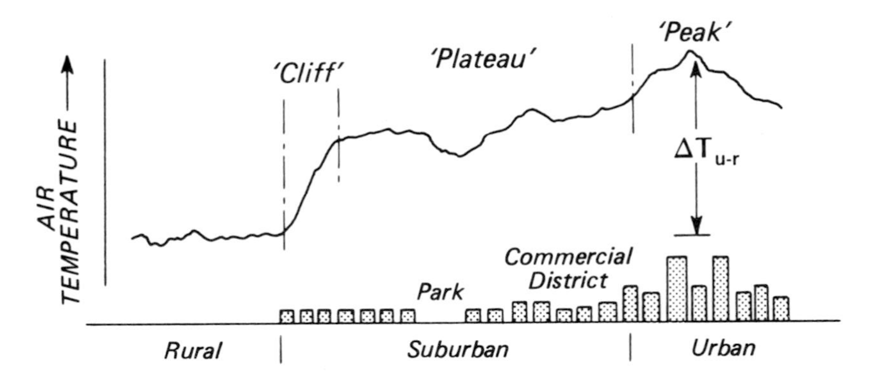

Figure 8.13 Generalized cross-section of a typical urban heat island (after Oke, 1976b).

city, with cloudless skies and light winds, just after sunset. Later we will add the temporal dimension, the effects of different meteorological conditions, and city size.

Figure 8.13 shows the characteristic variation of air temperature with distance whilst traversing from the countryside to the centre of an urban area under the conditions set out above. It demonstrates the aptness of the geomorphic analogy with an island, since the relative warmth of the city protrudes distinctly out of the cool 'sea' of the surrounding landscape. The rural/urban boundary exhibits a steep temperature gradient, or 'cliff' to the urban heat island. In this area the horizontal gradient may be as great as 4°C km^{-1}. Much of the rest of the urban area appears as a 'plateau' of warm air with a steady but weaker horizontal gradient of increasing temperature towards the city centre. The uniformity of the 'plateau' is interrupted by the influence of distinct intra-urban land-uses such as parks, lakes and open areas (cool), and commercial, industrial or dense building areas (warm). Especially in North American cities the urban core shows a final 'peak' to the heat island where the urban maximum temperature is found. The difference between this value and the background rural temperature defines the *urban heat island intensity* ($\Delta T_{u\text{-}r}$).

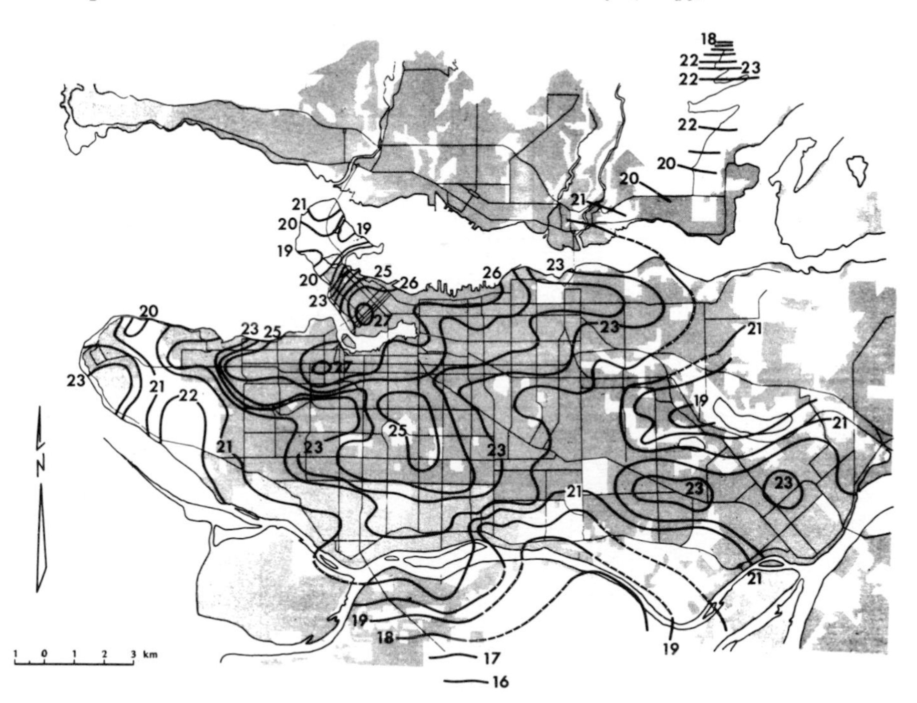

Figure 8.14 Two-dimensional distribution of air temperature in Vancouver, B.C. at 21 h on 4 July 1972. Cloudless skies, wind speed 2·0 m s^{-1} from the West. Isotherms (°C) are time-corrected potential temperatures (after Oke, 1976a).

Figure 8.14 is an example of an intense heat island (ΔT_{u-r} greater than 10°C) in Vancouver, British Columbia. Despite the complexity of the spatial pattern the general structure is in agreement with Figure 8.13. At the urban periphery in the south, west and north-west where the built-up margin is abrupt the 'cliff' is clearly evident. The approximately concentric isotherms converge on the dual 'peaks' located in the areas of greatest building density. In the Montréal example (Figure 8.19a) there are a number of such 'peaks' associated with areas of dense development but the centre of the city is occupied by a cool 'basin' due to the existence of a large park. These cases also serve to illustrate the fact that the urban heat island morphology is strongly controlled by the unique character of each city.

Given reasonably constant weather conditions the heat island intensity shows a diurnal variation. It is a maximum a few hours after sunset, and smallest in the middle of the day. In some cities ΔT_{u-r} may even be negative

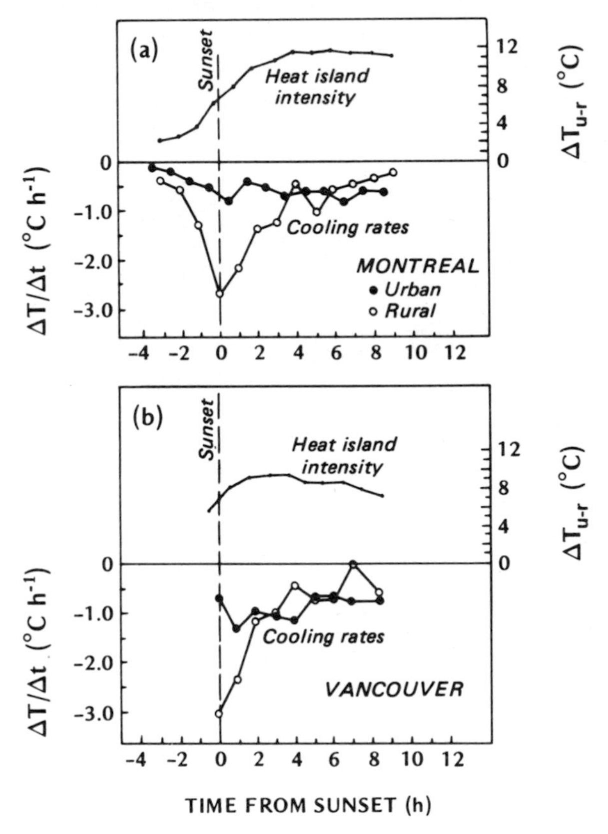

Figure 8.15 Mean hourly cooling rates ($\Delta T/\Delta t$), and urban heat island intensities (ΔT_{u-r}) for (a) Montréal and (b) Vancouver on calm, cloudless summer nights (after Oke and Maxwell, 1975).

(city cooler than its environs) around midday. The growth of $\Delta T_{u\text{-}r}$ is particularly rapid following sunset because of the difference between the urban and rural cooling rates (Figures 8.15 and 8.17b). During this period the rural area rapidly expends its energy store by long-wave radiation, but the urban area cools at a slower and more uniform rate. After a few hours the two areas cool at about the same rate and $\Delta T_{u\text{-}r}$ remains constant or slightly declines through the rest of the night. After sunrise (Figure 8.17b) the rural area also heats up more rapidly causing $\Delta T_{u\text{-}r}$ to decline.

This simplified diurnal picture is considerably modified by changes in weather conditions. At night when the heat island is best developed, $\Delta T_{u\text{-}r}$ is inversely related to wind speed ($\bar{u}$) and cloud cover. Thus $\Delta T_{u\text{-}r}$ is greatest under the 'ideal' conditions of weak winds and cloudless skies. This is to be expected since these conditions promote the differentiation of micro-climates between surfaces. In the present context this means that urban/rural cooling differences will be best exhibited. The heat island intensity is most sensitive to wind speed, probably indicating the importance of turbulent and advective activity. It appears that for a given city with no cloud, the value of $\Delta T_{u\text{-}r}$ near sunset is approximately related to $\bar{u}^{-1/2}$ (equation 8.3). The influence of cloud is related to its effectiveness in reducing radiation losses (L^*). Thus, a given coverage of low, thick Stratus is very much more important than a similar amount of high, thin Cirrus cloud (p. 25 and Table A2.3).

The heat island intensity is also related to the size of the city. Using population (P) as a surrogate of city size, $\Delta T_{u\text{-}r}$ is found to be proportional to log P (Oke, 1973). In the 'ideal' case with calm winds and cloudless skies the maximum heat island ($\Delta T_{u\text{-}r(max)}$) is very well related to log P for many

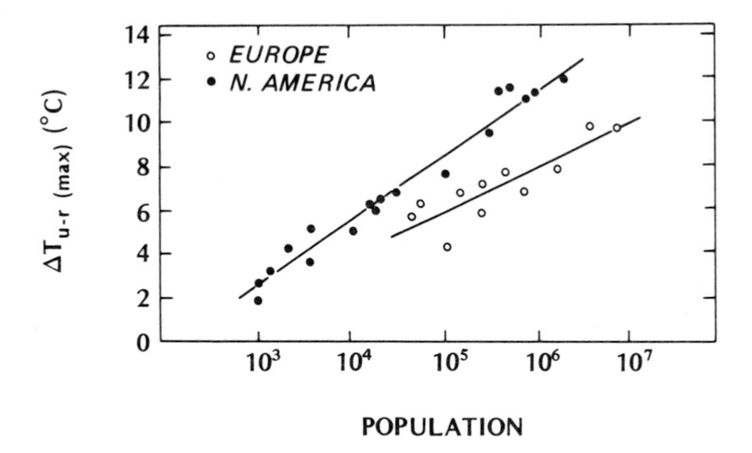

Figure 8.16 Relation between maximum observed heat island intensity ($\Delta T_{u-r(max)}$) and population (P) for North American and European settlements (modified after Oke, 1973).

North American and European settlements (Figure 8.16). This shows that even villages (population 1,000) have a heat island, indeed shopping centres and small groups of buildings are also warmer than their surroundings. At the other end of the scale it shows that the maximum thermal modification is about 12°C. The reason for the difference in slope between the North American and European relationships has not yet been identified but may be related to the fact that P is a surrogate index of the central building density. Urban heat islands have also been observed in low and high latitude settlements but there are insufficient data to establish relationships similar to those in Figure 8.16. If winds are included the relation between $\Delta T_{u\text{-}r}$ and city size (Oke, 1973) is given by:

$$\Delta T_{u\text{-}r} = P^{0.27}/4.04\bar{u}^{0.56} \simeq P^{1/4}/4\bar{u}^{1/2} \qquad (8.3)$$

where, $\bar{u}$ – regional (non-urban) wind speed at a height of 10 m. This equation applies to the heat islands of North American cities near sunset with cloudless skies. Obviously with very strong winds urban/rural thermal differences are obliterated. The form of equation 8.3 does not easily allow the identification of the critical wind speed at which this happens in a given city, but based on observation it appears that this value is approximately 10 m s^{-1} (measured at a height of 10 m at a rural site) in the case of a city with one million inhabitants. It is lower for smaller settlements.

The existence of the urban heat island has a number of biological economic and meteorological implications. Urban warmth is responsible for the earlier budding and blooming of flowers and trees in the city; a generally longer growing season; and the attraction of some birds to the thermally more favourable urban habitat. Humans however find the added warmth to be stressful if the city is located in an already warm climate. From an economic standpoint the heat island is beneficial in reducing the need for winter space heating, but disadvantageous in conversely increasing the demands on summer air-conditioning, and in speeding up the process of chemical weathering of building materials. Some of the atmospheric side-effects will be dealt with later.

The existence of the urban heat island has been attributed to various causes as listed in Table 8.4. All of these hypothesized 'causes' have been verified to operate in the right direction so as to make the urban area warmer, but the relative roles of each within the canopy is not yet certain. In the summer it would seem possible that 'causes' 3 and 4, and to a lesser extent 5 and 6, may combine to make the canopy a store of sensible heat by day, and that after sunset 2 and 7 prevent its rapid dissipation and hence keep urban temperatures higher than in the countryside. In the winter the role of 5 is likely to become more important. The cause of daytime negative heat islands awaits more work but their occurrence may be restricted to cities with deep and narrow canyons in their centre. This would mean that

TABLE 8.4 Commonly hypothesized 'causes' of the canopy layer urban heat island; after Oke, 1976b.

1. Increased counter radiation ($L\downarrow$) due to absorption of outgoing long-wave radiation, and re-emission by polluted urban atmosphere.
2. Decreased net long-wave radiation loss (L^*) from canyons due to a reduction in their sky view factor (SVF) by buildings.
3. Greater short-wave radiation absorption (K^*) due to the effect of canyon geometry on the albedo.
4. Greater daytime heat storage (ΔQ_S) due to the thermal properties of urban materials, and its nocturnal release.
5. Anthropogenic heat (Q_F) from building sides.
6. Decreased evaporation (Q_E) due to the removal of vegetation and the surface 'waterproofing' of the city.
7. Decreased loss of sensible heat (Q_H) due to the reduction of wind speed in the canopy.

the street-level is almost continually in shade and this would dictate that 'cause' 3 is only operative well above ground. It is unlikely that 1 is important within the canopy.

The above 'causes' relate to changes in the *surface* energy balance due to urbanization. But since the heat island is the difference between rural and urban *air* temperatures it is also helpful to consider the energy balances of

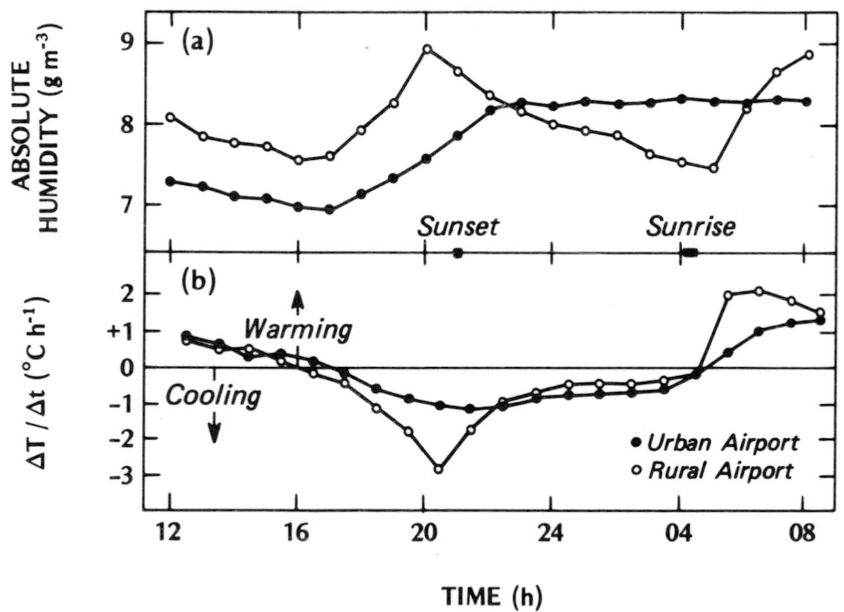

Figure 8.17 Diurnal variation of urban and rural (a) absolute humidities and (b) rates of temperature change ($\Delta T/\Delta t$), on 30 fine summer days in and near Edmonton, Alberta (modified after Hage, 1975). (Absolute humidity = ρq.)

the relevant rural and urban air volumes. Here we will investigate the case when the heat island is best developed (i.e. at night with near calm and cloudless skies). The energy balance of a rural air volume under these conditions was considered in Chapter 2 (p. 36), and it was shown that the cooling rate was determined by strong net long-wave radiative flux divergence ($-\Delta L^*$) with the convergence of sensible heat ($+\Delta Q_H$) by turbulence acting as a weak retarding (warming) influence.

In the urban canopy the appropriate body of air is the canyon-air volume (Figure 8.11). Measurements in such a volume under 'ideal' conditions show that here also the cooling is dominated by the divergence of net long-wave radiation, but that it is numerically very much weaker than in the rural case (Nunez and Oke, 1977). The reason for this has not yet been identified, but it may be related to the reduced sky view factor in the canyon. It is appealing to attribute the evening differences in urban/rural cooling rates (Figures 8.15 and 8.17b) to urban/rural differences in radiative flux divergence. Increases in wind speed and cloud, which have been noted to be associated with a decrease of $\Delta T_{u\text{-}r}$, might then be viewed in terms of enhancing turbulence as a cooling process relative to that of radiation, and thus reducing or eliminating urban/rural flux divergence differences.

(iii) *Humidity, fog and wind*

Urban/rural humidity differences are rather small, and the spatial pattern is often complex. The consensus of mid-latitude studies suggests that the urban canopy air is usually drier by day, but slightly more moist by night. This pattern is most evident during fine summer weather, as illustrated by Figure 8.17a. The rural data show the characteristic double wave of humidity which was explained in connection with Figure 2.16b, but the urban time-trend is quite different at night. During the day rural humidities are higher and this may be attributed to the greater rural evapotranspiration. In the early evening the rural air cools more rapidly and becomes more stable than the canopy air. Moisture therefore converges in the lower layers of the rural atmosphere because the evapotranspiration from the surface exceeds the loss to higher layers by the dampened turbulence. Thereafter the rural humidity decreases through the night, a vapour inversion forms, and the moisture content of the lower layers is depleted by dewfall. On the other hand, in the city weak evaporation, reduced dewfall, anthropogenic vapour, and the stagnation of airflow all combine at the same time to maintain a more humid atmosphere in the canyon-air volume. After sunrise the evaporation of dewfall and other surface water (including guttation and distillation) rapidly replenishes moisture in the rural atmosphere because convective transport is slow to develop. Later in the day instability promotes the mixing of vapour from the surface layer with that above and vapour concentrations in both areas are diluted.

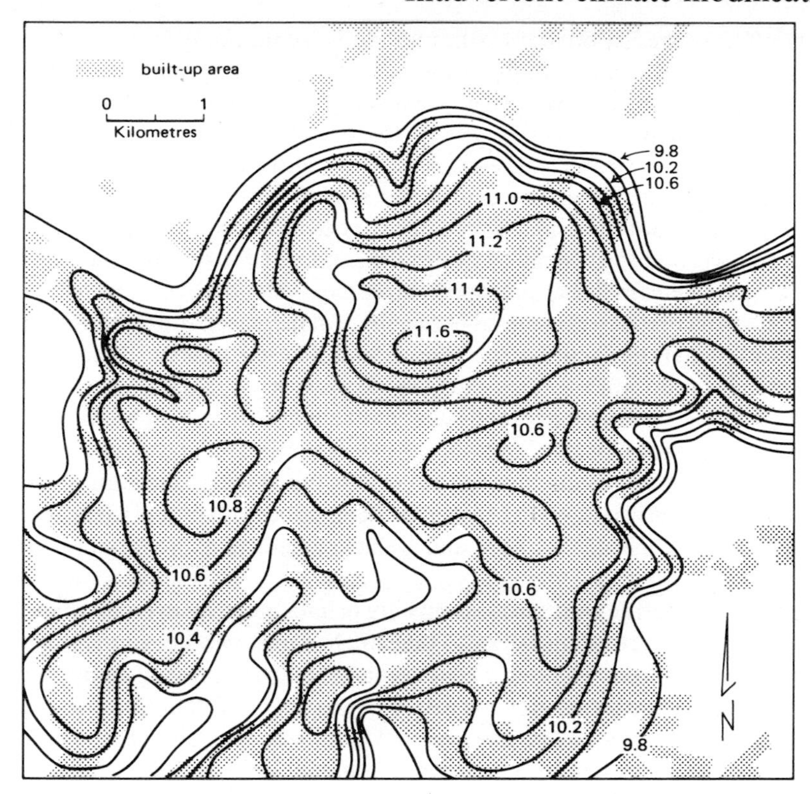

Figure 8.18 Two-dimensional distribution of vapour pressure ($\times 10^2$ Pa) in Leicester, U.K. at 2345 h on 23 August 1966 with near calm and virtually cloudless skies (after Chandler, 1967).

The night-time humidity excess in the city exhibits a moisture 'island' similar to that of temperature. The example shown in Figure 8.18 is from a summer night with almost 'ideal' conditions. The centre of the city is about $1{\cdot}8 \times 10^2$ Pa moister than the nearby rural areas at a time when the heat island intensity was 4·4°C. In this case the humidity 'island' also has a 'cliff' which corresponds well with the urban/rural boundary. In the daytime when the city air is drier the spatial pattern is more amorphous (Kopec, 1973). Increased wind speeds result in smaller urban/rural humidity differences, as with the heat island.

In cold climates the city during the winter can also be more humid during the day. Under these conditions the rural source of vapour (i.e. evapotranspiration) is virtually eliminated because the ground may be covered with snow or frozen, and vegetation is dormant, but in the city the anthropogenic releases from combustion (especially space-heating) provide a significant vapour input.

The effect of the city on fog is not as simple as is commonly assumed. The city is not always 'more foggy'. The situation is compounded by the definition of fog (p. 210) which only considers reduction in visibility and makes no distinction as to the type of fog or whether it is composed of pollutants or water droplets. Overall visibilities in the centre of large cities are indeed relatively low, and conditions tend to improve with distance from the centre. But the frequency of occurrence of thick fog (visibility less than 200 m) is often less in the city than the suburbs or rural surroundings. The improvement may be due both to the heat island effect and to the abundance of condensation nuclei in the city. An increased supply of nuclei results in greater competition for vapour and a larger number of smaller droplets which do not produce the very dense type of fog.

On the other hand there is no doubt that urbanization at high latitudes can produce ice-fogs (p. 230). The release of vapour into air at a temperature of less than −30°C results in a fog of ice crystals because the saturation vapour pressure is very low (Figure 2.15). The combustion of fuel for space-heating, and industrial, aircraft and vehicle operation, are principally responsible.

Wind speeds within the urban canopy are usually reduced in comparison with rural winds at the same height. There are however two situations when this may not be true. The first occurs when the faster moving upper air layers are either deflected downwards by relatively tall buildings or are channelled into 'jets' along streets oriented in the same direction as the flow (see pp. 234–5). The second occurs when regional winds are very light or calm (e.g. with an anticyclone). With cloudless skies at night this gives almost ideal weather for heat island development. The horizontal temperature (and therefore pressure) gradient across the urban/rural boundary can then be sufficient to induce a low-level breeze from the country into the city in the same manner as a sea breeze. The flow converges upon the city centre from all directions. Theoretically this must result in uplift over the city core and counter flow from the city to the country aloft (as in the complete circulation of Figure 5.6). If the canopy portion of the inflow is strong enough to overcome the frictional drag of the canyon walls then winds may be slightly greater than in the surrounding rural areas.

(c) CLIMATE OF THE URBAN BOUNDARY LAYER

The urban boundary layer (Figure 8.6) is that portion of the planetary boundary layer above the urban canopy whose climatic characteristics are modified by the presence of a city at the surface. In comparison with the surrounding landscape the city usually provides a rougher, warmer, and perhaps drier set of surface conditions.

The rough elements of a city are mainly its buildings. These relatively tall, sharp-edged and inflexible objects make cities the roughest of all aerodynamic boundaries. Practical problems make it difficult to measure

TABLE 8.5 Estimated† values of the roughness length (z_0) based on idealized building configurations (after Lettau, 1970).

Building type	Assumed building height h (m)	Assumed lot areas A' (m^2)	Assumed silhouette A^* (m^2)	Roughness length z_0 (m)
Low	4	2,000	50	0·05
Medium	20	8,000	560	0·70
High	100	20,000	4,000	10·00

† $z_0 = 0{\cdot}5\, h\, (A^*/A')$

the value of the roughness length (z_0) but the estimates in Table 8.5 are probably approximately correct. As in the case of tall vegetation the wind profile equations should include a zero-displacement height to allow for the upward movement of the effective momentum sink.

The effects of increased roughness on moderate or strong air flow across a city are illustrated in Figures 2.9a and 5.3. The increased drag and turbulence results in a deeper zone of frictional influence within which wind speeds are reduced in comparison with those at the same height in the country. This local slowing of the air flow causes it to 'pile-up' (i.e. converge) over the city, and this is relieved by the tendency for uplift. The vertical motion induced is in addition to that brought about by the heat island effect.

Thermal modification of the urban boundary layer occurs as cooler rural air traverses across the warmer city. During the day the influence of a large city may extend up to 0·6–1·5 km (i.e. virtually extending throughout the entire planetary boundary layer). This is possible because the normal daytime convection is augmented by both mechanical and thermal convection from the rougher, warmer city.

At night the urban heat island contracts to a depth of only 0·1 to 0·3 km because the bulk of the planetary boundary layer is stable and this suppresses vertical transfer. Nevertheless the combination of urban warmth at the surface and increased forced convection is capable of eroding the stability of rural air as it advects over the city. This concept first put forward by Summers (1964) is illustrated by the results from Montréal in Figure 8.19. The map shows that the canopy heat island intensity was about 4·5°C, and the profiles show the progressive development of the thermally modified boundary layer in an along-wind traverse from the rural surroundings to the city centre. Site 1 is in the upwind 'rural' area and the profile shows an inversion of potential temperature (i.e. stable) based at the ground extending up to at least 600 m. Just across the urban/rural boundary at site 2 surface temperatures rise by 2°C at the heat island 'cliff', but the bulk of the layer remains strongly stable. (The warming in the 100–300 m layer is probably due to heat released from the stacks of oil refineries

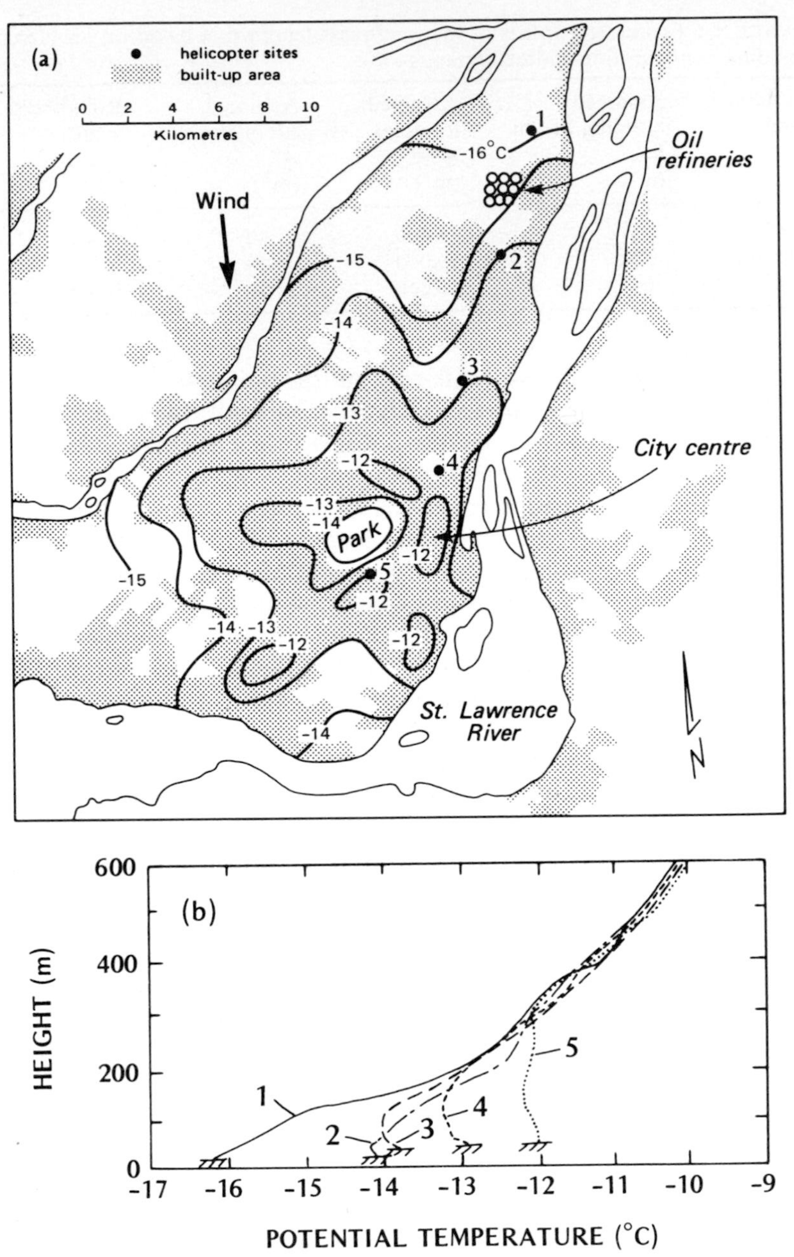

Figure 8.19 The urban heat island in Montréal on 7 March 1968 at 07 h with winds from the N at 0·5 m s^{-1} and cloudless skies. (a) Urban canopy temperature distribution (°C) from car traverses, and (b) vertical profiles of potential temperature at different along-wind distances into the city from helicopter measurements. Profile numbers correspond with location of sites in (a) (after Oke and East, 1971).

located between sites 1 and 2.) Moving further towards the city centre the amount of urban warming increases, and the depth of the modified boundary layer grows. At the centre (site 5) the depth of the heat island is about 300 m and the urban boundary layer is characterized by an almost isothermal (neutral) profile, capped by the stable 'rural' layer above. Downwind of a city under these conditions, the stable rural layer re-forms at the surface. This isolates the warmer urban layer aloft (Figure 8.6), giving what has been termed the urban 'plume' (Clarke, 1969b).

Thus unlike rural areas the urban atmosphere does not experience a strong diurnal change in stability. Both by day and by night the urban layer is well mixed. This tends to destroy strong temperature gradients and therefore both strong instability and stability are absent. The lack of stability at night also explains why winds are often greater in the urban boundary layer. In rural areas strong stability near the ground causes a partial de-coupling from the upper air flow. On the other hand the maintenance of a nocturnal mixing layer permits vertical exchanges of horizontal momentum to continue and thereby to maintain higher mean wind speeds in the city.

TABLE 8.6 Mechanisms hypothesized to cause the urban heat island in the urban boundary layer.

1. Entrainment of warm air scoured from the canopy layer heat island.†
2. Anthropogenic heat from roofs and stacks.
3. Downward flux of sensible heat from the overlying stable layer by penetrative convection.
4. Short-wave radiative flux convergence in the polluted urban air.

† See Table 8.4.

The four physical mechanisms thought to underly the thermal modification of the urban boundary layer are listed in Table 8.6. Some of the warming enters through the lower boundary (i.e. the plane ABCD of the building-air volume in Figure 8.7a) and this consists firstly of the canyon-warmed air, and secondly of the heat emanating from the building roofs, chimneys and industrial stacks. Thirdly, it is also likely that heat is added to the urban boundary layer from above. Both by day and by night the urban layer is capped by an inversion and hence there is warmer air aloft. The action of urban-generated turbulence may 'eat' away at the base of the inversion and mix this warmer air downwards in the process of *penetrative convection*. Such a re-distribution of heat in the vertical is conceptually the same as the practice of frost protection by mixing (p. 209), and also hastens the early morning growth of the mixed layer in rural areas (Figure 2.14). The detailed structure of urban temperature profiles (Figure 8.19b) tends to support this idea of dual heat convergence from below and above. The 'neutral' profile is actually composed of a slightly unstable lower layer and a slightly stable one above. Finally, it is probable that internal radiative exchanges play a role in urban boundary layer temperature changes.

There is only scanty information concerning humidity in the urban boundary layer. That available seems to suggest that urban/rural differences have a similar sign to those in the canopy, and may be detected up to 1 km above the city, and in the downwind 'plume'.

(d) URBAN EFFECTS ON CLOUD AND PRECIPITATION

Although this book does not consider convective precipitation phenomena, a brief statement of urban weather modification seems appropriate because the urban atmosphere provides an ideal natural laboratory for such studies. Convective precipitation requires a supply of water vapour and nuclei to form droplets, and uplift to carry these materials to sufficient heights so that cooling and condensation can occur. The urban 'plume' would seem to be characterized by these properties to a greater extent than the surrounding rural areas.

If urban precipitation enhancement does indeed occur it will take time for the materials to be carried up to cloud level and for the droplets to form and grow to a sufficient size to fall to the surface. Therefore any effects are

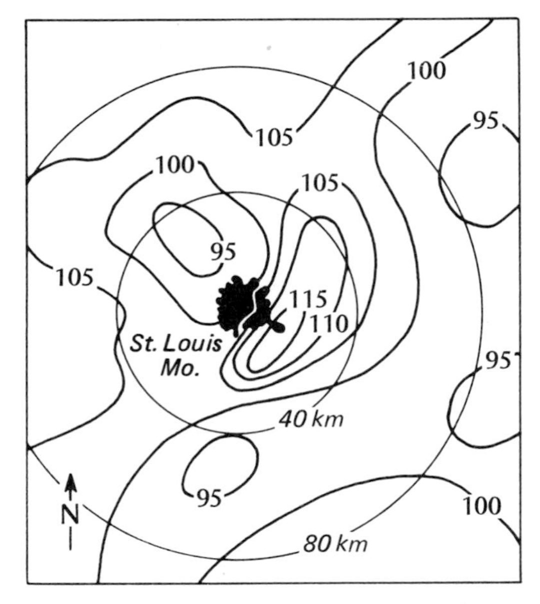

Figure 8.20 Average rural/urban ratios of summer rainfall in the St Louis area in the period 1949–1968 (after Changnon *et al.*, 1971). (Note – the rural/urban ratio is the ratio of the precipitation recorded at any station to that at two stations near the urban centre.)

likely to occur downwind of, rather than within, the city itself. Just such a pattern emerges from the results in Figure 8.20 which show the summer precipitation around St Louis compared with that in the city. Background fluctuations of ±5% are found all around the city except in the area centred about 20 km east of the urban area where precipitation is at least 15% greater. Since 90% of all rain systems move in a W to E direction at this location the zone of apparent enhancement lies downwind of the city. The effect of the St Louis urban area appears to be most marked in the case of summer convective rainfall involving high intensities. It is especially thought to increase the frequency and intensity of severe weather (thunder and hailstorms), and again a time lag is evident because although maximum thunder increases are experienced in the city, maximum hail excesses are recorded in an area 20 to 40 km downwind (Changnon, 1972).

As the body of such information on inadvertent weather modification develops a consensus of opinion seems to be emerging that a large city does indeed produce a downwind enhancement of certain types of precipitation. On an annual basis the total receipt at the position of the downwind maximum is from 5 to 30% above that of the surrounding region. The relative roles of the causative agents remain to be determined.

There is conflicting evidence regarding the influence of a city on snowfall. Climatological analyses have shown small increases in snow deposition and case studies report individual instances where snow amounts have increased downwind of urban and industrial areas. On the other hand other work suggests that if air temperatures are close to freezing the heat island effect may be sufficient to melt flakes over the city centre so that they are deposited as rain.

CHAPTER 9

Air pollution in the boundary layer

1 Introduction

Air pollutants are substances which, when present in the atmosphere under certain conditions, may become injurious to human, animal, plant or microbial life, or to property, or which may interfere with the use and enjoyment of life or property. This definition stresses the effects upon receptors, and includes modification due to both natural and anthropogenic sources. In the context of this book we will emphasize the physical factors leading to a more or less polluted atmosphere at a location rather

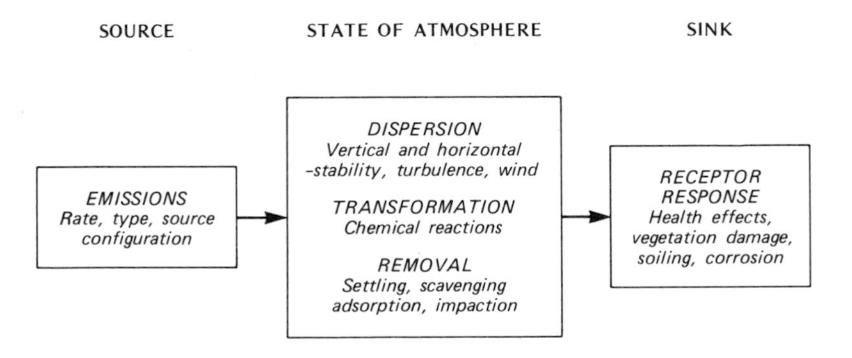

Figure 9.1 Progress of pollutants from sources to sinks (receptors) by way of the atmosphere.

than the effects evoked after receipt, and we will be concerned mainly with pollutants generated by human activities rather than by natural events.

Two classes of factors determine the amount of pollution at a site. They are (i) the nature of relevant emissions, and (ii) the state of the atmosphere. Figure 9.1 shows their relationship in the form of a flow chart, and this chapter is organized in the same sequence. First we examine the input of pollutants to the boundary layer. Second we consider the atmospheric mechanism governing the dispersal, transformation and eventual removal of pollutants from the air. Finally we outline a few examples of air pollution related to particular source-weather combinations.

A number of important characteristics of the emissions have a bearing upon the resulting air pollution. Obviously the rate of emission and the physical and chemical nature of the pollutants are central to the determination of the amount and type of pollutant loading. It is also important to know certain other characteristics of the source including the shape of the emission area, the duration of the releases and the effective height at which the injection of pollutants occurs.

After release, the dispersion of the pollutants is controlled by atmospheric motion (wind and turbulence) on many scales. The temperature stratification is important because it defines the atmospheric stability (Appendix A1) and this in turn controls the intensity of thermal turbulence (buoyancy) and the depth of the surface mixed layer. Together they regulate the upward dispersion of pollutants and the rate of replacement of cleaner air from above. The wind field is critical with respect to horizontal dispersion in the boundary layer. The wind speed determines both the distance of downwind transport and the pollutant dilution due to plume 'stretching' (p. 279), and in combination with the surface roughness it establishes the intensity of mechanical turbulence. The wind direction is important because it controls the general path the pollutants will follow, and its variability circumscribes the extent of cross-wind spreading.

Whilst suspended in the atmosphere, the pollutants may undergo physical and chemical transformation. These changes are related to meteorological characteristics such as the abundance of water vapour or droplets, the air temperature, the intensity of solar radiation, and the presence or absence of other atmospheric substances. Similarly the eventual removal of pollutants by precipitation-related processes (called *scavenging*), by gravitational settling, or by surface adsorption and impaction, is related to the state of the atmosphere.

It is salient to note that even if emissions in a given area remain fairly constant, the air quality can exhibit a wide range of conditions. The variability is introduced by the ever-changing state of the weather and therefore the ability of the atmosphere to transport, dilute, transform and remove pollutants. In general the atmosphere has a tremendous capacity for dispersal, but at certain times and locations this facility may be substantially curtailed. Under these conditions air pollution can pose severe problems.

This chapter does not treat the effects of air pollution in any detail. Some of the effects of pollution upon weather and climate are mentioned in Chapter 8 (e.g. the effects on transmission of radiation, on visibility and on the development of precipitation and fog). The effects on surface receptors are many and varied and rather than give an over-simplified account here, the reader is referred to the fuller texts in the supplementary reading section.

2 Emissions

Although it is a truism, it is worth stating that without sources there would be no emissions, and without emissions there would be no pollution. This does not mean that all emissions are harmful, indeed it should be appreciated that most natural releases are essential for the maintenance of life (e.g. carbon dioxide, nitrogen compounds) and it is only when their concentration becomes too high that they can be classed as pollutants. It does however highlight the fact that the only fundamental form of air pollution control is to curb emissions strictly at their source. This need is

TABLE 9.1 Types and sources of atmospheric pollutants (modified after Varney and McCormac, 1971).

Type	Source	
	Natural	Anthropogenic
Particulates	Volcanoes Wind action Meteors Sea spray Forest fires	Combustion Industrial processing
Sulphur compounds	Bacteria Volcanoes Sea spray	Burning fossil fuels Industrial processing
Carbon monoxide	Volcanoes Forest fires	Combustion engines Burning fossil fuels
Carbon dioxide	Volcanoes Animals Plants	Burning fossil fuels
Hydrocarbons	Bacteria Plants	Combustion engines
Nitrogen compounds	Bacteria	Combustion

further accented by the nature of the atmosphere as a medium for dispersal. Air is readily mixed and transported, therefore it could be said that in comparison with soil and water the atmosphere is an excellent diluting medium for effluents. On the other hand, the atmosphere's capability to disperse materials also means that after release it is virtually impossible to re-capture or contain air pollutants. Thus air pollution control must be a preventive rather than a remedial technology.

Table 9.1 gives a general listing of the most important types of air pollutants, and their major sources. We will use this as a basis for a brief description of the nature of emissions.

Particulates – about 90% of all atmospheric particles are derived from natural sources. Anthropogenic sources are mainly associated with combustion (domestic and power station coal and oil burning, automobiles, and refuse incineration), industrial processing (cement and brick works, iron foundries, metal-processing mills) and surface disturbances due to building activities (house and road building).

The term particulate matter includes both solid and liquid particles, and embraces a wide range of sizes varying from greater than 100 μm to less than 0·1 μm in diameter. Those greater than 10 μm consist of dust, grit, fly ash and visible smoke, and because of their weight tend to settle out relatively rapidly after emission. They are therefore of greatest nuisance close to the source (e.g. soiling laundry, houses, cars, vegetation, etc.). Particles of less than 10 μm remain suspended longer and hence the state of the atmosphere becomes more important in their dispersion. If they are less than 1 μm in diameter they may remain in the boundary layer for several days, and if ventilation is weak lead to the build-up of a smoky haze.

Particulates most commonly consist of carbon or silica but may also include iron, lead, manganese, cadmium, chromium, copper, nickel, beryllium and asbestos (U.S. DHEW, 1970a).

Sulphur compounds – sulphur enters the atmosphere in many forms especially as the gases sulphur dioxide (SO_2) and hydrogen sulphide (H_2S), but also as sulphurous (H_2SO_3) and sulphuric acid (H_2SO_4), and as sulphate salts on particulate matter. About two-thirds of all atmospheric sulphur comes from natural sources of which H_2S from bacterial action is the most important. Of the anthropogenic sulphur releases SO_2 is by far the most significant. It is mainly expelled as a waste product in the combustion of sulphur-bearing fuels such as coal and oil which are used for space heating in cold-winter climates, and to generate electricity by power plants. Ore smelters and oil refineries also give off SO_2 in large quantities, and Kraft pulp mills and oil refineries emit H_2S with its disagreeable smell.

Oxides of carbon (CO_x) – carbon monoxide (CO) is generated in any process involving the incomplete combustion of carbonaceous materials. Natural sources are relatively small, and the main anthropogenic source is the internal combustion engine. Emissions are therefore usually concentrated along major highways and especially near congested urban streets. In such locations the vehicle density is high, engine efficiency is low, and ventilation is restricted. Other sources include metal processing industries, gasoline refineries, and paper processing factories, but by far the most important source of CO to which human receptors are exposed is cigarette smoke (Bates, 1972). When inhaled into the lungs CO can be absorbed into the blood haemoglobin thereby reducing the ability to fix oxygen. In high concentrations CO is potentially lethal.

CO is a particularly stable gas and there is some discussion as to how it is removed from the atmosphere. It can oxidize to carbon dioxide (CO_2), or be absorbed by oceans, but soils are probably the main natural sink by virtue of the activity of soil micro-organisms.

CO_2 is often not considered to be a pollutant because it is such an essential element in life processes. It is evolved whenever a fuel is completely combusted in the presence of oxygen. Plants and animals are natural sources of CO_2, they respire (exhale) CO_2 after consuming carbohydrate fuels. The diurnal and seasonal variations of CO_2 emission from a soil-plant system have already been discussed (Chapter 4). Anthropogenic releases accompany the burning of fossil fuels and concentrations of CO_2 are therefore considerably higher in urban areas (Bach, 1972). Vegetation and the oceans are natural CO_2 sinks, but they do not seem to be keeping pace with the increasing rate of anthropogenic production so that global concentrations are generally increasing. The main concern with regard to this trend is its possible effect upon the total Earth-Atmosphere long-wave radiation budget.

Hydrocarbons (H_C) – most atmospheric H_C originates from the natural decomposition of vegetation. Despite being rather small compared with natural releases the anthropogenic production is very important because not only are some of the compounds potentially harmful to vegetation and humans, they are also highly reactive and can facilitate the formation of photochemical smog (p. 283). Hydrocarbons are produced in the combustion of fossil fuels, and evaporation from gasoline. Vehicles are the primary source and H_C emissions (like those of CO) are closely related to traffic density.

Oxides of nitrogen (NO_x) – natural emission of nitrogen compounds arises from organic decomposition in the soil and oceans. Most anthropogenic releases accompany the combustion of fuel under pressure and heat resulting in the fixation of nitrogen and oxygen to form nitric oxide (NO) which is a relatively harmless gas. However, in the atmosphere NO readily oxidizes to give nitrogen dioxide (NO_2) which is a yellow-brown coloured gas, and an irritant. The principal sources of oxides of nitrogen are vehicles, coal and natural gas burning, and fertilizer and explosives factories.

Minor and secondary pollutants – the above review only mentions the most common pollutants. There are other substances which may be classed as *minor pollutants* because they are emitted in small quantities, or are restricted to small areas. On the other hand, in other respects (e.g. toxicity) they may be significant and should not be neglected. In this class are such pollutants as hydrogen fluoride (from fertilizer factories), toluene (from paint solvents), radioactive substances (from reactors and weapons testing), and ammonia (from fertilizer and chemical plants). In addition there are a wide range of *secondary pollutants* created by chemical reactions between two or more pollutants, or between pollutants and natural atmospheric constituents. The most notable examples are the products of the photochemical chain reactions, such as ozone (O_3), peroxyacetyl nitrate (PAN) and aldehydes. In urban air where there is an almost haphazard mixture of a very wide range of chemical substances from many sources the detailed chemistry remains largely unexplained.

The scenario of anthropogenic emissions is continually changing. In terms of their sheer amount they are increasing almost everywhere, but the relative roles of individual pollutant types are also shifting. In many places there has been a significant decline in large particulate pollution associated with the phasing-out of steam locomotives, the use of filters in industrial chimney stacks, and the switch in heating fuel away from coal and wood, to oil and natural gas. In many countries this change has been hastened by enforcement of legislation setting limits on industrial emissions, and the establishment of smoke-control zones in urban areas where domestic coal burning or garden bonfires were a problem. The net result has been to reduce the average size of particulate pollutants. The switch in fuels has also tended to reduce the concentration of SO_2 especially in Europe where open-hearth coal-burning was until recently the traditional means of home heating. In North America the car is the largest source of pollutants except in highly industrialized regions. This means that the pollution-mix is dominated by CO, NO, NO_2, H_C and small particulates (including lead from gasoline and asbestos from brake linings). If sufficient sunlight is

TABLE 9.2 Typical source configurations.

Shape	Duration	Height	Example
Point	Continuous	Elevated	Chimney stack
		Ground	Bonfire
	Instantaneous	Elevated	Shell burst
		Ground	Explosion
Line	Continuous	Ground	Busy highway
	Instantaneous	Elevated	Crop spray; Aircraft exhaust
Area	Continuous	Elevated	City; Forest or field fire

available this leads to the possible development of photochemical smog and its secondary pollutants. As the vehicle population of other countries grows this type of contamination is becoming more widespread.

A complete knowledge of emissions also requires information on the *source configuration* including its shape, duration and height. These criteria are utilized in the classification scheme of Table 9.2 which includes most of the common source arrangements in the boundary layer. The most important *point source* is the chimney stack which can be the originator of very concentrated and harmful or exotic materials. A considerable amount of work has been conducted on the dispersion of effluent from point sources and is discussed later (p. 285). A busy highway is the most common example of a *line source* where it is assumed that the integration of the exhaust emissions from many separate vehicles constitutes a continuous output along its length. Cities are the paramount *area sources* wherein it is convenient to lump together the individual emissions from a multiplicity of small sources (e.g. houses etc.) to give an areal average. In the detailed modelling of an urban region it is often the practice to divide it into grid squares and to assign area emission strengths to each. The major highways are treated as line sources with a strength dependent upon the traffic density, and the major point sources (e.g. large power or industrial plants) are also treated separately. It should be pointed out however that these designations depend upon the scale of the study. For example, on a continental scale cities might appear as point sources.

3 Atmospheric controls

(a) THE EFFECT OF STABILITY ON DISPERSION

The vertical movement of pollutants in the boundary layer is largely controlled by the prevailing stability conditions, and therefore the air temperature stratification (Appendix A1). Free convection is an important means of diffusing material into a larger volume, and the depth of the mixed layer sets an upper limit to the vertical dimension of this volume. Therefore from a convective standpoint the best conditions for pollutant dispersion usually occur with strong instability and a deep mixed layer. This is characteristic of sunny, daytime conditions, especially in summer. Conversely, the worst conditions for dispersion occur when there is a temperature inversion and the boundary layer is stable. Turbulence is then suppressed and upward motion is effectively eliminated. There are a number of exceptions to these general remarks and some of these are outlined later. In view of the importance of inversions to dispersion, we will give special attention to their characteristics and modes of genesis.

By definition an inversion exists when warm air overlies cooler air. This could be brought about by cooling (usually radiative) from below; by warming (usually adiabatic) from above; or by the advection of warmer or cooler air.

Inversions due to cooling – we have already discussed the formation of the simple nocturnal *radiation inversion* (p. 53). The driving-force is surface long-wave radiative cooling and the inversion is based at the ground (or other active surface) and may extend up to heights of 50–100 m. Radiation inversions are characteristic of the lower atmosphere on nights with little or no cloud, and light winds or calm. Their strength and depth may be locally increased by katabatic air drainage (p. 154). During the period of 'polar night' in the winter at high latitudes they may remain intact for several weeks at a time. Elevated radiation inversions can also be formed within the atmosphere. In these cases the active cooling surfaces may be cloud tops, or polluted layers.

Evaporation cooling of the surface can also give rise to a surface-based inversion during the day in fine weather. For example, we have already noted that the 'oasis-effect' is accompanied by an inversion because the moist cool surface is overlaid by hot air. A summer rain shower can also cool the ground's surface by evaporation and lead to a similar temperature profile.

Inversions due to warming – when air sinks within the atmosphere it

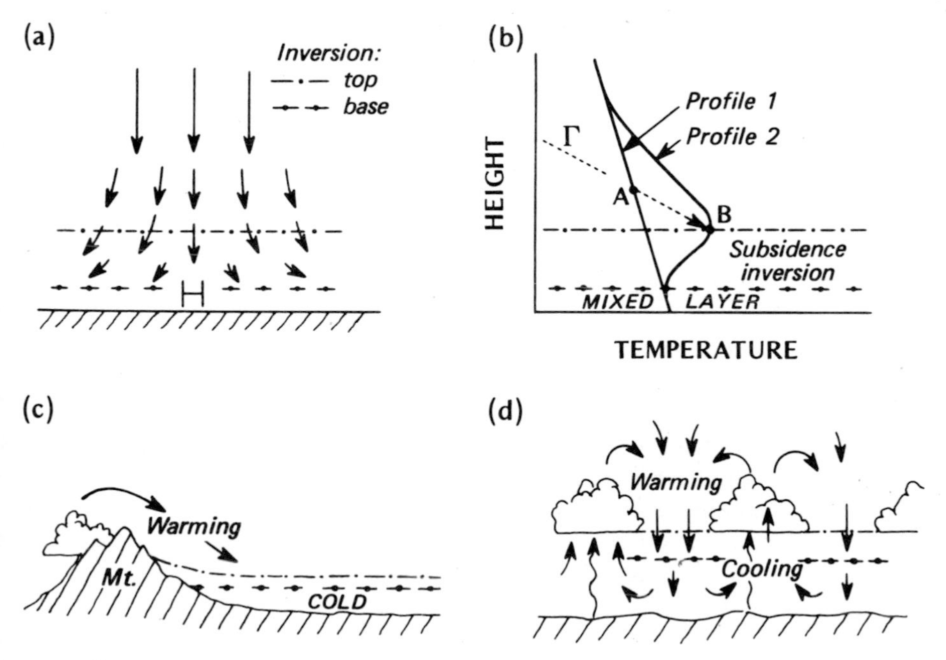

Figure 9.2 Inversions formed by adiabatic warming due to subsidence. (a) and (b) in an anticyclone, (c) in the lee of mountains, and (d) in convection cells between clouds.

encounters greater pressure, is compressed, and warms. In anticyclones (areas of high atmospheric pressure) it is quite common for air in the middle troposphere (0·5 to 5 km, Figure 1.2) to be gently subsiding at a rate of about 1 km day^{-1} (e.g. from A to B in Figure 9.2b). If this air is unsaturated it will warm at the dry adiabatic lapse rate ($\Gamma = 9{\cdot}8$°C km^{-1}) and therefore the temperature of a sinking parcel will increase approximately at the rate of 10°C day^{-1}. This results in a characteristic elevated 'bulge' in the temperature profile (Figure 9.2b) so that the lowest portion of the warm layer exhibits a *subsidence inversion* which forms a very effective 'lid' to the underlying mixed layer over an extensive area. In the mid-latitudes anticyclones tend to migrate but can stagnate over a region for up to 2–3 weeks. In the sub-tropics they are semi-permanent allowing pollutant concentrations to build-up over a period of time. In either case stagnation creates a thoroughly murky mixed layer with a sharp upper boundary. Viewed from above the scene is vividly described as 'anti-cyclonic gloom'.

Subsidence warming also accounts for the development of inversions in the lee of mountain ranges (Figure 9.2c). A good example occurs in the lee of the Rocky Mountains in winter. Radiative cooling over the northern interior of the North American continent creates a very cold air mass which 'ponds' itself up against the mountains. The descending warm air spreads over the top of this stagnant pool and provides an almost impenetrable cap to upward dispersion.

On a smaller scale adiabatic warming accounts for the *cloud-base inversion* found on summer afternoons when the sky is dotted by a patchwork of small cumulus clouds. The cloud pattern and density is not haphazard, it is the visual expression of an organized multi-cellular circulation system (Figure 9.2d). The upward limbs of these cells are buoyant thermals carrying heat and moisture away from the heated ground. During ascent they cool adiabatically and their moisture condenses to give cumulus cloud. Unless there is a low-level convergent supply of fresh air (e.g. a sea breeze) this upward displacement will be compensated for by subsidence from above. The downward limbs carry drier air from above between the clouds. During its descent this air is adiabatically heated and thereby creates a discontinuous layer of warm air just below cloud-base level.

Inversions due to advection – the weather fronts seen on weather maps are the boundary zones between contrasting air masses. The full frontal surface extends from the surface (where it is plotted) up into the atmosphere as a sloping plane with the warmer (less dense) air invariably overlying the colder air. Fronts are therefore always characterized by an inversion. In general, fronts are in motion and if the passage of a front causes cold air to replace warm air at a given station it is a cold front (Figure 9.3a), and the reverse case defines a warm front (Figure 9.3b). Due to their motion *frontal*

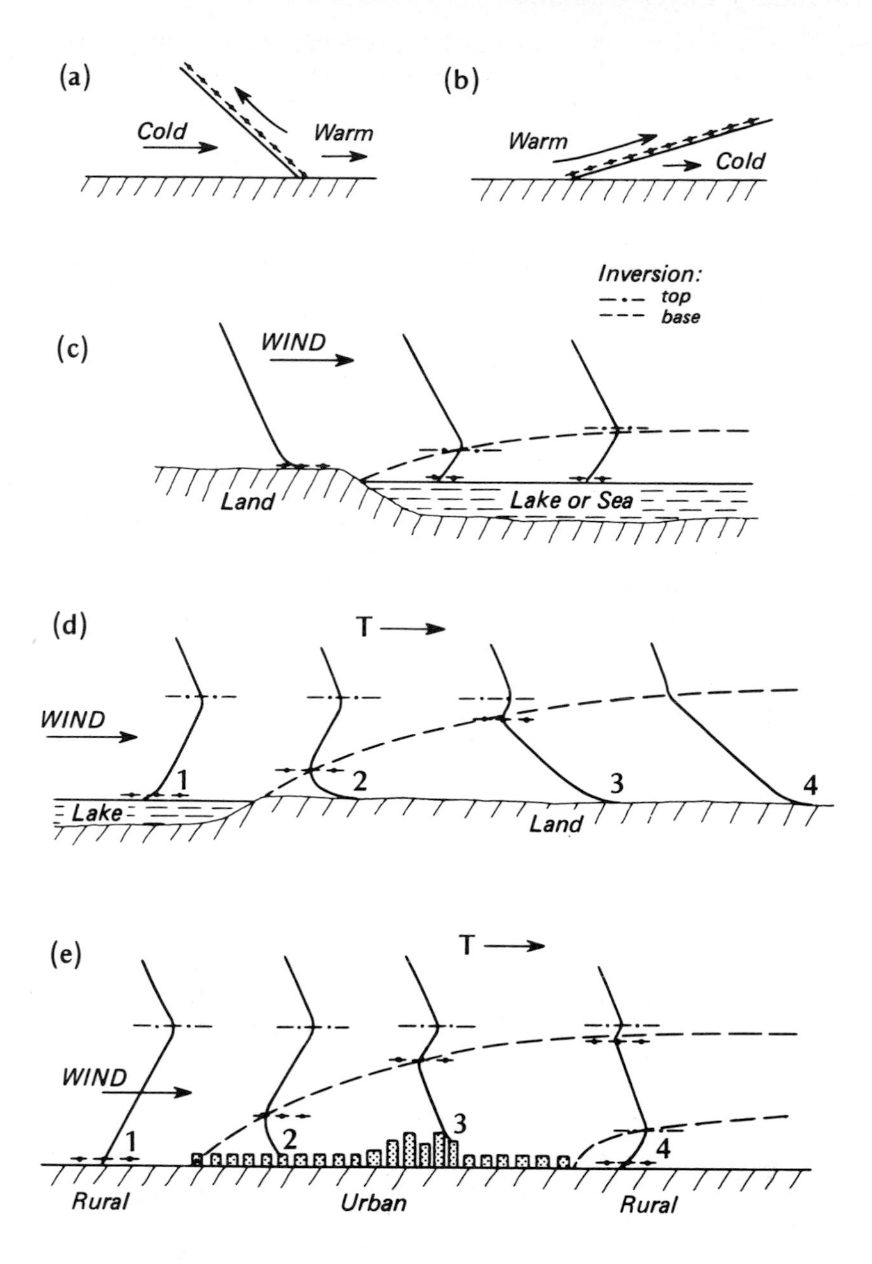

Figure 9.3 Inversions due to advection. A frontal inversion caused by (a) cold air wedging under warmer air (cold front), and (b) warm air over-riding colder air (warm front). (c) The modification of an unstable temperature (T) profile to give a surface-based inversion over a cool water body. Elevated inversions due to the advection of stable air across warmer surfaces, (d) near a lake on a spring afternoon and (e) over a city at night.

inversions are usually short-lived and not very important in air pollution considerations. But problems can occur with slow-moving warm fronts because the frontal slope is usually rather slight (averaging 1:200) and thus the elevated warm air is relatively close to the ground over a considerable area. Also since the frontal surface (inversion base) slopes towards the colder air its approach causes a progressive thinning of the mixed layer. Dispersion conditions therefore become increasingly poorer until the front has passed. Even then the danger may not be over if pockets of cold air remain trapped at the bottom of valleys because the warm air behind the front may ride over these cold pools and act as a 'lid' thereby inhibiting the diluting action of vertical mixing.

Advection inversions also arise when warm air flows across a cold surface such as a cold land surface, water body or snow cover. The cooling of the underside of the air mass creates an inversion with its base at the surface (e.g. Figure 9.3c).

Existing inversion structures can also be modified by advection across warmer surfaces. Figure 9.3d illustrates modification of the temperature profile as air is advected from a cool lake surface to a heated land mass on a spring afternoon, and corresponds to the lake breeze circulation shown in Figure 5.6a. Over the cool lake (profile 1) the lowest layers exhibit an advection inversion with a normal daytime lapse rate above. As the air moves inland (profiles 2 and 3) heating progressively erodes the inversion from below causing it to become elevated. At profile 4 the erosion is complete and the inversion is eliminated in favour of a deep, unstable mixed layer.

An almost completely analogous situation applies to the advection of stable rural air across an urban area at night (Figure 9.3e). This is the same case as illustrated in Figure 8.19b (using potential temperatures). Over the city (profiles 2 and 3) the urban boundary layer shows a weak lapse rate surmounted by the remnants of the upwind rural radiation inversion (profile 1). Downwind of the city (profile 4) the rural inversion re-establishes itself in the surface layer and is surmounted by the urban 'plume' (p. 265), and the last vestige of the elevated inversion.

(b) THE EFFECT OF WIND ON DIFFUSION AND TRANSPORT

Atmospheric motion serves both to *diffuse* (dilute) and to *transport* air pollutants. If the size of the eddies is smaller than the pollutant cloud or plume they will diffuse it; if they are larger they will transport it.

When a wind is blowing, pollution is *diffused* both in the along-wind direction and by turbulent eddy diffusion in the across-wind and vertical directions. This is shown in Figure 9.4 where the stack is emitting smoke at the constant rate of one puff (shown as a bubble) every second. If the wind speed ($\bar{u}$) is 2 m s^{-1} there will be 2 m between puffs; but if $\bar{u}$ is 6 m s^{-1} they will be spaced every 6 m. Thus the higher the wind speed, the greater is the

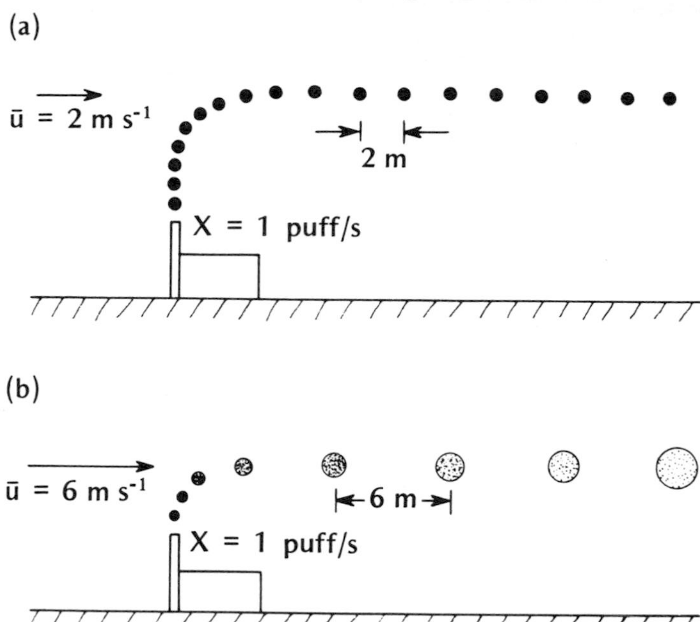

Figure 9.4 The effect of wind speed upon pollution concentration. In both cases the rate of emission (X) is 1 puff per second, but in (a) the wind speed ($\bar{u}$) is 2 m s^{-1}, whereas in (b) it is 6 m s^{-1}.

volume of air passing the stack exit per unit time, and the smaller the concentration per unit volume. This is the concept of dilution by forward 'stretching' and is directly related to wind speed.

Wind speed also governs the amount of forced convection generated in the boundary layer due both to internal shearing between air layers, and between the air and the surface roughness elements. Greater speeds mean greater turbulent activity. The eddies spawned in this manner are characteristically small and their action on a plume is to dilute it by rapidly mixing it with surrounding cleaner air. This effect is represented in Figure 9.4b where the size of the bubbles increases with distance downwind. Turbulence involves fluctuations in direction as well as speed, and these small-scale horizontal eddy motions act to diffuse a plume sideways.

Wind direction is also important in the *transport* of effluent. The perpetual variation of wind direction accounts for the sinuous outline of a plume when instantaneously viewed in plan (Figure 9.5). Over periods of an hour or more these fluctuations may cover an arc of 30–45 degrees centred upon the mean wind direction. This 'spraying' action is similar to that from an oscillating garden hose and serves to widen the time-averaged plume width.

The wind direction also determines the path followed by pollutants after emission. From the standpoint of an individual receptor this could clearly

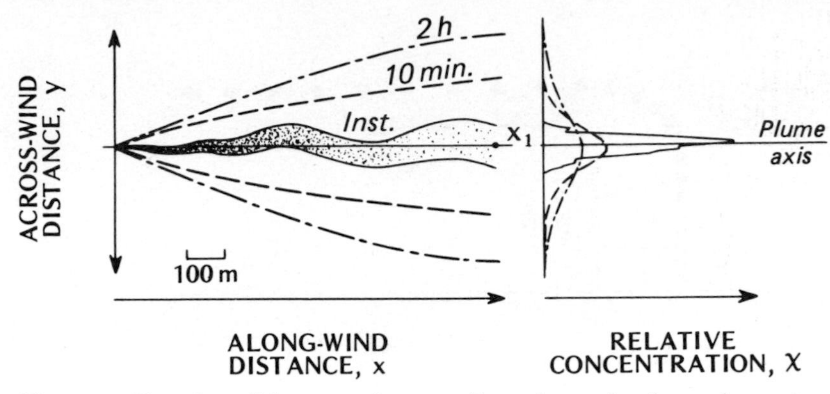

Figure 9.5 Plan view of the approximate outlines of a smoke plume observed instantaneously, and averaged over periods of 10 min and 2 h. At the right are the corresponding cross-plume concentration distributions at the distance x_1 downwind (after U.S. AEC, 1968).

mean the difference between a high pollution concentration or none at all. A particular wind direction may also result in multiple pollutant inputs due to the coincident alignment of sources. In addition to the problem of cumulative loading this might result in the assemblage of a particularly reactive set of chemicals leading to the development of secondary pollutants downwind.

In addition to forward 'stretching' the wind speed is also responsible for the distance of transport. In strong winds the effluent may be transported long distances but the concentration becomes so weak that it is of relatively little consequence. The greatest potential for pollution therefore often exists with weak winds because both horizontal transport and turbulent diffusion are curtailed. Unfortunately these are also the conditions under which local wind systems tend to develop, and these are difficult to predict with any accuracy. This places great value on a knowledge of local climates, but even then the complexity of the wind field often rules out any detailed understanding of dispersion.

Local circulation systems (e.g. land and sea breezes, mountain and valley winds and city winds) are not good pollution ventilators for three reasons. First, the speed of these breezes is usually rather low (less than 7 m s^{-1}); second, they are closed circulation systems (see Figures 5.6 and 5.10); and third, they exhibit a diurnal reversal in direction of flow. The latter two factors mean that there is little true air *exchange*. Instead of the flow replacing dirty air with 'clean' air there is a back-and-forth 'slurrying' movement involving a rather limited volume of contaminated air. This situation is illustrated by the trajectory of a balloon released near the shore of Lake Michigan (Figure 9.6). After release in the morning the balloon moved about 4 km inland to a position near the lake breeze front where it rose to a height of about 1 km, and was carried lakeward in the counter

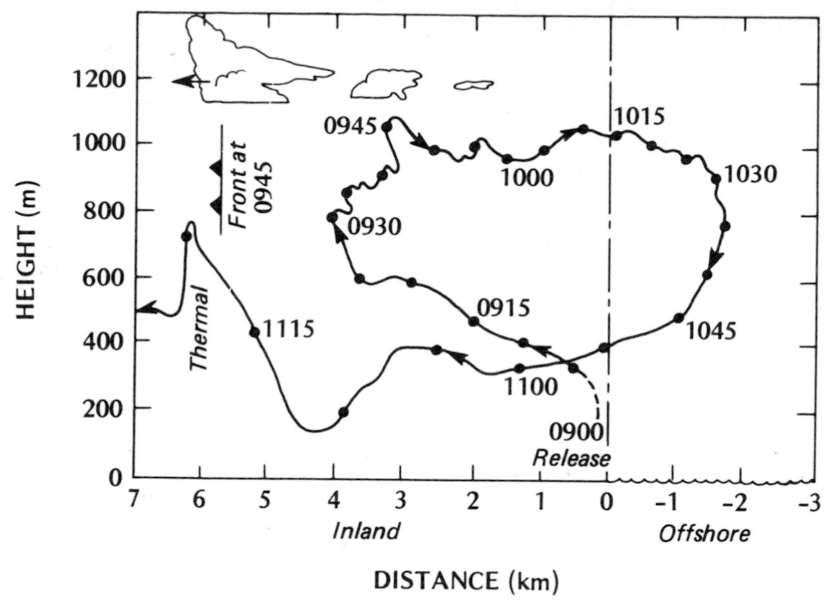

Figure 9.6 Side view of the trajectory of a balloon launched at 09 h on 12 August 1967 at the Chicago shoreline of Lake Michigan. Positions of the balloon are plotted every 5 min. Also shown are the positions of the lake breeze front at 0945 h, and of prevailing clouds (after Lyons and Olsson, 1973).

flow. It descended again about 2 km offshore and returned inland to complete a full cycle in less than 2 h. Thereafter the balloon travelled inland encountering strong thermal convection (e.g. at 6 km), but was unable to penetrate the elevated stable layer near the top of the inflow layer (see temperature profile at position 3 in Figure 9.3d). By this time the lake breeze front had penetrated about 13 km inland and the balloon was lost to sight moving towards it. Localities within the inflow zone find themselves in a situation analogous to being 'locked in a closed room with a fan running' (Lyons and Olsson, 1973). This leads to a re-circulation of pollutants and a progressive increase in pollutant loading with time. The onset of the nocturnal land breeze may merely reverse the flow and true ventilation must await the flushing action of winds from a stronger synoptic system.

The city thermal wind system, which is thought to exist with large scale stagnation (p. 262), is particularly dangerous because the system is totally self-contained over an area of densely packed sources. The low-level (mainly canopy) flow converges upon the city centre from all directions, rises, diverges aloft, and then moves outward to subside on the urban/rural fringe and rejoin the inflow.

(c) PROCESSES OF POLLUTANT TRANSFORMATION

Atmospheric chemistry is not as well understood as atmospheric physics.

In fact until relatively recently meteorologists tended simply to avoid the field by considering the dispersion of inert or elemental substances. Two problems which have attracted attention are the atmospheric transformation of the oxides of sulphur which are at the root of *sulphurous (or London-type) smogs*, and of the oxides of nitrogen and hydrocarbons which are involved in the production of *photochemical (or Los Angeles-type) smogs*.

Sulphur dioxide (SO_2) is a primary pollutant mainly generated in the combustion of fuels, but even in air it may oxidize to sulphur trioxide (SO_3) which reacts with water vapour (H_2O) in the presence of catalysts to form sulphuric acid mist (H_2SO_4). Subsequently the acid reacts with other substances to form sulphate particles which settle out. The ideal conditions for this chain of events occur in urban areas where sulphur-bearing fuels are burned, and when poor dispersion conditions coincide with fog (or at least high humidity). This was the scenario in London in December 1952 from which this smog type gained its name. Dense smog remained stationary for 4–5 days and was at least partially responsible for the deaths of approximately 4,000 people (Wilkins, 1954). The precipitation of acid droplets and sulphate particles is also responsible for the so-called 'acid rain' which can lead to contamination of lakes and rivers downwind of major industrial regions. Acidification has been observed in Scandinavia, and emissions from Britain, the Ruhr, Bohemia and elsewhere are suspected to contribute.

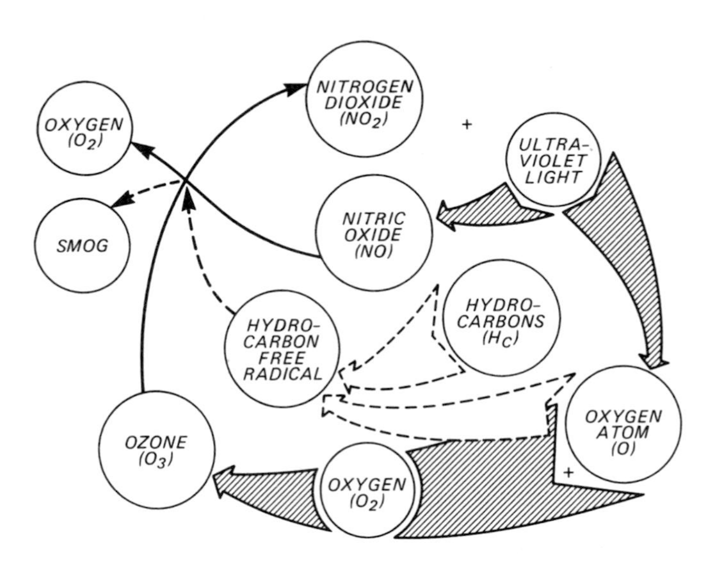

Figure 9.7 The basic NO_2 photolytic cycle and its disruption by reactive hydrocarbons to form photochemical smog (modified after U.S. DHEW, 1970b).

TABLE 9.3 Simplified scheme of the chemical reactions involved in the formation of photochemical smog (modified from ACS, 1969).

1. NO_2 + Ultra-violet light $\longrightarrow$ $NO + O$	NO_2 Photolytic cycle
2. $O + O_2 \longrightarrow O_3$	
3. $O_3 + NO \longrightarrow NO_2 + O_2$	
4. $O + H_C \longrightarrow H_CO^*$	
5. $H_CO^* + O_2 \longrightarrow H_CO_3^*$	
6. $H_CO_3^* + NO \longrightarrow H_CO_2^* + NO_2$	
7. $H_CO_3^* + H_C \longrightarrow$ Aldehydes, ketones, etc.	
8. $H_CO_3^* + O_2 \longrightarrow O_3 + H_3O_2^*$	
9. $H_CO_x^* + NO_2 \longrightarrow$ Peroxyacetyl nitrate	

Photochemical smog is initiated by the action of solar radiation upon nitrogen oxides and is aided by the presence of hydrocarbons. These primary pollutants are thereby transformed through a complex series of chemical reactions to spawn a wide range of secondary pollutants (especially oxidants such as ozone, oxygen, nitrogen dioxide and peroxyacetyl nitrates).

The photochemical smog sequence is centred around the naturally occurring *NO_2 photolytic cycle* (reactions 1–3 in Table 9.3 and the solid arrow portion of Figure 9.7). This set of photochemical reactions results in the rapid cycling of nitrogen dioxide (NO_2) as follows. In the presence of ultra-violet radiation, in the 0·37–0·42 μm waveband, NO_2 can be photodissociated into nitric oxide (NO) and atomic oxygen (O). The latter being highly reactive combines with the ambient molecular oxygen (O_2) to form ozone (O_3), which then reacts with NO to yield NO_2 and O_2.

The NO_2 cycle alone however cannot account for the high O_3 concentrations observed in photochemical smog because O_3 and NO are formed and destroyed continuously with no *net* production. What is required is a process to unbalance the cycle by converting NO to NO_2 without consuming an equivalent amount of O_3. This function is performed by reactive hydrocarbons (H_C) released by vehicle exhausts. Reactive hydrocarbons are oxidized to form organic radicals (reaction 4 in Table 9.3 where O combines with H_C to form the radical H_CO^*, and the dotted arrow portion of Figure 9.7). Further reactions produce other radicals (reaction 5) which in turn react with NO to produce the required extra NO_2 (reaction 6); with H_C to form aldehydes and ketones (reaction 7); and with O_2 to form more O_3 (reaction 8). Yet further reactions between H_C radicals and NO_2 yield peroxyacetyl nitrates (PAN), and even some particulates.

Photochemical smog is accompanied by a characteristic odour (partially due to aldehydes and formaldehyde), a brownish haze (due to NO_2 and light scattering by particulates), eye and throat irritation (due to O_3, aldehydes and PAN), and plant damage (due to NO_x, O_3, PAN and ethylene).

Photochemical smog finds its fullest expression on the Californian coast, and especially in the Los Angeles Basin after which it is named. In this area the vehicle density is very high (about 1,500 vehicles/km^2 in Los Angeles County in 1970) and so therefore are the emissions of NO, NO_2 and H_C. At certain times of the year the meteorological setting is dominated by a sub-tropical anticyclone with clear skies, weak winds and a subsidence inversion, and the geographic setting favours the operation of local wind systems. Together these features provide the ideal photochemical framework, viz: large vehicular emissions, air stagnation, and strong solar radiation. These criteria can also be met in other locations around the world, and hence photochemical smogs have been reported in other areas of the United States and in Canada, Mexico, Brazil, Britain, Holland, Greece, Israel, Iran, Japan and Australia. Outside the tropics their occurrence tends to be restricted to the summer season because of the importance of strong sunshine. Solar radiation also imposes a marked diurnal cycle upon oxidant pollution (Figure 9.14b).

(d) PROCESSES OF POLLUTANT REMOVAL

The atmosphere rids itself of pollutants in a number of ways. These may be conveniently classified under the headings of gravitational settling, dry deposition, and precipitation scavenging.

Gravitational settling is responsible for the removal of most of the particulate matter greater than 1 μm in diameter. That greater than 10 μm settles out relatively close to the source and in a matter of minutes after release. Even strong turbulence is unable to hold these particles in suspension for long, and their rate of removal is simply related to their size and density, and to the strength of the wind. Smaller particles become influenced by turbulent diffusion which slows their descent, and those less than 1 μm in diameter could theoretically remain aloft almost indefinitely. In practice most are eventually removed by dry deposition, by precipitation scavenging, or by settling after their aggregation into larger particles. Gaseous pollutants can be adsorbed onto particles and removed with them, and in photochemical smog it is even possible for gases to react to form small particles and thence be expelled by settling.

Surface adsorption, or *dry deposition*, is a truly turbulent transfer process similar to that involved in the vertical transfer of heat, water vapour, momentum and carbon dioxide (Chapter 2). The vertical flux of pollution (F) is given by the flux-gradient expression:

$$F = -K_F \frac{\Delta\bar{\chi}}{\Delta z} \qquad (9.1)$$

where, K_F – atmospheric diffusivity (m^2 s^{-1}), and $\Delta\bar{\chi}/\Delta z$ – mean vertical gradient of pollution concentration ($\mu g\ m^{-3}/m$), so that F has the flux density units, $\mu g\ m^{-2}\ s^{-1}$. The deposition process involves a downward flux where the underlying surface acts as a pollutant sink. For a given atmospheric pollution loading the rate at which pollution is delivered to the surface is governed by the value of K_F (i.e. the degree of turbulence present) and this is highly sensitive to the nature of the surface roughness and configuration as well as to the vertical temperature profile. The actual adsorption rate is further complicated by a number of pollution and/or surface attributes. Over vegetation the rate is affected by stomatal aperture, over soil by bacterial activity, and over water by the surface tension. The uptake can also be affected by electrostatic attraction, and chemical reactions between the surface materials and the pollution.

Precipitation is the best cleanser of gaseous and small particulate pollutants. Particles and gases are delivered to clouds by convective updrafts. Some of the particles, particularly the hygroscopic ones, become nuclei around which water or ice is condensed to form a droplet or ice crystal. During their life in the cloud (or fog) these droplets capture or adsorb other particles and gases. Eventually, if the droplet becomes large enough, it is precipitated and thereby transports the pollutants with it to the surface. This in-cloud scavenging process is called *rainout* (or *snowout*). Below the cloud precipitation is also active in cleansing the boundary layer by 'sweeping-up' the materials in the air through which it falls. This *washout* process is in fact much more efficient than rainout. The efficiency and rate of washout scavenging depends upon the rainfall rate, and the sizes and electrical charges of the droplets (flakes) and the pollutants. A rainstorm following a period of smog build-up can transform a murky haze into visibly cleaner environment. Munn and Bolin (1971) however warn that the improvement in air quality following rain may in some cases be due not to precipitation scavenging but to the passage of a front which can bring with it a shift in wind direction and greater instability.

4 Dispersion in the boundary layer

(a) INDIVIDUAL PLUME CHARACTERISTICS

In order to illustrate air pollution dispersion we will first consider the relatively simple situation of the plume from a single stack (i.e. from an elevated continuous point source).

The first important feature is the height of emission. Other things being equal the higher the point of injection the smaller will be the ground-level concentrations downwind, because eddy diffusion will have had a longer time to dilute the plume contents. Tall stacks are therefore usually of help in combating poor air quality near the point of release. The effective height of release can also be increased if the effluent emerges at a high velocity, and at

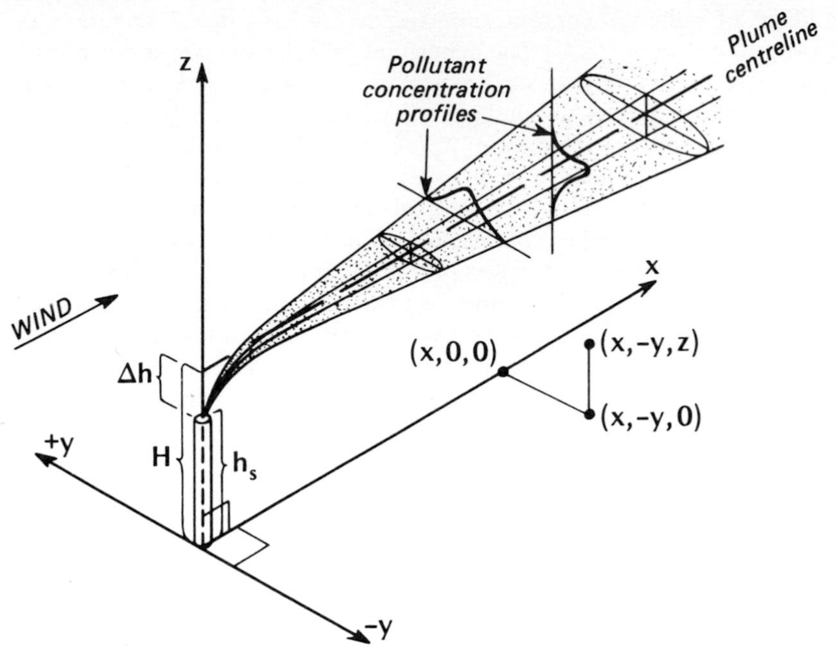

Figure 9.8 A typical plume from an elevated point source illustrating plume rise (Δh), and the normal (Gaussian) distribution of pollutant concentrations in the horizontal and vertical. Also includes the three-dimensional co-ordinate system used in the Gaussian plume model, see text (after Turner, 1969).

a temperature well above that of the environmental air so that it possesses buoyancy. The plume then ascends well above the stack exit before bending-over and proceeding downwind (Figure 9.8). The *effective stack height* (H) is then composed of the stack height (h_s) plus the additional height due to the *plume rise*, Δh so that:

$$H = h_s + \Delta h \tag{9.2}$$

The height H depends upon the stack dimensions (height and exit diameter), the effluent (exit velocity and temperature) and the prevailing meteorological conditions (wind speed and lapse rate). The non-meteorological factors can to some extent be controlled by carefully engineering the stack. On the meteorological side the effect of increasing wind speed is to 'push' the plume over progressively and to diminish the amount of plume rise (see Figure 9.4); and the environmental lapse rate is important because instability encourages upward penetration whereas stability produces a restraining influence.

After its initial rise the form of the plume downwind is largely governed by the prevailing structure of turbulence. Using atmospheric stability as a

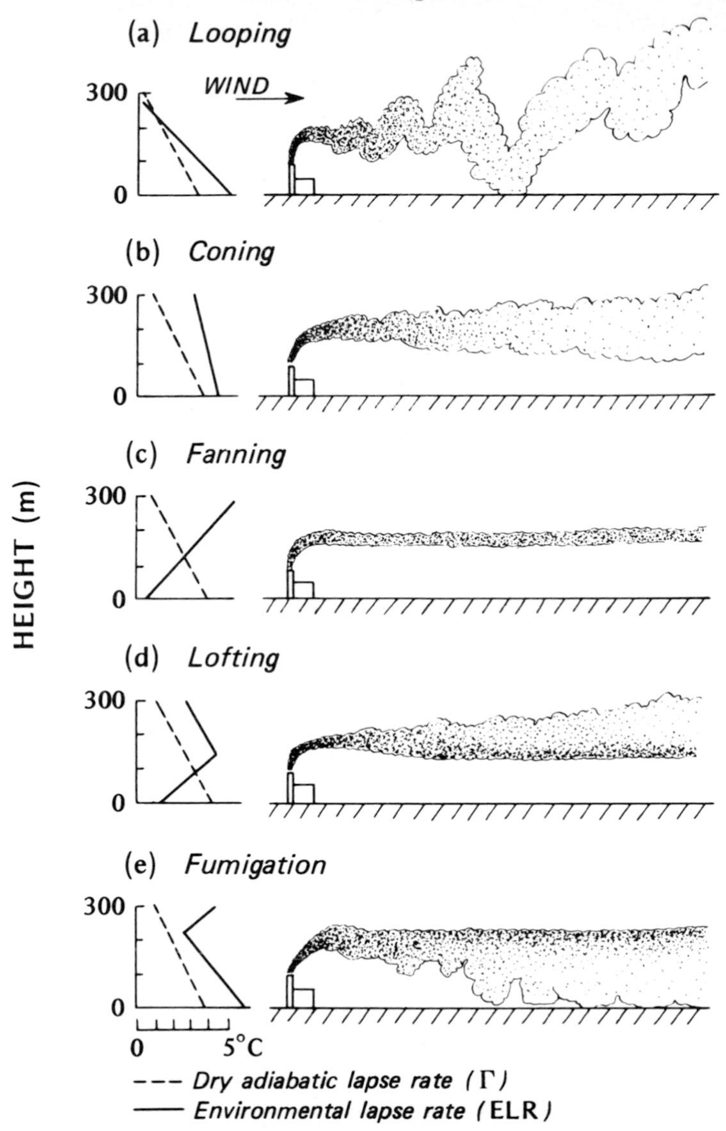

Figure 9.9 Characteristic plume patterns under different stability conditions (modified after Bierly and Hewson, 1962).

surrogate for turbulence it is possible to classify the most characteristic plume patterns into the five basic types illustrated in Figure 9.9.

Looping – is typical of daytime conditions on a fine summer day with strong instability. The atmosphere is then dominated by the relatively large eddy structures associated with free convection (e.g. upper trace Figure 2.10). Since the eddies are larger than the plume diameter their main effect is to transport it up and down in a sinuous track. The 'loops' travel with the wind and grow in size as they go. This erratic transport is capable of bringing the relatively undiluted plume in contact with the ground at quite short distances downwind from the stack resulting in high instantaneous concentrations at these points. (Therefore this is one of the exceptions noted on p. 274.) Over longer periods the ground-level pattern of concentrations assumes the smoother downwind distribution shown in Figure 9.11. Plume dilution is accomplished by the action of small forced convection eddies which 'eat' away at the plume edges, and break it up into increasingly smaller portions. The rate of diffusion is dependent upon the wind speed, and the terrain roughness.

Coning – can occur both by day or night, and in all seasons. It is characteristic of windy and/or cloudy conditions with stability close to neutral (i.e. an adiabatic or slightly lapse temperature profile). In the absence of buoyancy turbulence is composed mainly of the smaller frictionally-generated eddies of forced convection. Without the vertical amplification of instability (or conversely the compression of stability) the vertical and lateral spreading of the plume are about equal so that it forms a cone shape symmetric about the plume centreline. Since the plume diameter only grows by diffusion, and there is little vertical transport, the plume intersects the ground at a greater distance downwind of the stack than is the case with looping.

Fanning – is characteristic in a strongly stable atmosphere (i.e. with an inversion). The ideal conditions are likely to occur with anticyclonic weather and especially at night. At these times turbulence is weak or almost absent, and there is little motion to act upon the plume. Stable air actively suppresses any buoyant stirring and with very light winds forced convection is unlikely to be significant at the height of an elevated plume such as that in Figure 9.9c. The lack of vertical diffusion keeps the plume thin in that direction, but the erratic behaviour of wind direction in stable conditions may allow a V-shape to develop (i.e. resembling a fan when viewed in plan). At other times the outline may appear as a straight or meandering ribbon. Plumes of this type can remain essentially unchanged up to 100 km from the point of emission. But because there is no means of vertical transport the ground-level pollutant concentrations are close to zero unless (i) the stack is very short, or (ii) there are downwind changes of terrain causing the plume to intersect with the ground. In these cases there can be a serious problem because the effluent concentration is almost equal to that at the point of exit. This plume type is commonly the precursor of the potentially more dangerous fumigation type (see Figure 9.9e and discussion below).

Lofting – is the most favourable dispersal condition. It is found in the early evening during the period when the nocturnal radiation inversion is building up from the surface (e.g. the near sunset profile in Figure 2.13).

(a)

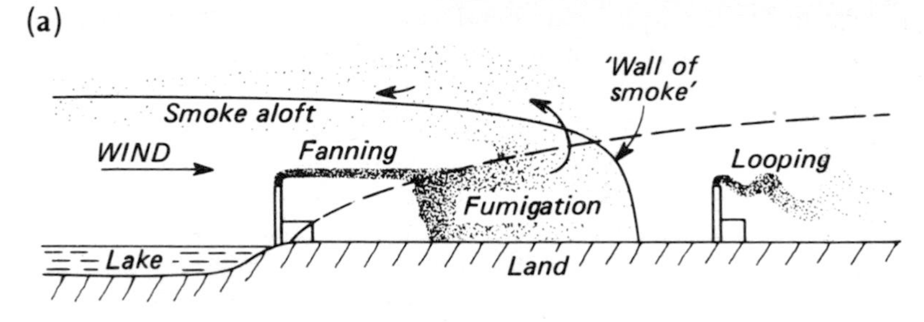

(b)

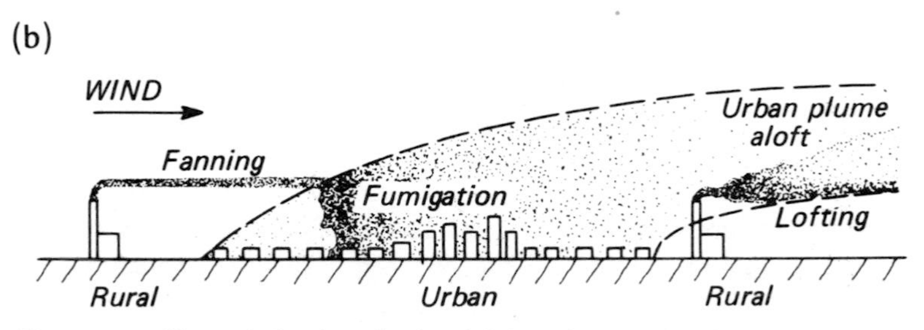

Figure 9.10 Plume behaviour in the vicinity of (a) a coastline on a fine spring day (based on Lyons and Olsson, 1973), and (b) a city at night with clear skies and light winds.

The stable layer beneath the plume prevents its transport downwards but the moderately unstable layer aloft allows the plume to disperse upwards. Unfortunately this condition is often only transitory because when the inversion depth exceeds the effective stack height the plume changes to the fanning type.

Fumigation – is the reverse situation to that of lofting. In this case there is an inversion 'lid' above the plume which obstructs upward dispersion, but there is a lapse temperature profile beneath so that there is ample buoyant mixing capable of bringing the plume contents to the surface. This unfavourable temperature structure can arise in a number of ways. For example in rural areas it occurs during the period after sunrise when the nocturnal surface inversion is being eroded by surface heating (Figure 2.13). During the previous night the plume may have assumed a fanning form and this will continue until the depth of the developing mixed layer reaches the plume, whereupon its contents will be carried downwards in the descending limbs of convection cells. Ground-based receptors will then receive high pollution concentrations all along the line of the plume at about the same time. This very unpleasant situation may persist for 30 minutes, and could affect locations many kilometres from the source. A particularly clearcut example of this process was observed in the Columbia

River valley at Trail, British Columbia (Dean *et al.*, 1944). A lead and zinc smelter was releasing large quantities of SO_2 to the atmosphere in a deep mountain valley. It was found that at night the effluent travelled down-valley with the drainage winds, and within the valley radiation inversion there was little dilution. Soon after sunrise heating of the valley floor and slopes eroded the inversion, and produced a valley fumigation which occurred almost simultaneously along the valley for distances up to 55 km from the source. Temperature profiles conducive to fumigation can also occur along coastlines and in cities, as illustrated by profiles 2 and 3 in Figures 9.3d and e. The aerodynamic 'downwash' of pollutants in the lee of topographic features (escarpments and hills, Figure 5.15) and buildings (Figure 8.5) can create equally severe but different types of fumigation.

Figure 9.10 shows how four of the above plume types may be encountered in relation to the temperature distributions shown in Figure 9.3. In (a) a shoreline source is emitting effluent into the stable air of the off-shore portion of a sea (lake) breeze. The associated fanning plume drifts inland until it encounters the developing unstable boundary layer of the warmer land at which point it fumigates. The resulting murky mixture is advected further inland and the sea breeze front is seen to approach as a 'wall of smoke'. Notice also that a portion of this polluted mass is carried aloft and back out over the water by the sea-breeze counter current (Figures 5.6a, and 9.6) to form an elevated smoke pall. At the same time, ahead of the breeze front, the plume from a similar source is seen to exhibit looping because it exists in the as yet unmodified, and highly unstable air inland. The left-hand side of (b) reveals a situation having much in common with that in (a). A rural plume is fanning in a stable nocturnal inversion layer, but upon entering the urban boundary layer the increased turbulence caused by the heat island and the building roughness causes it to fumigate along with the pollution from urban sources. On the other hand downwind of the city is a more favourable arrangement, where an elevated source is able to loft its output into the slightly adiabatic urban 'plume', whilst the surface is buffered from contamination by the newly developing rural stable layer.

Observations reveal that the horizontal distribution of pollutants in a plume is as shown in Figure 9.5. On an instantaneous basis the concentration is very peaked with a maximum in the plume centre. A 10-minute average shows a smoother and wider plume envelope which contains all of the short-period plume fluctuations. The wider spread results in a flatter concentration curve with a lower peak value. Over a 2 h period the plume fluctuates over an even wider arc, and the concentration curve is even flatter, but remains centred on the time mean axis of the plume (i.e. the mean wind direction). Entirely analogous bell-shaped concentration profiles are also found in the vertical plane through the plume (e.g. Figure 9.8). There may however be a difference between the width of

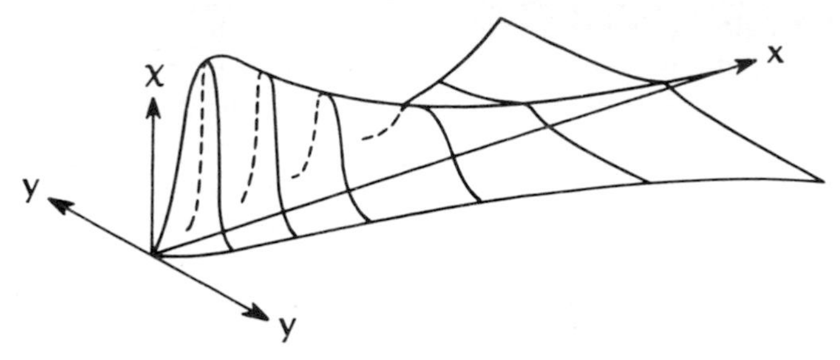

Figure 9.11 Representative ground-level concentration (χ) distribution downwind of an elevated continuous point source.

the horizontal and vertical profiles indicating that turbulence, and therefore plume spreading, is greater in one or other direction. Note also that since the amount of effluent passing through any vertical cross-section at any moment must equal the emission rate at the source, and since the plume width is continually expanding, then mean concentrations must decrease with distance downstream. Put another way this means that the *area* under the bell-curves remains constant at all points downwind, but since their width is increasing their peak value must decrease. These characteristics are also evident in the along-wind pattern of ground-level concentrations found downwind from an elevated source whose plume intersects with the ground (Figure 9.11). Concentrations are zero near the stack, rise sharply to a peak at some distance downstream, and thereafter tail-off to increasingly smaller values. At all distances the concentration at the plume centreline is greater than on either side.

The bell-curve is known in statistics as the normal or Gaussian distribution, and conforms to the pattern that a series of completely random errors would assume about the correct value of a measurement. In our case we may translate this to mean that the almost random nature of atmospheric turbulence serves to mix pollutants so that their concentration is distributed bi-normally about the plume's central axis (i.e. with a normal distribution in both the horizontal, y and vertical, z planes). The mathematical description of such curves then provides a means of modelling the dispersion of plumes. For example the following equation can be utilized to calculate the concentration of pollution (χ) at any point in a plume (see Figure 9.8 for the three-dimensional co-ordinate system):

$$\chi_{(x,y,z,H)} = \frac{\mathrm{X}}{2\pi\sigma_y\sigma_z\bar{u}} \exp\left[-\frac{y^2}{2\sigma_y^2}\right] \times \left[\exp\left(-\frac{(z-H)^2}{2\sigma_z^2}\right) + \exp\left(-\frac{(z+H)^2}{2\sigma_z^2}\right)\right] \tag{9.3}$$

where, X-rate of emission from the source (kg s^{-1}), σ_y, σ_z – horizontal and vertical standard deviations of the pollutant distribution in the y and z directions (m), $\bar{u}$ – mean horizontal wind speed through the depth of the plume (m s^{-1}), and H – effective stack height given by equation 9.2 and another formula to calculate the plume rise (m). The units of χ are therefore kg m^{-3} (more reasonably μg m^{-3}), or units of mass concentration. Alternatively pollution can be expressed as a concentration by volume in which case the units are parts per million (ppm).[1] This refers to the number of pollutant molecules per million molecules of air, and would require that X be given as a volumetric rate (m^3 s^{-1}).

Equation 9.3 may look rather formidable, but upon inspection includes some simple physical principles which we have already described. First, it is obvious that pollution concentrations must be proportional to the source strength (i.e. $\chi \propto X$). Second, we noted in relation to Figure 9.4 that concentration is inversely related to the wind speed, due to forward plume 'stretching' in the x direction (i.e. $\chi \propto 1/\bar{u}$). Third, we know that the concentration behaves in a Gaussian manner inside the plume and this is incorporated in equation 9.3 by the use of the standard deviations. These are related to the dimensions of the plume as it grows by turbulent diffusion, and hence are functions of downwind distance and stability. The concentration is inversely related to σ_y and σ_z because larger values indicate better diffusion. Fourth, as pointed out earlier, the concentration at a given distance downwind is decreased by raising the effective stack height (e.g. by providing good plume rise). The last term on the right-hand side is included to account for the increased concentration at positions downwind of the point at which the plume first reaches the ground. In this formulation it is assumed that all of the material is 'reflected' back up into the atmosphere and none is deposited.

The basic form of equation 9.3 can be simplified greatly if for example one only requires to know the values at ground-level (i.e. $z = 0$):

$$\chi_{(x,y,0,H)} = \frac{X}{\pi\sigma_y\sigma_z\bar{u}} \exp\left[-\left(\frac{y^2}{2\sigma_y^2} + \frac{H^2}{2\sigma_z^2}\right)\right]$$

or if only plume centreline values at ground-level are required (i.e. $y = 0$):

$$\chi_{(x,0,0,H)} = \frac{X}{\pi\sigma_y\sigma_z\bar{u}} \exp\left[-\frac{H^2}{2\sigma_z^2}\right]$$

or if in addition the source were not elevated with plume rise but were at the surface with no plume rise (i.e. $H = 0$):

$$\chi_{(x,0,0,0)} = \frac{X}{\pi\sigma_y\sigma_z\bar{u}}$$

[1] To convert ppm vol. to μg m^{-3} at 25°C, and a pressure of 101 kPa, multiply by the following factors: SO_2 – 2620; H_2S – 1390; CO – 1150; CO_2 – 1800; NO – 1230; NO_2 – 1880; O_3 – 1960; PAN – 4950.

However, it should be pointed out that this Gaussian plume model only applies under certain limiting conditions. For example, it only applies to continuous emissions from a point source; to inert almost weightless pollutants (e.g. gases and particles less than 20 μm in diameter); over time periods greater than 10 minutes; and to distances in the range from a few hundred metres to 10 km downwind from the source. The model can be modified to cope with special conditions such as an elevated inversion which restricts upward dispersion, or topographic constrictions, and it can be manipulated to handle most of the source configurations listed in Table 9.2.

(b) URBAN AND REGIONAL POLLUTION

Over distances of 10 to 50 km most of the individual plumes discussed above lose their identity and contribute to a more general contamination of the boundary layer. This is accomplished even sooner in the case of small, low-level sources with little plume rise. In urban areas the sheer multiplicity of sources produces the same effect. The efficiency of turbulence then tends to produce a rather homogeneous melange of contaminants occupying the entire depth of the mixed layer. This is the hazy type of smog so characteristic of urban regions, but which is also observed to some degree over almost all settled land areas, and is most visible when viewed from above in an aircraft, or from mountain lookouts.

The pollutant concentration in such circumstances can be related to a simple model of mass input and output such as that in Figure 9.12. The box represents an air volume located above a region with a large number of sources emitting pollutants at the rate X ($kg\ m^{-2}\ s^{-1}$). The upward

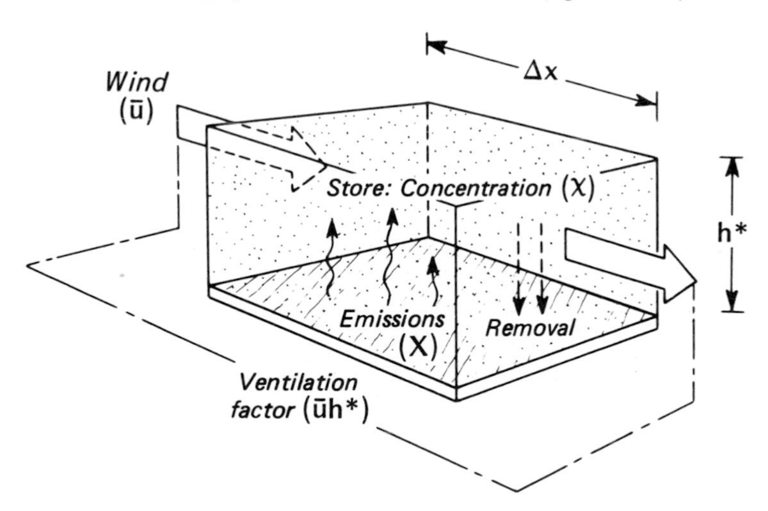

Figure 9.12 A simple input/output 'box' model of pollution in the boundary layer.

dispersion of these materials is restricted by an elevated inversion layer at the height h^* which is therefore also the depth of the mixed layer. Pollutant output from the volume is either via vertical removal processes (settling, deposition or scavenging), or through the flushing action of the mean wind $\bar{u}$ ($m\ s^{-1}$) averaged over the depth h^*. Except during periods of precipitation it is probably acceptable to assume that vertical removal is negligible by comparison with that by the wind. Further – if we assume that air entering the box is 'clean'; that the emission rate is uniform across the area; that the effluent is thoroughly mixed over the depth h^*; and that lateral mixing does not produce a decrease in concentration – then we can write that the average concentration $\bar{\chi}$ ($kg\ m^{-3}$) at any distance Δx (m) from the upwind border is given by:

$$\bar{\chi}_{(x)} = \frac{X\Delta x}{\bar{u}h^*} \qquad (9.4)$$

This shows that the concentration should be directly related to the strength and distance of travel over the sources, and inversely related to the wind speed and the depth of the mixed layer. It also shows that at any given location (value of Δx) $\bar{\chi}$ depends on the amount of emissions and the meteorological conditions, both of which vary with time. The rate X depends on the daily and seasonal pattern of human activity, and the product $\bar{u}h^*$ (known as the *ventilation factor*) similarly exhibits diurnal and seasonal variations related to boundary layer and synoptic meteorological controls. Other things being equal $\bar{u}$ is characteristically stronger by day than at night (p. 67), and during the passage of weather disturbances (low pressure systems). Similarly h^* is larger by day due to the convective growth of the mixed layer, and is smaller at night when it shrinks again (Figure 2.14). In rural areas h^* is essentially zero at night if the weather is good for radiative inversion development, but in the city the heat island effect is capable of maintaining a mixing layer throughout the night (p. 265). Thus we can see that on a diurnal basis both components of the ventilation factor are linked in such a way that the best environment for dispersion occurs by day and the worst by night.

The temporal variations of emissions and of the ventilation factor combine to produce characteristic pollution concentration cycles at a given location. Figure 9.13 shows typical examples of the diurnal and seasonal cycles of smoke (COH[1]) in central Montréal. On a seasonal basis (Figure 9.13a) pollution is clearly greatest in the winter. This can be attributed to both higher emissions (especially from fuel burning for space heating), and to a reduced capacity for atmospheric dispersal (especially as a result of the very stable atmosphere) in this season. The daily curves (Figure 9.13b)

[1] COH (coefficient of haze) is a measure of small particulate pollution based on light transmittance through a filter paper exposed to the ambient air for a given period.

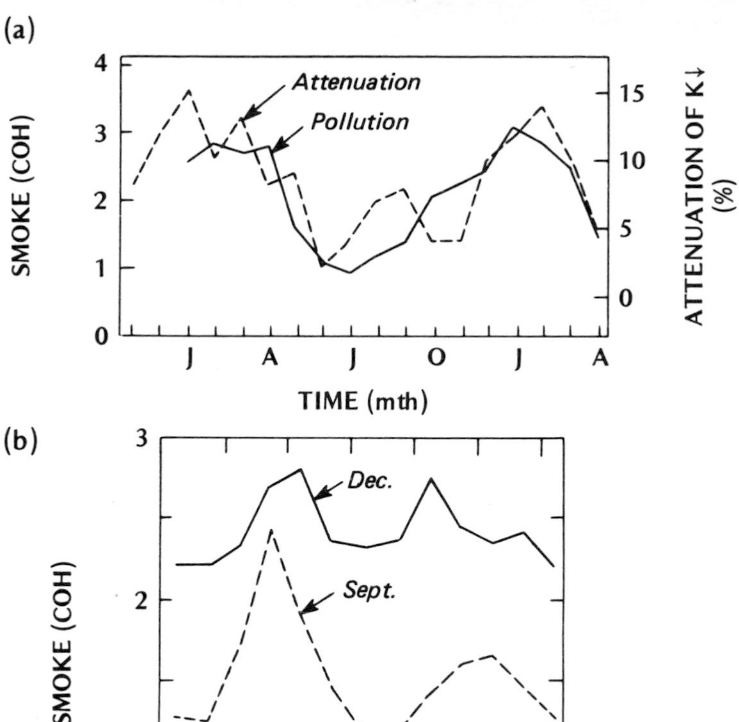

Figure 9.13 Temporal variations of smoke concentration (COH) in Montréal. (a) The seasonal pattern for 1960–61 including measurements of solar radiation attenuation (%) by the polluted atmosphere based on a comparison of data from a city, and a rural station over the period 1965–67 (after East, 1968). (b) The average diurnal pattern for weekdays in December and September 1960 (after Summers, 1962).

characteristically show two maxima, one in the early morning and one in the evening, which coincide with peaks in activities which generate pollution. For example, the morning peak occurs at the same time as the commuter traffic surge and when industries are re-starting; and the evening one coincides with the return home and the need for energy for cooking, lighting and heating. But they also occur at times when winds tend to be light and atmospheric stability is either breaking down (perhaps causing a morning fumigation), or developing. Notice also that there is a seasonal difference between the two diurnal curves, with the summer pattern showing a greater

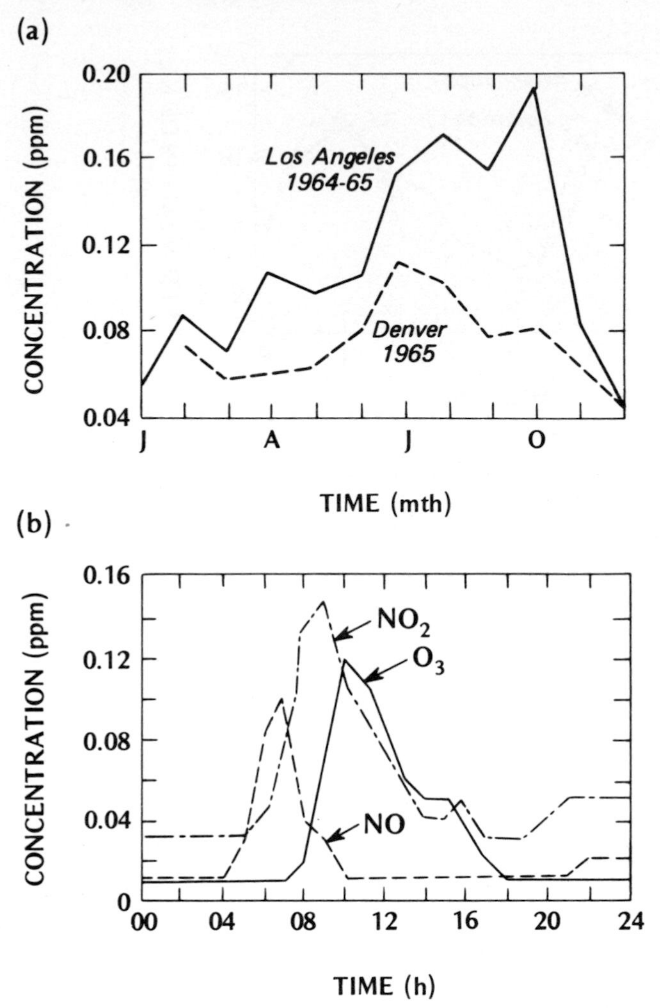

Figure 9.14 Annual and diurnal cycles of photochemical smog. (a) Annual variation of mean daily maximum 1 h average oxidant concentrations for Denver and Los Angeles. (b) Diurnal variation of NO, NO_2 and O_3 concentrations in Los Angeles on 19 July 1965 (after U.S. DHEW, 1970b).

range as well as being at a lower absolute level. The increase in range is a reflection of the fact that there is a greater diurnal *change* in stability during the summer period. The general form of these particulate concentration cycles also holds for other major pollutants such as SO_x, CO, NO_x and H_C in urban atmospheres. These patterns can however be upset by the passage of weather disturbances, or if there is photochemical activity because that involves the destruction of some materials and the production of others.

Photochemical smog variations are particularly geared to the diurnal and annual cycles of solar radiation because high intensity short-wave radiation is necessary to initiate the NO_2 photolytic cycle. The annual patterns of oxidant concentration in Los Angeles and Denver are shown in Figure 9.14a. These values are mean daily maximum 1 h averages, which means they are a measure of the worst conditions averaged firstly over 1 h periods and then over a month. Denver exhibits a simple mid-summer peak of oxidants which can be largely attributed to the availability of sufficient sunlight. Los Angeles shows a peak skewed towards the late summer and autumn because this is the period with least cloud and when flushing by winds is rather weak.

The diurnal sequence of oxidant concentration (Figure 9.14b) embodies the interaction between the temporal variations of (i) the relevant emissions

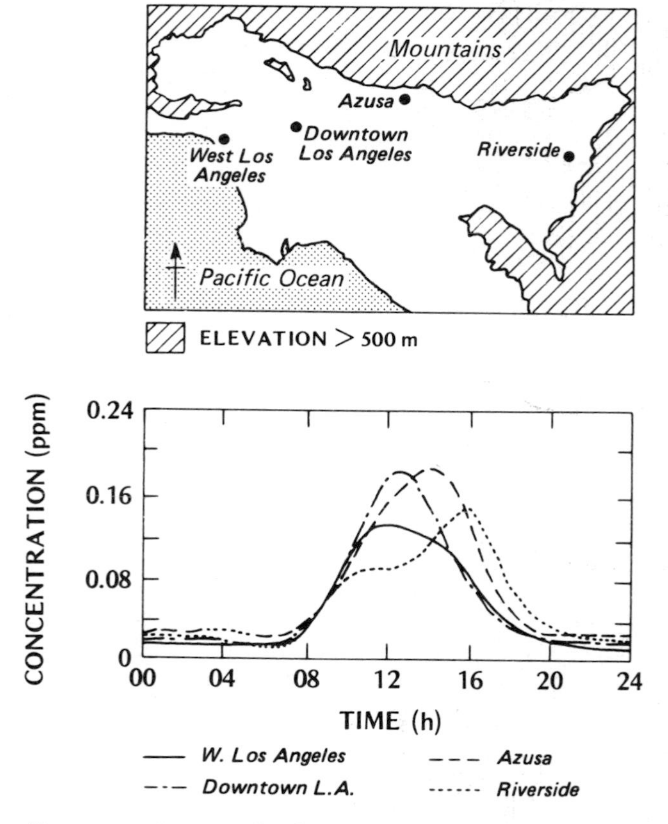

Figure 9.15 An example of smog build-up due to accumulation during eastward drift in the Los Angeles Basin. Data are monthly mean hourly average concentrations of oxidants during October 1965 (compiled from data in Pitts, 1969).

(mainly from automobiles), (ii) the atmospheric dispersion capacity, and (iii) the solar radiation intensity. Peak emissions occur in the early morning traffic build-up when dispersion and solar intensity are weak. This allows the rapid accumulation of exhaust products such as NO and H_C (not shown), but at this time there is no appreciable effect upon levels of O_3. By about 07 h NO begins to decrease as it reacts with hydrocarbon radicals and boosts the production of NO_2. After 08 h the high NO_2 levels in association with increasing radiation intensity allows photodissociation of atomic oxygen (O) and the rapid increase of O_3 to a peak at midday. Meanwhile primary emissions drop, increasing instability aids dilution, and other secondary reactions alter the nature of the essential smog ingredients (especially NO_2 and H_C) so that the sequence passes its most active phase. In the afternoon O_3 concentrations decrease as the radiation intensity declines, dilution continues, and O_3 is removed by reactions with other atmospheric constituents and with surface receptors such as plants.

The simple model embodied in equation 9.4 and illustrated in Figure 9.12 is not valid in detail because the large number of simplifying assumptions required are never fully met, but it has pedagogic value in alerting us to situations likely to produce particularly unfavourable urban or regional problems. For example it shows that the most unfavourable conditions occur when the ventilation factor is smallest (i.e. weak air flow and a shallow mixing depth). Anticyclonic weather can often provide this unfortunate combination due to weak horizontal pressure gradients and a subsidence inversion. In an extreme case it is possible for the air to become virtually stagnant resulting in almost no pollution flushing and leading to an almost continuous increase of pollutant concentration with time. It has been pointed out (p. 280) that even the thermal breeze systems which develop in these conditions are not helpful because they do not result in a *net* transport, rather they circulate the air around inside an almost closed box. The problem is further aggravated if the source area is topographically confined so as to restrict lateral spreading, and especially if the inversion lid lies at a lower elevation than the valley or basin sides.

Equation 9.4 also shows that $\bar{\chi}$ is related to the distance of travel (Δx) so that the area can become increasingly more polluted as a result of the accumulation of materials as the air advects across the source region. This is partially responsible for the observation that oxidant pollution in the Los Angeles Basin is greatest in downwind communities, and in the late afternoon (Figure 9.15). The West Los Angeles oxidant curve peaks at midday and is consistent with the idea of local smog production in phase with solar radiation intensity. The pollutants drift eastward, and augment the local production in downtown Los Angeles thereby causing the maximum to occur there about 1 h later. Similarly at Azusa further eastward the peak occurs 2 h after that at West Los Angeles, and despite lower local emissions the concentration is even higher than that at either of the

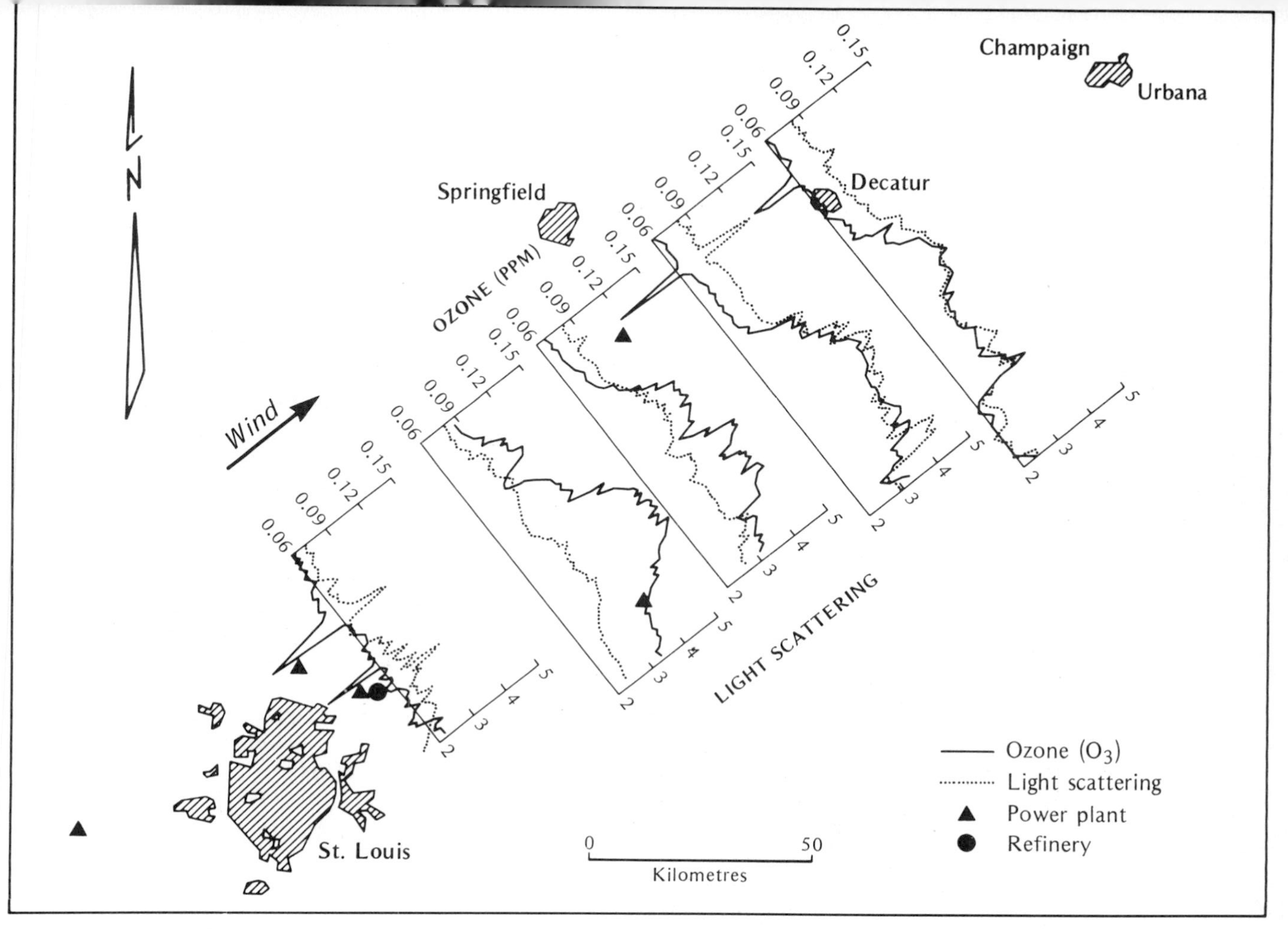

Figure 9.16 The pollution plume of St Louis on 18 July 1975 as defined by the concentrations of ozone and the light scattering produced by the aerosol content. Data gathered from an aircraft traversing the plume as it progressed downwind (after White *et al.*, 1976).

Los Angeles stations. Even further downwind at Riverside there are two distinct maxima: one close to midday which is interpreted to be due to local smog development, and a later higher peak which corresponds to the arrival of the accumulated metropolitan Los Angeles smog. Williamson (1973) reports that the arrival is sometimes seen as a well-defined 'smog front' (i.e. like the 'wall of smoke' in Figure 9.10a).

On the meso-scale (Figure 1.1) cities appear as large point sources with their plume extending many kilometres downwind. Figure 9.16 shows the plume of St Louis out to 160 km downwind; it was further tracked to at least 240 km, at which distance it was about 50 km wide. Note that the O_3 concentrations *increase* for quite a distance indicating that this gas is a secondary pollutant due to photo-chemistry in the plume. It will also be seen that O_3 concentrations dip when the plume passes over a major power plant. This is thought to be due to the scavenging of O_3 by the NO released from these plants (i.e. reaction 3 in Table 9.3). The aerosol content also increased with distance from St Louis indicating that it too included secondary production of particulates (especially sulphates) in the photo-chemical process (p. 283).

Urban plumes are subject to fumigation in much the same manner as stack plumes. For example the elevated nocturnal plume from a city may be fumigated next day by the development of the normal rural mixing layer. The situation is particularly acute if the urban plume is confined in a valley. Fumigation can also result if the plume from one city passes over another (cf. the upwind stack in Figure 9.10b). The plume from a coastal city may become caught up in the sea breeze circulation. Thus, for example, emissions from the morning rush-hour may be carried offshore, some may be transformed to new products, and the mixture returned to the city in the afternoon sea breeze whereupon it fumigates in the urban boundary layer.

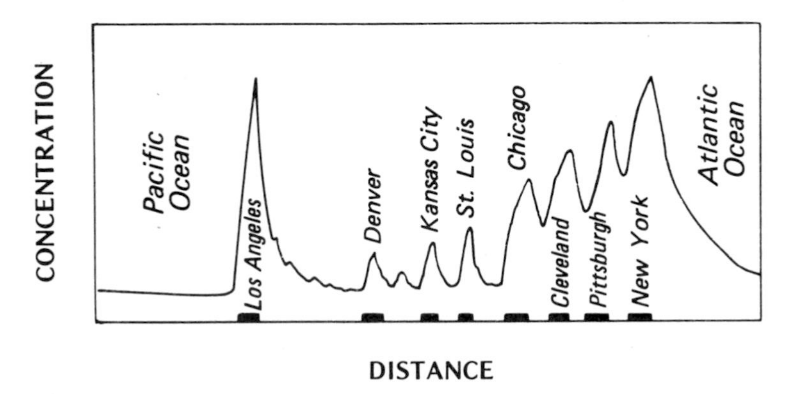

Figure 9.17 Hypothetical pollution concentration in an air mass traversing North America (after Neiburger, 1969).

Figure 9.17 further develops this theme to the continental scale. It hypothetically traces the concentration of pollutants in an air mass forced to traverse North America from west to east. In crossing the Los Angeles Basin the pollutant loading increases dramatically. As it is carried further eastward, over areas with little or no emissions, the pollutants are diluted and the air partially cleansed by removal processes. It then encounters a series of city sources some of which are only separated by short distances. Insufficient cleansing then means that the air arrives at the next city in an already polluted state leading to an escalation of concentrations. In this way urban plumes merge to form megalopolitan plumes of giant proportions. It now seems likely that such plumes are capable of transporting pollutants thousands of kilometres.

APPENDIX A1

Atmospheric lapse rates and stability

Consider a discrete parcel of air moving up through the atmosphere, and assume that it neither receives nor gives out heat to the surrounding air (such a parcel is said to be moving *adiabatically*). As it rises it encounters lower atmospheric pressure because the mass of air above it becomes progressively less. Thus the internal pressure of the parcel becomes greater relative to its surroundings and the parcel will tend to expand. To push away the surrounding air requires work and therefore energy. But the only energy available is the thermal energy of the parcel itself (since we are assuming no exchange with the surroundings), thus as the parcel rises it cools. In a dry (unsaturated) atmosphere the rate of cooling with height is the constant value of $9{\cdot}8 \times 10^{-3}$ °C m^{-1} called the *dry adiabatic lapse rate* (Γ). If the parcel becomes saturated some vapour condenses into droplets thereby releasing latent heat (L_v) which reduces the rate of cooling, but the value is not constant. In most of the applications in this book (i.e. in the boundary layer below cloud base height) the dry adiabatic assumptions are approximately valid. Eventually of course the parcel will cease to rise and will impart its heat by mixing with the air at that level.

It is important however not to confuse the dry adiabatic lapse rate (Γ) with the *environmental lapse rate* (ELR). The former is the rate at which a dry parcel will cool if it is moved upward through the atmosphere, and also the rate at which it will warm if it moves down towards the ground (i.e. when it encounters increased atmospheric pressure and becomes compressed). The ELR, on the other hand, is a measure of the actual temperature structure

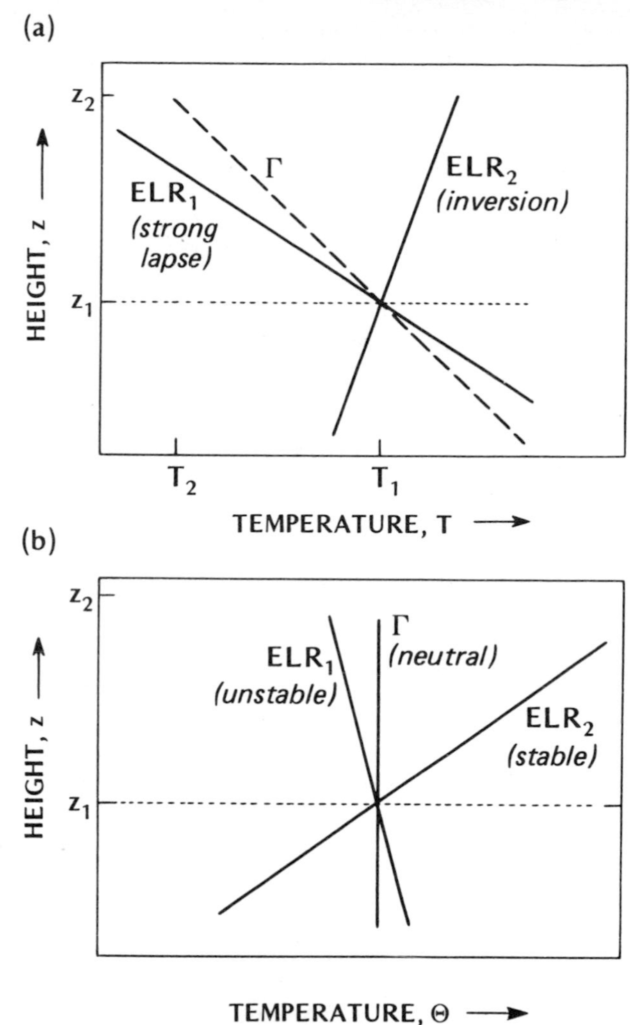

Figure A1.1 Temperature/height graphs illustrating atmospheric stability using (a) the observed, and (b) the potential air temperatures.

existing above a given location (i.e. ELR $= \Delta\overline{T}/\Delta z$). For example, close to the surface on a sunny day the ELR as sensed by thermometers on a mast, or attached to a balloon or an aircraft, may well be much greater than Γ. Conversely, at night the ELR may show an increase of temperature with height (i.e. an inversion) and hence have the opposite sign to Γ. Therefore the complete temperature structure above a location is quite likely to exhibit changes in the ELR in different layers, some being lapse, some inversion, and some isothermal.

Atmospheric stability may be viewed as the relative tendency for an air parcel to move vertically, and can be evaluated in a dry atmosphere by comparing the values of the ELR against Γ. In Figure A1.1a consider a parcel of dry air at rest at the level z_1 with a temperature T_1. If the parcel is forcibly displaced upwards it will cool at the constant rate Γ so that at the level z_2 its temperature has dropped to T_2. If at any level the displacing force were removed the parcel would tend to continue to rise or would start to sink depending on its density relative to that of the air at the same level. Basically, if warm air is embedded in an otherwise cool (more dense) air layer it possesses *buoyancy* and tends to rise like the heated water at the bottom of a kettle; conversely if a parcel is colder (denser) than its surroundings it will tend to sink; and if it has the same temperature it will tend to remain static.

Three possibilities exist depending upon the value of the ELR, viz:

(a) ELR greater than Γ – this is the situation described by ELR_1 in Figure A1.1a, and is typical of sunny days near the ground when surface heating gives rise to a strong lapse rate. If the parcel at z_1 is displaced upwards its temperature (which follows Γ) is always higher than that of the environmental air, therefore if the displacing force is removed its buoyancy will ensure that it continues to rise. The parcel is therefore said to be *unstable*. The greater the value of ELR_1 the greater is the divergence between the two lines and hence the greater the instability. Notice also that had the parcel originally been displaced below z_1 it would always find itself colder than the environmental air, and therefore if let alone would tend to continue sinking. Its instability is therefore independent of the direction of displacement.

(b) ELR less than Γ – if the ELR is less than $9{\cdot}8 \times 10^{-3}$°C m^{-1} then the parcel is said to be *stable*. These conditions are most obviously met in an inversion such as that described by ELR_2 in Figure A1.1a. In this case a displaced parcel above z_1 always finds itself colder than the environmental air and hence tends to sink back towards z_1. Equally if displaced below z_1 it would find itself warmer and tend to rise back up to its equilibrium position. The greater the difference between the ELR and Γ the greater is the damping tendency.

(c) ELR equal to Γ – in this situation (not shown in Figure A1.1a) the parcel is said to be *neutral*. After displacement to any level above or below its initial position z_1 the temperature of the parcel and of the air are the same. Hence there will be no relative tendency for the parcel to rise or sink, and if the displacing force is removed the parcel remains stationary. This situation occurs in the boundary layer under cloudy, windy conditions. Cloud restricts surface heating and cooling thereby minimizing the development of any horizontal temperature stratification, and wind helps to homogenize the temperature structure by vigorous mechanical convection.

With fine weather it is normal for the planetary boundary layer to be unstable by day, and stable by night (Figure 2.14). Exceptions occur over high latitude snow surfaces in winter where the boundary layer is stable for long periods, and over tropical ocean surfaces where it may be unstable for equally long spells. It is also important to realize that the atmosphere is commonly made up of a number of layers of different stability. For example, the daytime unstable boundary layer is usually capped by an inversion layer – which is almost impenetrable to air parcels rising from beneath. It therefore halts their vertical motion and hence the heat and water vapour they transport is largely retained in the boundary layer. Layering can also be caused by the advection of relatively warm or cold air across a site. This will alter the stability conditions and therefore enhance or dampen vertical motion.

When considering atmospheric stability it is often useful to use *potential temperature* (θ) instead of the observed air temperature (T). We have seen that the temperature of parcels behaving adiabatically is related to pressure. In order to compare parcels existing at different pressures (levels in the atmosphere) it is therefore useful to standardize conditions to a common pressure. The potential temperature of a parcel is the value it would have if it were at the arbitrary pressure value of 100 kPa. This is tantamount to correcting the observed temperature to allow for Γ, so that replotting Figure A1.1a using θ involves rotating all lapse rates to the right by the slope of Γ (Figure A1.1b). Interpretation of stability now becomes straightforward. If the ELR plotted as θ is constant with height (i.e. a vertical line) the layer is neutral, if the ELR slopes to the left it is unstable, and to the right stable. It should also be pointed out that equation 2.12 could be more simply written using θ instead of T, viz:

$$Q_{\mathrm{H}} = -\rho c_{\mathrm{p}} K_{\mathrm{H}} \frac{\Delta \overline{\theta}}{\Delta z}$$

APPENDIX A2

Evaluation of energy and mass fluxes in the surface boundary layer

Much of modern micro-meteorology has been concerned with the development of suitable techniques to determine the fluxes of energy and mass in the surface layer. Here we can only deal with a few examples in each category but this should be sufficient to give an idea of the basic instrumentation and methodological approaches employed to evaluate many of the climatic features dealt with in the body of the text. Some of the theoretical background to flux evaluation is also included with the aim of helping the reader to bridge the gap between the largely explanatory approach to atmospheric systems that is incorporated in this book, and the more analytical treatment embodied in most micro-meteorological texts.

Initially we consider the measurement of the standard atmospheric variables (air temperature, humidity, wind speed and carbon dioxide). This is followed by a review of the methods used to determine the vertical fluxes of energy and mass which comprise the surface radiation budget, and the surface energy, water and carbon dioxide balances. Instruments employed for routine climatological observations at weather stations are not considered here since they are well covered in most introductory texts or manuals. Included in this category are the instruments used to measure air temperature (standard thermometers in a weather screen, thermographs, and grass-minimum thermometers); soil temperature (mercury-in-glass or mercury-in-steel thermometers); air humidity (wet- and dry-bulb thermometers and hygrographs); wind speed and direction (cup anemometer and wind vane); precipitation (rain and snow gauges); atmospheric pressure (barometer); and duration of bright sunshine (sunshine recorder).

1 Evaluation of temperature, humidity, carbon dioxide and wind

(a) TEMPERATURE

In most boundary layer studies it is necessary to have remote reading instruments so as to avoid interference with the environment being sensed. This explains the popularity of electrical methods whose sensors can be manufactured to give minimal interference and whose signals can be monitored at a distant location by standard electronic recording equipment.

Thermocouples. If two dissimilar metals are joined to give a circuit, and the junctions are at different temperatures, an electromotive force (emf) will be generated. The value of the emf (voltage difference, ΔV) is proportional to the temperature difference (ΔT) so that:

$$\Delta V = a_1 \, \Delta T + a_2(\Delta T)^2 \qquad \text{(A2.1)}$$

In practice, for typical temperature ranges found in the boundary layer, the second term on the right-hand side can be neglected. The constant a_1 depends upon the nature of the metals used, for the common combination of copper and constantan (an alloy of copper and nickel) $a_1 \simeq 40 \, \mu V \, °C^{-1}$. To obtain absolute temperature values one junction must be referenced against a known, usually constant, temperature such as that provided by a mixture of ice and water in equilibrium. To measure the temperature of a substance using equation A2.1 it is necessary to know a_1 and to measure ΔV by a voltmeter or potentiometric recorder. Then knowing ΔT and the temperature of the reference junction it is possible to solve for the temperature of the other junction (the sensor). The ΔV signals are small (typically 10^{-3} to 10^{-6} V) thereby requiring high quality monitoring equipment, but the problem can be lessened by connecting a number of junctions in series so that their outputs are added arithmetically. Then neglecting the quadratic term equation A2.1 becomes:

$$\Delta V = [a_1(\Delta T)]_1 + [a_1(\Delta T)]_2 + \cdots + [a_1(\Delta T)]_n \qquad \text{(A2.2)}$$

where, n – number of junctions. Such a device is called a *thermopile.*

Other approaches. The principles of thermocouples have been outlined because they are a good example of electric thermometry and because of their use in the thermopile format as radiation and soil heat flux transducers (p. 315), but they only represent one of many ways of sensing temperature. The fact that the electrical resistance of metals and semiconductors depends on temperature is utilized in the case of resistance wire and thermistor thermometers respectively. Similarly the resonant frequency of quartz crystals, the speed of sound in dry air, the refraction of light, and the behaviour of transistors, are all temperature-dependent and have been utilized in thermometry.

Air temperatures

The temperature registered by a thermometer is the result of its energy balance, and this is determined by the net heat exchanges to and from the thermometer by radiation, convection and conduction. Ideally only a convective balance is required, so that the thermometer approaches the temperature of the air passing over it. Most thermometer systems are especially designed to minimize radiative exchanges between the instrument, the Sun and its surroundings by encasing the sensor in a radiation shield whose temperature is close to that of the air, or by reducing its dimensions to the point where radiative exchange is minimal (e.g. using very fine wires). On the other hand, the convective exchange can be maximized by artificially aspirating the sensor with forced ventilation, or if this is unacceptable by making the sensor so small that even the slightest air movement is sufficient to prevent air from stagnating around it. An example of a shielded and aspirated thermometer for use in boundary layer studies is shown in Figure A2.1a.

When constructing thermometers attention should also be paid to the required response. A very fine wire thermometer will follow the rapid turbulent fluctuations of temperature (e.g. Figure 2.11). For certain applications this may be ideal but if average values are required some form of integration is made necessary. Alternatively, if the mass of the sensor is increased it provides a lagged response to temperature fluctuations and a

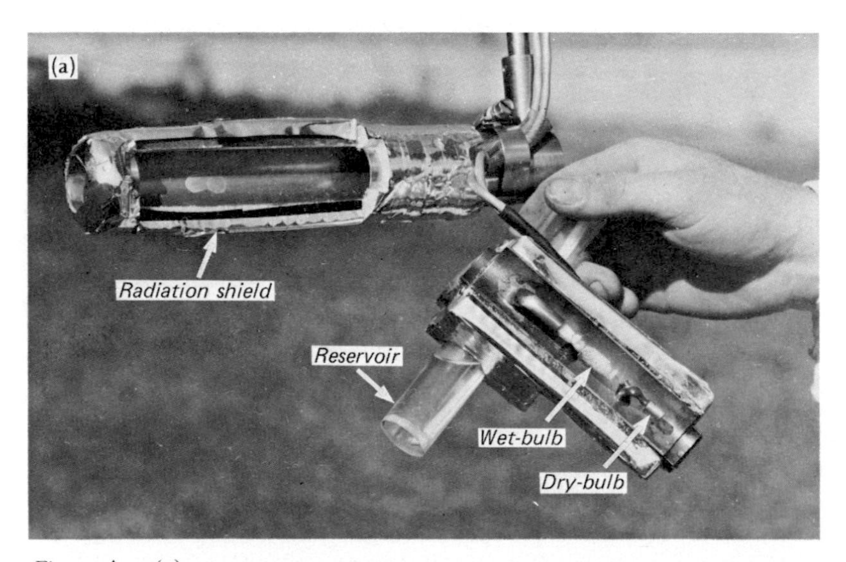

Figure A2.1(a).

more easily readable output. The desired output characteristics can usually be achieved by careful choice of the gauge of the wire used in construction.

In many applications *differences* of air temperature are more important than *absolute* values (e.g. in evaluating the vertical transfer of entities via flux-gradient approaches, p. 325). Differences of the order of 0·01°C are often required and this cannot be achieved by taking the difference between two absolute measurements. It is better to measure the difference directly

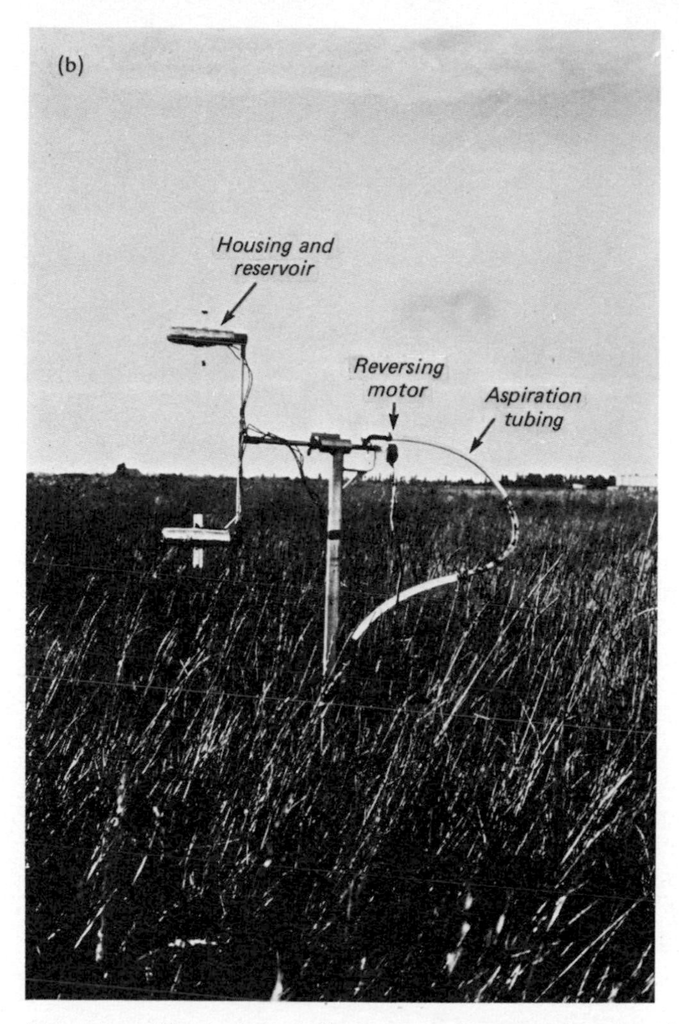

Figure A2.1 Wet- and dry-bulb thermometer system for measuring vertical differences of air temperature and humidity. (a) Wet- and dry-bulb sensors and water reservoir detached from the housing (photograph courtesy T.A. Black, Dept. Soil Science, Univ. British Columbia). (b) Difference system showing the complete sensor housings (aspirated radiation shields) and the reversing motor which rotates the horizontal arm at regular intervals.

Figure A2.2 Hand-held infra-red radiation thermometer (Barnes PRT-10) for surface radiation temperatures.

and to reference one level to an absolute measurement. This can be achieved using a thermopile with the junction pairs forming a difference (i.e. what would normally be the reference junctions become sensors as part of a difference-pair). Similar difference systems can be constructed from resistance wires or thermistors if matched resistance elements are placed in the opposite arms of a bridge.

Soil temperatures

Soil temperatures vary less rapidly than air temperatures, and since radiative and convective exchanges are virtually absent in the soil, thermometer requirements and errors are less than in the air. The types of thermometer outlined for use in the air are also applicable for the soil. Care must be taken to cause as little disturbance to the soil structure as is possible when installing the thermometers. Ideally they should be inserted horizontally from a pit excavated to one side so that heat conduction along the leads is minimized, and soil moisture flow is not overly disrupted.

Surface temperatures

The surface temperatures of leaves or the ground are difficult to measure. Very fine-wire electrical thermometers can be attached to leaves or appressed to the ground, but even using a large number may not give an adequate spatial sample. Probably the best approach is to sense the surface temperature remotely by a radiation thermometer (Figure A2.2). The instrument measures the long-wave radiation (limited to the 8–14 μm waveband) emitted by surfaces placed in its field-of-view. The radiation 'seen' by the instrument is that emitted by the surface ($L\uparrow = \varepsilon\sigma T_0^4$, equation 1.4) plus any radiation in the same waveband from the sky which is reflected ($L\downarrow(1 - \varepsilon)$). Since most natural surfaces are close to full radiators ($\varepsilon \simeq 1{\cdot}0$) in this waveband the reflected term can often be ignored, and the apparent surface radiative temperature (T_k) can be equated with the true surface temperature (T_0) so that:

$$T_0 \simeq T_k = (L\uparrow/\sigma)^{1/4} \qquad \text{(A2.3)}$$

The approach is very helpful because no contact with the surface is involved, and the radiation 'seen' is an integration of that emitted from an area. The effect of neglecting the variation of emissivities is of the order of 1°C for most natural surfaces.

(b) HUMIDITY

Atmospheric humidity is also a difficult quantity to measure with any high degree of accuracy. There are at least thirty different instruments (hygrometers) designed to measure humidity but most can be classified under the following five categories: psychrometric approaches (thermodynamic methods involving the measurement of air temperatures); absorption methods (based on changes in the physical dimensions of substances due to moisture absorption); condensation approaches (determining the dew-point temperature at which a water film forms on a cooled surface); chemical and electrical approaches (based on changes in the chemical or electrical properties of substances due to moisture absorption); radiation absorption approaches (utilizing the fact that water vapour absorbs radiation in specific wavebands).

In practice the number of methods actually used in the field is much smaller, and of these the simple but well-tested approach of wet- and dry-bulb psychrometry remains the most popular. The method consists of exposing two identical thermometers, one to measure the actual air temperature (T_a), and the other (covered with a wetted wick) to measure the wet-bulb temperature (T_w) which is lower than T_a due to evaporative cooling. In the absence of external energy it is assumed that all of the energy used to evaporate the water is supplied by cooling the air. When equilibrium evaporation is achieved it can be shown that:

$$e = e^*_{(T_w)} - \frac{c_p P}{0{\cdot}622 L_v}(T_a - T_w) \qquad \text{(A2.4)}$$

where, $e^*_{(T_w)}$ – saturation vapour pressure at the wet-bulb temperature. The term ($c_p P/0{\cdot}662 L_v$) is known as the psychrometric constant and is equal to 66 Pa °C^{-1} at sea level. Therefore, since $e^*_{(T)}$ is a unique function of T

(Figure 2.15) measurements of T_a and T_w allow calculation of e (usually from tables), and as shown in Chapter 2 this term can be related to a wide range of other humidity measures.

Wet- and dry-bulb measurements are subject to the same errors as for T_a alone, and similar radiation shielding and aspiration are in order (Figure A2.1a). In addition it is important to ensure that the wick is properly wetted and clean, and that only distilled water is used (the $e^*_{(T)}$ vs T relation only holds for pure water). Determination of humidity differences to the accuracy required in flux-gradient equations is particularly difficult. As with air temperature differences it is advisable to measure the differences rather than absolute values, and any systematic errors in sensors can be eliminated by interchanging sensors at regular intervals during a measurement period (Figure A2.1b).

Other approaches commonly in use include dew-point hygrometers (condensation category); lithium chloride dew-cells (chemical category); and infra-red hygrometers (radiation absorption category, see next section).

(c) CARBON DIOXIDE CONCENTRATION

The absorption of certain wavelengths of infra-red radiation by CO_2 can be used to measure the concentration of this gas in the air. Air samples are drawn through a tube in an infra-red gas analyser and subjected to radiation from dull-red filaments. The degree of depletion of the radiation is then related to the amount of absorbing gas in the sample. To measure absolute CO_2 concentrations the depletion in the sample is compared

Figure A2.3(a).

Figures A2.3 Anemometers suitable for surface layer wind speed observations. (a) Cup anemometers (right) installed to measure the profile of horizontal wind speed over a crop. Mast at the left is instrumented to measure wet- and dry-bulb temperature profile. (b) Propeller anemometer mounted to measure vertical wind speeds.

against that in a reference tube containing CO_2-free air. When determining profile differences the air from two different levels is compared directly in the two tubes. In the absolute mode changes of 1 to 2 ppm can be detected; in the differential mode precision of $\pm 0{\cdot}1$ ppm is possible with normal background concentrations (approximately 300 ppm).

Infra-red gas analysers can also be fitted with detectors to measure water vapour instead of, or as well as, CO_2.

(d) WIND SPEED

Wind speed can be measured by rotating cup anemometers, propeller anemometers, heat-transfer devices, differential pressure devices, and acoustic anemometers. For average horizontal wind speed the first three types are the most used, and of these the cup anemometer is by far the most popular (Figure A2.3a). The rotation of the vertical shaft supporting the cup arms may be utilized to provide voltage pulses, or a continuously variable voltage signal. Friction in the bearing plus any in the electrical contacts causes the shaft to stop rotating before the horizontal wind actually becomes zero. Typical stall speeds for sensitive micrometeorological anemometers are about 0·1 to 0·3 m s^{-1}. In profile studies where small differences of horizontal wind speed are important the anemometers must be accurately matched and calibrated, usually in a wind tunnel.

Propeller-type anemometers can be used to measure horizontal wind speeds if the propeller is continually orientated into the wind by a vane. When placed with the axis of rotation in the vertical (Figure A2.3b) the output is proportional to the vertical wind speed. In confined spaces where cups or propellers cannot rotate (e.g. crop canopy) hot-wire anemometers are sometimes used. The sensor is an electrically-heated wire or junction whose temperature is dominated by convective heat exchange and this can be related to the wind speed. Unfortunately such devices respond to almost the complete wind field and not just the horizontal or vertical components. Differential pressure devices and acoustic anemometers are usually confined to work on the fine structure of turbulence and not for average winds.

(e) PROFILE MEASUREMENTS

When making measurements of air temperature, humidity, carbon dioxide and wind speed to estimate the vertical fluxes of heat, water vapour, carbon dioxide and momentum it is usual to mount the instruments on a mast thereby obtaining vertical profiles from which vertical gradients ($\Delta\bar{T}/\Delta z$, $\Delta\bar{q}/\Delta z$, $\Delta\bar{c}/\Delta z$ and $\Delta\bar{u}/\Delta z$) can be computed (Figures A2.3a, A2.4). On such masts it is common to space the instruments logarithmically. This is based on the knowledge that the profiles of most properties vary approximately in this fashion, therefore greater sampling is needed near the surface where properties are changing most rapidly with height. As a rule-of-thumb, however, the lowest measurement level is never placed below a height equal to five times the value of the surface roughness length (z_0). In addition, if measurements are to be considered representative of the local surface the upper observation level must lie within the local internal boundary layer (see Figure 5.2c and discussion).

2 Evaluation of radiative fluxes

(a) MEASUREMENT

The principal instruments used to measure component fluxes of the radiation budget are listed in Table A2.1. They represent a range of

Figure A2.4 Meteorological tower instrumented to measure profiles of air temperature and humidity over a forest at Thetford, England (photograph courtesy Director, Institute of Hydrology, Wallingford, England).

instrumental configurations, but most modern designs are united by the use of a multi-junction thermopile (p. 309) as the method of transducing the radiation flux into a thermal response, and thence into a voltage signal suitable for electronic monitoring. The receiving surface of the thermopile is often covered by a dome of glass, quartz, polyethylene etc. which acts as: a protection from weather damage; a spectral filter to distinguish short- from long-wave radiation fluxes; and a means of standardizing convective heat exchange at the thermopile surface so as to reduce the effects of wind speed on the energy balance of the instrument.

Short-wave radiation

Figure A2.5a shows a typical pyranometer used to measure incoming short-wave radiation on a horizontal surface ($K\downarrow$). The thermopile is

TABLE A2.1 Radiation instrument terminology.

Instrument	Definition
Radiometer	Instrument measuring radiation.
Pyrradiometer	Measures total radiation from the solid angle 2π incident on a plane surface ($Q\downarrow$ or $Q\uparrow$).
Pyranometer (solarimeter)	Measures short-wave radiation from the solid angle 2π incident on a plane surface ($K\downarrow$ or $K\uparrow$).
Net pyranometer	Measures net short-wave radiation (K^*).
Pyrheliometer	Measures direct-beam short-wave radiation at normal incidence
Diffusometer	Pyranometer and shade device used to measure diffuse-beam short-wave radiation (D).
Pyrgeometer	Measures long-wave radiation on a horizontal blackened surface at the ambient air temperature ($L\downarrow$ or $L\uparrow$).
Net pyrradiometer	Measures net all-wave radiation from above and below (Q^*).

covered by double glass domes whose radiative properties are such as to only allow radiation in the band from 0·3 to 3·0 μm to pass through to the receiving surface. In this example the receiving surface is painted with a special optical black paint so that it has a very high absorptivity. Half of the thermo-junctions are attached to thin strips whose temperature fluctuates rapidly as $K\downarrow$ varies, the others are attached to a large brass block whose

Figure A2.5(a).

Figure A2.5 (a) Pyranometer (Kipp) and, (b) pyranometer and shade ring used to measure total incoming and diffuse-beam short-wave radiation, respectively. (c) Net pyrradiometer (Swissteco) for net all-wave radiation. The ports for external ventilation are visible, but not connected.

temperature varies slowly. The difference can be related to the short-wave receipt. In another design the junctions are alternately in contact with white- and black-painted surfaces.

An *inverted* pyranometer senses the short-wave radiation reflected from the underlying surface ($K\uparrow$). Therefore using equation 1.11 the surface albedo (α) can be obtained as $\alpha = K\uparrow/K\downarrow$. A pyranometer can become a diffusometer by adding a shade ring set at an angle to obscure the sensing surface from direct-beam radiation at all times (Figure A2.5b). The instrument therefore measures only diffuse-beam short-wave radiation (after correction has been made for the amount of diffuse-beam cut out by the ring itself). If $K\downarrow$ from an un-shaded pyranometer is available at the same time, then the direct-beam radiation (S) can be obtained by difference from equation 1.10 (i.e. $S = K\downarrow - D$). Alternatively S can be gained from a pyrheliometer which focuses only upon the solar disc and measures S at normal incidence to the beam. To convert this value to that for a horizontal surface resort must be made to the cosine law of illumination (equation 5.2).

In vegetation canopies where radiation fluxes vary spatially, a single, fixed pyranometer of the usual pattern is insufficient. Sampling can be improved either by moving the instrument along a trackway, or by constructing an instrument with a long tubular thermopile.

All-wave and long-wave radiation

The receiving surface of a net pyrradiometer is a blackened plate across which there is a thermopile with one set of junctions in contact with the upper face and the other set attached to the lower face. With the plate aligned parallel to the surface the thermopile output is related to the temperature difference across the plate, and this is proportional to the difference between the total incoming ($Q\downarrow = K\downarrow + L\downarrow$), and outgoing ($Q\uparrow = K\uparrow + L\uparrow$) radiation fluxes at all wavelengths ($Q^* = Q\downarrow - Q\uparrow$). However, the temperature difference is really an expression of the difference in the energy balances of the two faces and these are affected by convective as well as radiative exchanges. To overcome the effects of wind differences on the two faces the plate is either forcefully ventilated at a constant rate, and/or protected by a hemispheric dome of polyethylene. This material is chosen because it is virtually transparent to radiation with wavelengths in the range of 0·3 to 100 μm. There are a few absorption bands in the infra-red but these can be allowed for in calibration.

The net pyrradiometer shown in Figure A2.5c is of the polyethylene dome type. The thin domes are kept inflated by a constant supply of nitrogen or dry air which circulates between the upper and lower domes and helps to equalize convective exchange. This model is also provided with exterior ventilation which further aids convective equalization and prevents the accumulation of dust or dew on the domes. Such substances must be removed because they absorb radiation and therefore reduce the transparency of the domes. For this reason no net pyrradiometer is reliable during rain.

There are no similarly accepted instruments to measure the long-wave radiation budget terms. Of course at night a net pyrradiometer measures L^* (i.e. it is a net pyrgeometer) but by day the situation is dominated by short-wave exchanges. One approach towards identifying long-wave radiation is to modify a net pyrradiometer by removing the polyethylene dome from one face and replacing it with a black body cavity. The cavity consists of an aluminium dome whose interior is coated with optical black paint, and whose interior temperature (T_{cav}) is sensed by means of a thermocouple. Thus with a cavity on the lower surface, by day the instrument output is due to the difference between $Q\downarrow$ on the upper face, and the black body output of the cavity interior (σT_{cav}^4) on the lower face, i.e.:

$$\text{Instrument output} = Q\downarrow - \sigma T_{cav}^4$$

therefore by re-arrangement:

$$Q\downarrow = \text{Instrument output} + \sigma T_{cav}^4$$

If $K\downarrow$ is available at the same time from a pyranometer then the incoming long-wave radiation from the atmosphere ($L\downarrow$) can be obtained by difference (i.e. $L\downarrow = Q\downarrow - K\downarrow$). Further if $K\uparrow$ from an inverted pyranometer and Q^* from a net pyrradiometer are also available the radiation budget (equation 1.14) can be solved for the long-wave radiation emitted by the surface ($L\uparrow$), i.e.:

$$L\uparrow = K\downarrow - K\uparrow + L\downarrow - Q^*$$

At night when $Q\downarrow = L\downarrow$ all of the long-wave terms can be obtained using two net pyrradiometers, one of which is equipped with a cavity.

Given proper installation and maintenance of radiometers, and appropriate recording equipment, all of the radiation budget terms can be measured to an accuracy of better than 5% (Latimer, 1972). However, in the absence of such equipment, or the lack of a sufficiently dense network of observations, there is often the need for estimation procedures.

(b) ESTIMATION

Approaches to the estimation of radiation components fall into two broad categories: those based on theory, and those relying on statistics.

(i) *Theoretically-based approaches*

A number of numerical models have been developed to estimate the short-wave radiation incident upon the surface (Houghton, 1954; Monteith,

1962; Davies *et al.*, 1975). Their basic approach is to compute the extra-terrestrial radiation (K_{Ex}) above a station (a purely geometric problem) and using known or assumed relationships for the absorption, scattering and reflection by atmospheric constituents (water vapour, dust and air molecules) to calculate the proportion of K_{Ex} transmitted to the surface. Initially this is undertaken assuming cloudless skies, but later this value can be modified to incorporate the effects of cloud using the observed cloud distribution, and coefficients for absorption and reflection. The approach relies heavily upon the values of the coefficient for transmission but is now capable of giving estimates of $K\downarrow$ with an accuracy approaching that of measurement with cloud-free conditions ($\sim 5\%$). The effects of cloud are more difficult to model. Obviously with a knowledge of the surface albedo these models can estimate the reflected flux and the net short-wave absorption.

A similar array of models are available to calculate the incoming long-wave radiation at the surface (Yamamoto, 1952; Elsasser and Culbertson, 1960). These models are used to calculate long-wave radiation exchanges in the atmosphere due to the absorption and emission by water vapour, carbon dioxide and ozone. The complexity of the absorption spectra for these gases (Figure 1.9) is usually simplified by considering bulk absorptivities and emissivities as a function of the temperature and a calculated path length (based on observed profiles of temperature and water vapour from radiosonde balloon ascents).

(ii) *Empirical formulae*

An alternative, but less rigorous approach is to seek statistical relationships between the required radiation term and a surrogate atmospheric variable which is more readily available.

Short-wave radiation ($K\downarrow$) has been related to the extra-terrestrial input (K_{Ex}) and the number of sunshine hours. The value of K_{Ex} places an upper limit on the available solar radiation, and the sunshine term (from a sunshine recorder) adjusts this to give the amount penetrating to the surface in proportion to the sunniness/cloudiness of the day. Similarly, net all-wave radiation (Q^*) can be statistically related to $K\downarrow$ or better K^*, at a given site by a simple linear equation:

$$Q^* = a + b(K\downarrow \text{ or } K^*) \qquad \text{(A2.5)}$$

where, a, b – constants derived from a linear regression analysis. Such relationships however are only used as a last resort and then only for long-term averages such as daily radiation totals, not hourly values. The constants in equation A2.5 are likely to be site-specific because of the important role of surface radiative properties in determining Q^*, and this seriously restricts the use of such a relationship.

Empirical formulae have also been developed for the long-wave

TABLE A2.2 Formulae used to compute atmospheric and net long-wave radiation at the surface with cloudless skies ($L{\downarrow}_{(0)}$ and $L^*_{(0)}$).† Net long-wave equations suitable for daily estimates only.

Author (year)	Equation	Remarks
1. Brunt (1932)	$L{\downarrow}_{(0)}=\sigma T_a^4[a+b(e_a)^{1/2}]$ $L^*_{(0)}=\sigma T_a^4[a+b(e_a)^{1/2}-1]$	Constants (a, b) vary with location, Budyko (1958) gives average values: $a=0{\cdot}61$ and $b=0{\cdot}05$ [T_a (K); e_a (mb)]
2. Swinbank (1963)	$L{\downarrow}_{(0)}=1{\cdot}20\sigma T_a^4-171$ $L^*_{(0)}=0{\cdot}20\sigma T_a^4-171$	[T_a (K)] Limited to temperatures above $\sim 0°C$
3. Monteith (1973)	$L{\downarrow}_{(0)}=208+6T_a$ $L^*_{(0)}\simeq T_a-107$	Valid over the range of T_a from -5 to $+25°C$ [T_a (°C)]
4. Idso and Jackson (1969)	$L{\downarrow}_{(0)}=\sigma T_a^4\{1-c\exp[-d(273-T_a)^2]\}$ $L^*_{(0)}=\sigma T_a^4\{-c\exp[-d(273-T_a)^2]\}$	Constants: $c=0{\cdot}261$, $d=7{\cdot}77\times10^{-4}$ [T_a (K)]

Note: Units of $L{\downarrow}_{(0)}$ and $L^*_{(0)}$ are $W\,m^{-2}$ if $\sigma=5{\cdot}67\times10^{-8}\,W\,m^{-2}\,K^{-4}$.
† The subscript (0) indicates cloudless skies.

radiation terms that are so awkward to measure. The choice of surrogate variables depends upon two facts. First, with cloudless skies air temperature and water vapour are the most important atmospheric characteristics influencing long-wave radiation exchange. Second, these two variables show their greatest variability, and have their largest values, close to the ground because of the role of the active surface. As a result most of the long-wave radiation from the atmosphere originates within 100 m of the surface. These two facts explain why most long-wave radiation formulae use screen-level ($\sim 1{\cdot}5$ m) values of the air temperature (T_a) and vapour pressure (e_a).

Table A2.2 lists some of the most commonly used equations. The Brunt, Swinbank and Idso-Jackson atmospheric long-wave radiation formulae were derived by statistical regression between measured $L{\downarrow}_{(0)}$, T_a and e_a (in Brunt). The corresponding net long-wave formulae are an extension of those for $L{\downarrow}_{(0)}$ by using equation 1.13 and assuming that the surface emission ($L{\uparrow}$) can be approximated by that from a full radiator at the screen air temperature (i.e. assuming that the surface emissivity is unity, and that the air temperature at screen-level is not greatly dissimilar to that of the surface). The Monteith equations are a simplification of the Swinbank relationships by assuming linearity over the range of air temperatures from -5 to $+25°C$. There are few clear guidelines governing the choice of formula to be used in a given situation except that the Brunt relation requires an additional variable (e_a), and that of Swinbank is unsuited to low temperature conditions.

Clouds have a strong influence on long-wave exchange because they are almost full radiators. The most common approach to estimating the effect of cloud upon $L\downarrow$ is to modify the cloudless sky value by a non-linear cloud term:

$$L\downarrow = L\downarrow_{(0)}(1 + an^2)$$

and for L^*:

$$L^* = L^*_{(0)}(1 - bn^2)$$

where the constants a and b allow for the decrease of cloud-base temperature with increasing cloud height (Table A2.3), and n is the fraction of sky covered with cloud (expressed in tenths on a scale from zero to unity).

TABLE A2.3 Values of the coefficients used to allow for decreasing cloud temperature with height (modified after Sellers, 1965).

Cloud type	Typical cloud height (km)	Coefficients	
		a	b
Cirrus	12·20	0·04	0·16
Cirrostratus	8·39	0·08	0·32
Altocumulus	3·66	0·17	0·66
Altostratus	2·14	0·20	0·80
Stratocumulus	1·22	0·22	0·88
Stratus	0·46	0·24	0·96
Fog	0	0·25	1·00

3 Evaluation of conductive fluxes

Ideally the conduction of heat in the soil (or other solid substance) can be calculated from equation 2.2 if the thermal conductivity and vertical temperature gradient are known. In practice the variability of the conductivity usually renders this approach impractical. A better method is the use of a heat flux plate similar to that used to measure net radiation (p. 318). The plate consists of a disc of known conductivity, and the temperature difference between its upper and lower faces is measured by a thermopile. When placed horizontally within the soil or other medium its output is proportional to the temperature gradient across, and the heat flux through, the plate (and therefore the soil, if the plate is in good thermal contact). To avoid radiative and convective errors the plate must be buried at least

10 mm below the surface. This however creates the possibility that heat flux convergence (or divergence) in the overlying layer will cause the plate to under- (or over-) estimate the value of the conductive partitioning (Q_G) in the *surface* energy balance. This can be corrected by adding (or subtracting) the change of heat storage in the overlying layer (ΔQ_S) calculated from:

$$\Delta Q_S = C(\Delta \overline{T}/\Delta t)\,\Delta z$$

where, $(\Delta \overline{T}/\Delta t)$ is the measured average rate of temperature change with time in the layer Δz between the surface and the plate, and C is the heat capacity of the soil or other medium.

4 Evaluation of convective fluxes

Much of the field of micro-meteorology has been concerned with attempts to characterize the state of the turbulent atmosphere and in devising methods to evaluate the vertical transfer of entities by convective motion. Here it is neither appropriate nor possible to consider this work in detail so we will concentrate upon an explanation of the methodological principles involved.

There are two basic approaches towards the measurement of vertical fluxes in the surface boundary layer. First, there is the *eddy fluctuation* method which seeks to measure the flux directly by sensing the properties of eddies as they pass through a measurement level on an instantaneous basis. Second, there are the *profile* (or *flux-gradient*) methods which seek to infer the flux on the basis of average profiles of atmospheric properties and the degree of turbulent activity.

(a) EDDY FLUCTUATION METHOD

All atmospheric entities show short-period fluctuations about their longer term mean value. This is the result of turbulence which causes eddies to move continually around carrying with them their properties derived elsewhere. Therefore we may write that the value of an entity (s) consists of its mean value ($\bar{s}$), and a fluctuating part (s'), so that:

$$s = \bar{s} + s' \qquad \text{(A2.6)}$$

where the overbar indicates a time-averaged property and the prime signifies an instantaneous deviation from the mean. The vertical wind trace in Figure 2.11 illustrates these two components: the horizontal line at $w = 0$ is the value $\overline{w}$ since at an extensive site mass continuity requires that as much air moves up as moves down over a reasonable period of time (e.g. 10 min); and the detail of the fluctuating trace gives the value of w' at any instant as a positive or negative quantity depending upon whether it is above the mean (an updraft) or below it (a downdraft).

The properties contained by, and therefore transported with, an eddy are its mass (which by considering unit volume is given by its density, ρ), its vertical velocity (w) and the volumetric content of any entity it possesses (s). Since each one can be broken into a mean and a fluctuating part the mean vertical flux density of the entity (S) can therefore be written:

$$S = \overline{(\bar{\rho} + \rho')(\bar{w} + w')(\bar{s} + s')} \qquad \text{(A2.7)}$$

which upon full expansion yields:

$$S = (\overline{\bar{\rho}\bar{w}\bar{s}} + \overline{\bar{\rho}\bar{w}s'} + \overline{\bar{\rho}w'\bar{s}} + \overline{\bar{\rho}w's'} + \overline{\rho'\bar{w}\bar{s}} + \overline{\rho'\bar{w}s'} + \overline{\rho'w'\bar{s}} + \overline{\rho'w's'}) \qquad \text{(A2.8)}$$

Although equation A2.8 looks rather formidable it can be greatly simplified. First, all terms involving a *single* primed quantity are eliminated because by definition the average of all their fluctuations equals zero (i.e. we lose the second, third and fifth terms). Second, we may neglect terms involving fluctuations of ρ since air density is considered to be virtually constant in the lower atmosphere (i.e. we lose the sixth, seventh and eighth terms). Third, if observations are restricted to uniform terrain without areas of preferred vertical motion (i.e. no 'hotspots' or standing waves) we may neglect terms containing the mean vertical velocity (i.e. we lose the first term). With these assumptions equation A2.8 reduces to the form of the relation underlying the eddy fluctuation approach, viz:

$$S = \rho\overline{w's'} \qquad \text{(A2.9)}$$

where the bar over the ρ has been dropped since it is considered to be a constant. At first glance it might appear as though this term also could be ignored since both w' and s' averaged over time will be zero. However, the overbar denotes the time average of the *instantaneous covariances* of w and s (i.e. the time average of their instantaneous *product*) and this will only rarely be negligible. Note that this technique is also called the eddy correlation method.

In terms of the fluxes and entities with which we are concerned, equation A2.9 can be written:

$$\tau = -\rho\overline{u'w'} \qquad \text{(A2.10a)}$$

$$Q_H = \rho c_p\overline{w'T'} \qquad \text{(A2.10b)}$$

$$Q_E = \rho L_v\overline{w'q'} \qquad \text{(A2.10c)}$$

$$C = \overline{w'c'} \qquad \text{(A2.10d)}$$

and, Figure 2.11 very clearly illustrates how the product of w and T fluctuations combine to produce an instantaneous sensible heat flux (Q_H). The time average of this heat flux is the value given by equation A2.10b.

To obtain the fluxes given by these equations it is necessary to have instruments which can very rapidly sense virtually every variation in the vertical wind velocity and in the entity under study, and processing and recording equipment must be capable of integrating and/or quickly recording very large amounts of information. The response of the instruments should be matched, and sufficiently fast to sense the properties of the smallest eddies capable of contributing to the transport. Since the size of eddies increases with height (Figure 2.9) these requirements become increasingly harder to meet closer to the ground.

Typical w sensors for eddy fluctuation instruments include hot-wire anemometers, differential pressure devices and acoustic anemometers. At heights greater than about 4 m the vertical propeller anemometer is adequate (Figure A2.3b). Temperature is measured by fine-wire resistance elements or acoustic thermometers, and both water vapour and carbon dioxide by infra-red gas analysers. Vapour can also be sensed by very fine wet- and dry-bulb thermometers and chemical hygrometers.

The instrumental requirements still keep the eddy fluctuation method from being widely used but continued technological advances promise to make it more practical. The method has the great advantages of being based on an essentially simple theory; of measuring the fluxes directly; and of requiring no additional specifications of the nature of the surface (such as roughness) or of the atmosphere (such as stability).

(b) PROFILE METHODS

Under this heading there are two basic approaches: the aerodynamic method, and the energy balance or Bowen's ratio method. Both rely upon the principle of similarity (p. 59) regarding the equivalence of the diffusion coefficients for momentum, heat, water vapour and carbon dioxide (i.e. $K_M = K_H = K_W = K_C$). If this assumption is valid then we may use the flux-gradient relationships (i.e. equations 2.8, 2.12, 2.15 (2.16) and 2.18 respectively) to obtain a series of ratios such as:

$$\frac{Q_E}{\tau} = \frac{L_v \, \Delta\bar{q}}{-\Delta\bar{u}}, \qquad \frac{Q_H}{Q_E} = \frac{c_p \, \Delta\bar{T}}{L_v \, \Delta\bar{q}}, \qquad \frac{C}{\tau} = \frac{\Delta\bar{c}}{-\rho \, \Delta\bar{u}} \quad \text{etc.}$$

where the diffusion coefficients (and in the first two cases the air density) have been cancelled. This is advantageous because measurement of two appropriate property differences, and the knowledge of one flux enables the other flux to be obtained. For example in the case of the first ratio listed a measure of τ plus the differences $\Delta\bar{q}$ and $\Delta\bar{u}$ allows determination of Q_E. Alternatively even if equality of the K's cannot be assumed a knowledge of the behaviour of their ratios (e.g. K_W/K_M etc.) allows the fluxes to be evaluated.

Aerodynamic approach

It should be pointed out that the un-modified aerodynamic method only applies under the following restricted conditions:

(i) neutral stability – buoyancy effects are absent;
(ii) steady state – no marked shifts in the radiation or wind fields during the observation period;
(iii) constancy of fluxes with height – no vertical divergence or convergence;
(iv) similarity of all transfer coefficients.

Under these conditions the logarithmic wind profile (equation 2.9) is valid, and the wind gradient ($\Delta\bar{u}/\Delta z$) is found to be inversely proportional to the height above the surface (z). Since the constant of proportionality can be equated to the slope of the neutral wind profile (i.e. k/u_*, equation 2.9) it follows that:

$$u_* = kz(\Delta\bar{u}/\Delta z) \qquad \text{(A2.11)}$$

and substitution of equation 2.10 ($u_*^2 = \tau/\rho$) in A2.11 gives an expression for the vertical flux of horizontal momentum (τ) in terms of wind speed differences alone:

$$\tau = \rho k^2 z^2 (\Delta\bar{u}/\Delta z)^2 \qquad \text{(A2.12)}$$

Now with a measurement of one flux (τ) and invoking the principle of similarity we can use ratios of fluxes involving τ to obtain:

$$Q_H = -\rho c_p k^2 z^2 \left(\frac{\Delta\bar{u}}{\Delta z}\cdot\frac{\Delta\bar{T}}{\Delta z}\right) \qquad \text{(A2.13a)}$$

$$Q_E = -\rho L_v k^2 z^2 \left(\frac{\Delta\bar{u}}{\Delta z}\cdot\frac{\Delta\bar{q}}{\Delta z}\right) \qquad \text{(A2.13b)}$$

and

$$C = -k^2 z^2 \left(\frac{\Delta\bar{u}}{\Delta z}\cdot\frac{\Delta\bar{c}}{\Delta z}\right) \qquad \text{(A2.13c)}$$

These are the neutral stability aerodynamic equations.

There are two major limitations to the use of this approach deriving from the necessary assumptions of neutral stability, and similarity of all coefficients. The former restricts its use to a very narrow range of natural conditions, and to periods when fluxes are likely to be small. However, even given a means to extend these equations for use in non-neutral stability, there is concern that the similarity principle does not apply (especially with regard to K_M). There is an extensive literature concerned with attempts to extend the aerodynamic method by incorporating adjustments which depend upon stability and which include empirical terms to account for non-similarity of the diffusion coefficients. Here we will review one simple approach.

The Richardson Number (Ri) is a convenient means of categorizing atmospheric stability (and the state of turbulence) in the lowest layers:

$$\mathrm{Ri} = \frac{g}{\tilde{T}} \cdot \frac{(\Delta \overline{T}/\Delta z)}{(\Delta \bar{u}/\Delta z)^2} \qquad \text{(A2.14)}$$

where g – acceleration due to gravity (m s^{-2}), $\tilde{T}$ – mean temperature in the layer $\Delta z(K)$, and Ri is a dimensionless number. In general terms equation A2.14 shows that Ri relates the relative roles of buoyancy (numerator) to mechanical (denominator) forces (i.e. free to forced convection) in turbulent flow. Thus in strong lapse (unstable) conditions the free forces dominate and Ri is a negative number which increases with the size of the temperature gradient but is reduced by an increase in the wind speed gradient. In an inversion (stable) Ri is positive, and in neutral conditions Ri approaches zero.

The neutral form of the aerodynamic equations can be generalized according to stability (as given by Ri) in the following manner. From equation A2.11 rearrangement gives the neutral wind gradient as:

$$\frac{\Delta \bar{u}}{\Delta z} = \frac{u_*}{kz} \qquad \text{(A2.15)}$$

and in the general case we may write:

$$\frac{\Delta \bar{u}}{\Delta z} = \frac{u_*}{kz} \Phi_{\mathrm{M}} \qquad \text{(A2.16)}$$

where, Φ_{M} – dimensionless stability function to account for curvature of the logarithmic wind profile due to buoyancy effects (Figure 2.9e). The value of Φ_{M} is unity in the neutral case (i.e. equation A2.16 collapses to A2.15), and is greater or less than unity in stable and unstable conditions respectively. Similarly the neutral temperature, humidity and carbon dioxide gradients can be generalized to read:

$$\frac{\Delta \overline{T}}{\Delta z} = -\frac{Q_{\mathrm{H}}}{\rho c_{\mathrm{p}} k u_* z} \Phi_{\mathrm{H}} \qquad \text{(A2.17a)}$$

$$\frac{\Delta \bar{q}}{\Delta z} = -\frac{Q_{\mathrm{E}}}{\rho L_{\mathrm{v}} k u_* z} \Phi_{\mathrm{W}} \qquad \text{(A2.17b)}$$

and

$$\frac{\Delta \bar{c}}{\Delta z} = -\frac{C}{k u_* z} \Phi_{\mathrm{C}} \qquad \text{(A2.17c)}$$

where Φ_{H}, Φ_{W} and Φ_{C} are dimensionless stability functions for heat, water vapour and CO_2. Hence equations A2.13a–c take the general form:

$$Q_H = -\rho c_p k^2 z^2 \left(\frac{\Delta \bar{u}}{\Delta z} \cdot \frac{\Delta \bar{T}}{\Delta z}\right)(\Phi_M \Phi_H)^{-1} \quad \text{(A2.18a)}$$

$$Q_E = -\rho L_v k^2 z^2 \left(\frac{\Delta \bar{u}}{\Delta z} \cdot \frac{\Delta \bar{q}}{\Delta z}\right)(\Phi_M \Phi_W)^{-1} \quad \text{(A2.18b)}$$

$$C = -k^2 z^2 \left(\frac{\Delta \bar{u}}{\Delta z} \cdot \frac{\Delta \bar{c}}{\Delta z}\right)(\Phi_M \Phi_C)^{-1} \quad \text{(A2.18c)}$$

Observations suggest that $\Phi_H = \Phi_W = \Phi_C = \Phi_M$ in moderately stable conditions, but that $\Phi_H = \Phi_W = \Phi_C = \Phi_M^2$ in the unstable case. Further empirical evidence leads to the following description of the stability functions used in equations A2.18a–c:

Stable case (Ri positive)

$$(\Phi_M \Phi_x)^{-1} = (1 - 5\,\text{Ri})^2 \quad \text{(A2.19a)}$$

Unstable case (Ri negative)

$$(\Phi_M \Phi_x)^{-1} = (1 - 16\,\text{Ri})^{3/4} \quad \text{(A2.19b)}$$

where Φ_x is the appropriate stability function for the property being transferred. These relationships are plotted on logarithmic co-ordinates in Figure A2.6, which also shows the types of flow regimes existing under different stability conditions. When the atmosphere is neutral (Ri between $\pm 0{\cdot}01$) thermal effects are minimal and only forced convection is present. Moving towards greater instability (to the right) buoyancy effects grow in importance through the mixed regime, and at values of Ri larger than $-1{\cdot}0$ only free convection is in operation (weak horizontal motion, very strong convective instability). Conversely moving from neutrality towards greater stability (to the left) negative buoyancy increasingly dampens turbulent motion so that beyond Ri values of about $+0{\cdot}25$ the flow is virtually laminar and vertical mixing is absent (weak horizontal motion, strong temperature inversion).

In summary, although the basic aerodynamic approach is only applicable in neutral conditions semi-empiric relationships can be used to extend its usefulness to a wide range of stability regimes (i.e. by substituting equations A2.19a, b in equations A2.18a–c). The evaluation of fluxes via this method requires the accurate measurement of a wind difference, and the difference of a related property (usually over the same height interval). Figures A2.1b, A2.3a and A2.4 illustrate instrumental arrays designed to obtain these differences in the field. Should the height of the surface roughness elements necessitate the inclusion of a zero-plane displacement (d) in the wind profile (equation 4.5), the z^2 term in equations A2.18a–c is

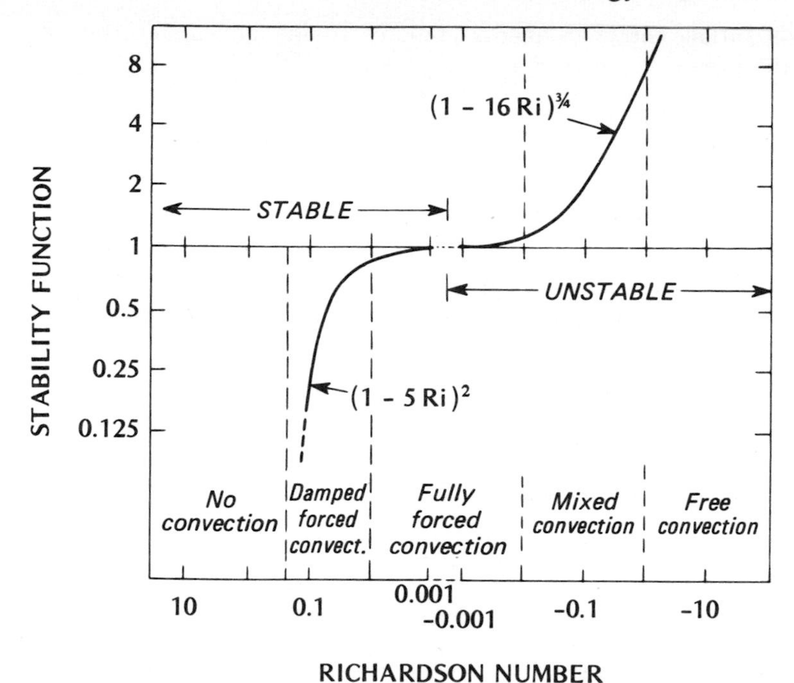

Figure A2.6 Non-dimensional 'stability factor' $(\Phi_M \Phi_x)^{-1}$ plotted logarithmically against the Richardson number stability parameter. Fluxes calculated in non-neutral conditions using flux-gradient equations valid for neutral conditions must be multiplied by this factor. Also showing the characteristic flow regimes at different stabilities (after Thom, 1975).

modified to read $(z - d)^2$. Averaging periods of about 30 minutes are normally appropriate.

Energy balance (Bowen's ratio) method

The energy balance approach to estimating convective fluxes seeks to apportion the energy available ($Q^* - Q_G$, or $Q^* - \Delta Q_S$) between the sensible and latent heat terms by considering their ratio, β (given by equation 2.17). Assumptions (ii) and (iii) necessary to the aerodynamic method also apply in this case. On the other hand, this method has the advantage of not being stability-limited because it only requires similarity between K_H and K_W, and not K_M. Since it can be shown that $K_H/K_W = \Phi_H/\Phi_W$ and that $\Phi_H = \Phi_W$ for all stability regimes (p. 328) it follows that K_H and K_W are similar, and therefore:

$$\beta = \frac{Q_H}{Q_E} = \frac{c_p \, \Delta \overline{T}}{L_v \, \Delta \bar{q}} \qquad \text{(A2.20)}$$

From the surface energy balance (equation 1.16) the individual turbulent fluxes are given in terms of β as:

$$Q_H = \beta(Q^* - Q_G)/(1 + \beta) \quad \text{(A2.21a)}$$

and

$$Q_E = (Q^* - Q_G)/(1 + \beta) \quad \text{(A2.21b)}$$

Therefore to evaluate Q_H and Q_E over an extensive surface all that is required are accurate measurements of Q^* (net pyrradiometer), Q_G (soil heat flux plate) and β from temperature and humidity differences over the same height interval (e.g. using the reversing wet- and dry-bulb system in Figure A2.1b). Averaging periods of about 30 min are found to be appropriate.

5 Evaluation of the water balance

The component terms of the water balance equation of a soil column (equation 1.18):

$$p = E + \Delta r + \Delta S$$

operate over different time scales (p is discrete, Δr can be semi-discrete, and E and ΔS are continuous variables). However, over relatively long periods (e.g. one week or more) all the terms can be evaluated to provide a budget estimate. This is commonly undertaken on the scale of a whole catchment basin, a lake or a glacier. Precipitation (p) is measured with standard rain gauges arranged in a suitable network for spatial sampling, and net run-off (Δr) by hydrologic stream gauging at the boundaries of the system. Therefore if either soil moisture change (ΔS) or evaporation (E) are evaluated the budget is obtained, providing deep drainage to or from the system can be neglected.

Soil moisture change can be measured by a regular programme of measurement using the simple gravimetric or tensiometer methods, or the neutron-scattering technique. The *gravimetric* method involves direct sampling using an auger to remove soil from the required depths. The weight loss after oven-drying gives the water content by weight, and knowledge of the soil's bulk density allows calculation of the soil water content by volume (S):

$$S = \text{Water content by weight} \times \text{Soil bulk density}$$

Tensiometers measure the soil moisture tension (m). The simplest form consists of a porous ceramic cup attached by a tube to a pressure sensor such as a mercury manometer. The cup is filled with water and placed at the required depth in the soil. As soil moisture is depleted water moves out of the cup and a suction is created which is directly related to the soil moisture tension. Values of tension are primarily of use in determining plant water availability but since m is related to S (Figure 2.7) these data can also be used to obtain ΔS. It should however be noted that the relationship depends upon soil texture (the texture classes in Figure 2.7 are only approximations), and is different for the same soil depending whether it is drying or wetting.

The *neutron-scattering* approach is based on the fact that fast neutrons are slowed by collision with hydrogen atoms, and in the soil hydrogen is most abundant in water. Neutron soil moisture meters operate by inserting a radioactive source of fast neutrons, and a detector of the density of slow neutrons into the soil by means of an access tube. The count rate of slow neutrons returning to the detector after being scattered by hydrogen atoms can be related to the amount of water in the soil within a sphere of radius about 0·10 m from the probe. The method is potentially accurate but great care must be exercised in the placement of the access tubes so as not to disturb the soil, and to avoid excessively stony sites. Ideally the neutron meter readings should be calibrated against gravimetric determinations for a particular soil.

In the case of the water balance of a lake the ΔS term relates to changes in the water volume which can be approximated by observations of the lake depth. For a glacier it relates to changes in the height of the glacier surface due to precipitation accumulation or ablation.

Evaporation on the basin, lake and glacier scale can be obtained by residual after evaluating p, Δr and ΔS as above, or from empirical equations, or from the aerodynamic and energy balance approaches outlined in the preceding section as long as there is appropriate spatial sampling. Evaporation from saturated surfaces (e.g. open water, melting ice, well-watered vegetation) can also be estimated from a simple bulk aerodynamic equation involving the mean wind speed and the difference of humidity between the surface and the air, viz:

$$E = -\rho C_{(z)} \bar{u}_{(z)} \Delta \bar{q} \qquad \text{(A2.22)}$$

where, $C_{(z)}$ –Dalton number which varies with stability and height but for water surfaces is usually approximately $1{\cdot}5 \times 10^{-3}$. Other approaches include attempts to relate E almost entirely to the available energy (i.e. $Q^* - Q_G$), and those which combine both bulk aerodynamic and energy factors.

At a level site where Δr is negligible the water budget approach can be used to estimate evaporation totals over periods as short as 1–3 days, as long as ΔS is sampled down to the depth where vertical water exchange is absent. In order to gain hourly measurements of E it is necessary to use micro-meteorological techniques (e.g. converting the latent heat transfer equations A2.10c, A2.18b and A2.21b to mass by dividing by L_v, equation 1.20), or the lysimetric approach. A *lysimeter* is a true water balance device which hydrologically isolates a volume of soil (and its plant cover). In the most accurate examples an undisturbed soil monolith is enclosed in a water-tight container (at least 1 m deep, and 1 to 6 m in diameter) with only its upper side open (Figure A2.7a). This arrangement allows complete specification of the water budget of the soil monolith because p can be measured, Δr is zero, and any deep drainage is either monitored in a sump or kept zero by providing no outlet. Therefore any changes in the mass of the tank must be due to evapotranspiration (including dewfall) or any irrigation applied (i.e. changes in storage, ΔS must be related to the mass flux of water to or from the atmosphere, E). Mass changes by the lysimeter are monitored either by a mechanical balance system installed under the monolith, or by changes in the hydrostatic balance of a fluid system which transmits pressure changes to a manometer as changes of fluid level. Both of the examples in Figure A2.7 are of the latter, 'floating' design. In the most accurate lysimeters E can be evaluated to within 0·02 mm water equivalent.

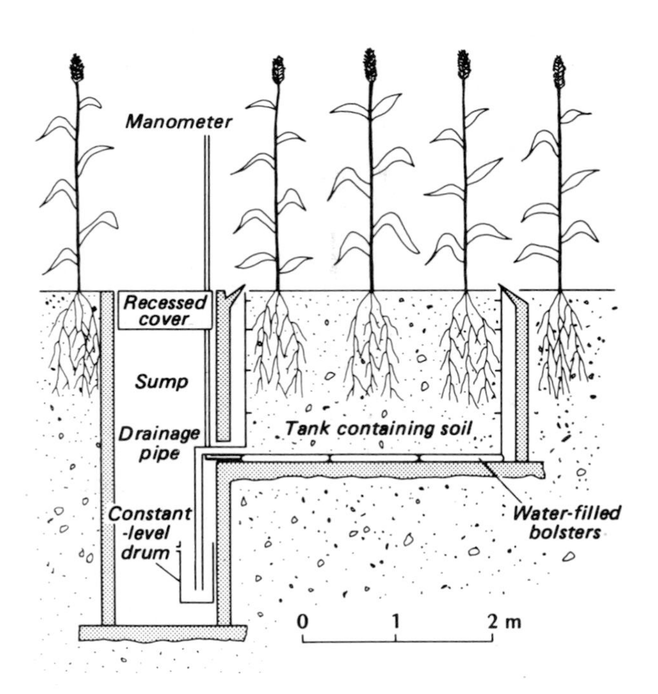

Figure A2.7
(a) Schematic side view of a lysimeter (modified after Forsgate *et al.*, 1965).

(b) A giant 'floating' lysimeter containing a mature Douglas fir tree at Cedar River, Washington. The observer (centre left) is reading the manometer which monitors mass changes by the soil-tree monolith. Tensiometers are installed both inside and outside the lysimeter to ensure that similarity of moisture content is maintained. The brace in the foreground prevents rotation, and the tree is lightly 'guyed' to surrounding trees to prevent it falling over in high winds.

APPENDIX A3 Temperature dependent properties of air, water, and water vapour.

Temperature		Density			Saturation vapour pressure	Latent heat of vaporization	Thermal conductivity of air	Molecular diffusion coefficients of air		
		Dry air	Saturated air	Water						
T	T	ρ_a	$\rho_{a*(T)}$	ρ_w	$e^*_{(T)}$	L_v	k_a	κ_H	κ_M	κ_W
(°C)	(K)	(kg m^{-3})	(kg m^{-3})	(kg m^{-3} × 10^3)	(Pa)	(J kg^{-1} × 10^6)	(W m^{-1} K^{-1} × 10^{-3})	(m^2 s^{-1} × 10^{-4})		
−5	268·2	1·316	1·314	0·9992	421	2·513	24·0	0·183	0·129	0·205
0	273·2	1·292	1·289	0·9999	611	2·501	24·3	0·189	0·133	0·212
5	278·2	1·269	1·265	0·9999	872	2·489	24·6	0·195	0·137	0·220
10	283·2	1·246	1·240	0·9997	1227	2·477	25·0	0·202	0·142	0·227
15	288·2	1·225	1·217	0·9991	1704	2·465	25·3	0·208	0·146	0·234
20	293·2	1·204	1·194	0·9982	2337	2·454	25·7	0·215	0·151	0.242
25	298·2	1·183	1·169	0·9971	3167	2·442	26·0	0·222	0·155	0·249
30	303·2	1·164	1·145	0·9957	4243	2·430	26·4	0·228	0·160	0·257
35	308·2	1·146	1·121	0·9941	5624	2·418	26·7	0·235	0·164	0·264
40	313·2	1·128	1·096	0·9923	7378	2·406	27·0	0·242	0·169	0·272

Sources: van Wijk, 1963; Monteith, 1973.

APPENDIX A4

Système International (SI) units and their equivalents

(a) SI base units

The SI units used throughout this book involve a small number of '*base*' units from which other '*derived*' units can be conveniently obtained. Derived units can be related to base units by the processes of multiplication or division using unity as the only multiplying factor.

The base units used in this book are:

Base unit	Dimensions	SI units
Length	L	m (metre)
Mass	M	kg (kilogram)
Time	T	s (second)
Temperature	θ	K (kelvin)†

† The *unit* temperature of one kelvin is the same as the *unit* degree Celsius (°C). The Kelvin and centigrade scales however differ in respect of the point at which they are based. The Kelvin scale has 0 K at absolute zero (theoretically where molecular motion ceases and a body contains no heat energy), whereas the centigrade scale has 0°C at the freezing point of water. The two are therefore linked by: °C = K − 273·15.

(b) Scientific notation

The scientific or exponential notation is a convenient 'shorthand' way of depicting very large or very small numbers without the use of many zeros. The notation x^n means that the number x is multiplied by itself n times (e.g. $2^3 = 2 \times 2 \times 2 = 8$) where n is called the *exponent*. Similarly x^{-n} involves a negative exponent, and is the reciprocal of x^n, that is $x^{-n} = 1/x^n$ (e.g. $2^{-3} = 1(2 \times 2 \times 2) = \frac{1}{8} = 0{\cdot}125$). It is often especially convenient to express large or small numbers as powers of 10 (i.e. 10^x or 10^{-x}) and certain of these are given prefixes as listed in Table A4.1. Examples used in this book include the micrometre (or micron, μm) for radiation wavelength; the kilowatt (kW) for power; and the megajoule (MJ) for heat energy.

TABLE A4.1 Prefixes used to describe multiples or fractions of ten.

Prefix			Scientific notation	Decimal notation
T	tera-	one trillion (US)	10^{12}	1,000,000,000,000
G	giga-	one billion (US)	10^9	1,000,000,000
M	mega-	one million	10^6	1,000,000
k	kilo-	one thousand	10^3	1,000
h	hecto-	one hundred	10^2	100
da	deka-	ten	10	10
d	deci-	one tenth	10^{-1}	0·1
c	centi-	one hundredth	10^{-2}	0·01
m	milli-	one thousandth	10^{-3}	0·001
μ	micro-	one millionth	10^{-6}	0·000001
n	nano-	one billionth (US)	10^{-9}	0·000000001
p	pico-	one trillionth (US)	10^{-12}	0·000000000001

Scientific notation is also very useful in multiplying or dividing large or small numbers. If the numbers have the same base of 10 then multiplication is achieved by adding the exponents so that $10^n \times 10^m = 10^{n+m}$ (e.g. $10^2 + 10^4 = 10^6$). Similarly for division the exponents are subtracted so that $10^n/10^m = 10^{n-m}$ (e.g. $10^6/10^4 = 10^2$, and $10^2/10^4 = 10^{-2}$, and $10^{-3}/10^2 = 10^{-3-2} = 10^{-5}$). Finally, to multiply or divide complete numbers in scientific notation it is necessary to operate on both parts of the number separately. For example to multiply $1{\cdot}4 \times 10^3$ by $2{\cdot}7 \times 10^{-4}$, first multiply $1{\cdot}4 \times 2{\cdot}7 = 3{\cdot}78$, then add the exponents $10^{3+(-4)} = 10^{-1}$, giving the final answer that the product of the two numbers is $3{\cdot}78 \times 10^{-1}$, or 0·378.

(c) SI derived units

Table A4.2 lists the important quantities used in this book. All are stated in terms of the dimensions of the SI base units (L, M, T and θ) and then they are given in the units of the SI system and with their equivalent values in the c.g.s. (centimetre, gram, second) and British (foot, pound, hour) systems. The following derived quantities may require further definition:

Velocity – is a vector quantity and hence for its complete specification should include a direction as well as a speed.

TABLE A4.2 SI units with c.g.s. and British system equivalents.

Quantity	Dimensions	SI	c.g.s.	British
Basic				
Length	L	1 m	$= 10^2$ cm	= 3·281 ft
Mass	M	1 kg	$= 10^3$ g	= 2·205 lb
Time	T	1 s (or min, h, day, yr)	= 1 s	$= 2{\cdot}778 \times 10^{-4}$ h
Temperature	θ	1 K (or 1°C)	= 1 K (or 1°C)	= 1·8°F
Derived				
Area	L^2	1 m^2	$= 10^4\ cm^2$	$= 10{\cdot}76\ ft^2$
Volume	L^3	1 m^3	$= 10^6\ cm^3$	$= 35{\cdot}31\ ft^3$
Density	ML^{-3}	1 kg m^{-3}	$= 10^{-3}\ g\ cm^{-3}$	$= 6{\cdot}243 \times 10^{-2}\ lb\ ft^{-3}$
Velocity	LT^{-1}	1 m s^{-1}	$= 10^2\ cm\ s^{-1}$	$= 3{\cdot}281\ ft\ s^{-1}$
Acceleration	LT^{-2}	1 m s^{-2}	$= 10^2\ cm\ s^{-2}$	$= 3{\cdot}281\ ft\ s^{-2}$
Force	MLT^{-2}	1 kg m s^{-2} = 1 N (Newton)	$= 10^5\ g\ cm\ s^{-2} = 10^5$ dynes	= 0·225 lb f (lb force)
Pressure	$ML^{-1}T^{-2}$	1 kg $m^{-1}\ s^{-2}$ = 1 Pa (Pascal)	$= 10\ g\ cm^{-1}\ s^{-2} = 10^{-2}$ mb	$= 0{\cdot}021\ lb\ f\ ft^{-2}$
Work, energy	ML^2T^{-2}	1 kg $m^2\ s^{-2}$ = 1 J (Joule)	$= 10^7\ g\ cm^2\ s^{-2} = 10^7$ ergs	= 0·738 ft lb f
Power	ML^2T^{-3}	1 kg $m^2\ s^{-3}$ = 1 W (Watt)	$= 10^7\ g\ cm^2\ s^{-3} = 10^7\ ergs\ s^{-1}$	$= 0{\cdot}738\ ft\ lb\ f\ s^{-1}$
Heat, energy	Q (or ML^2T^{-2})	1 J	= 0·2388 cal	$= 9{\cdot}478 \times 10^{-4}$ BTU
Heat flux	QT^{-1}	1 W	$= 0{\cdot}2388\ cal\ s^{-1}$	$= 3{\cdot}412\ BTU\ h^{-1}$
Heat flux density†	$QL^{-2}T^{-1}$	1 W m^{-2}	$= 2{\cdot}388 \times 10^{-5}\ cal\ cm^{-2}\ s^{-1}$	$= 0{\cdot}317\ BTU\ ft^{-2}\ h^{-1}$
Latent heat	QM^{-1}	1 J kg^{-1}	$= 2{\cdot}388 \times 10^{-4}\ cal\ g^{-1}$	$= 4{\cdot}29 \times 10^{-4}\ BTU\ lb^{-1}$
Specific heat	$QM^{-1}\theta^{-1}$	1 J $kg^{-1}\ °C^{-1}$	$= 2{\cdot}388 \times 10^{-4}\ cal\ g^{-1}\ °C^{-1}$	$= 2{\cdot}388 \times 10^{-4}\ BTU\ lb^{-1}\ °F^{-1}$
Thermal conductivity	$QL^{-1}\theta^{-1}T^{-1}$	1 W $m^{-1}\ °C^{-1}$	$= 2{\cdot}388 \times 10^{-3}\ cal\ cm^{-1}\ s^{-1}\ °C^{-1}$	$= 0{\cdot}578\ BTU\ ft^{-1}\ h^{-1}\ °F^{-1}$
Thermal diffusivity	L^2T^{-1}	1 $m^2\ s^{-1}$	$= 10^4\ cm^2\ s^{-1}$	$= 10{\cdot}8\ ft^2\ s^{-1}$

† Atmospheric scientists have also used the langley (ly) = 1 cal $cm^{-2}\ min^{-1}$ = 697·3 W m^{-2}.

Force – the SI derived unit is the Newton (N) defined as the force required to accelerate a body having a mass of 1 kg at 1 metre per second per second.

Pressure – the SI derived unit is the Pascal (Pa), defined as the pressure exerted by a force of 1 N evenly distributed over an area of one square metre.

Work, energy – the SI derived unit is the Joule (J), defined as the energy required to displace a force of 1 N through a distance of 1 metre.

Power – the SI derived unit is the Watt (W) defined as the power required to equal the rate of working of 1 Joule per second.

(d) Numerical constants

Quantity	Magnitude and units
Dry adiabatic lapse rate	9·8°C km^{-1}
Solar constant	1353 ± 20 W m^{-2}
Specific heat of air at constant pressure	1010 J kg^{-1} °C^{-1}
Specific heat of water vapour	1880 J kg^{-1} °C^{-1}
Specific heat of carbon dioxide	850 J kg^{-1} °C^{-1}
Standard atmospheric pressure†	101·33 kPa
Standard gravitational acceleration†	9·80665 m s^{-2}
Stefan–Boltzmann constant	5·67 × 10^{-8} W m^{-2} K^{-4}

† Based on the *US Standard Atmosphere* (Anon., 1962).

Atmospheric quantities that vary with temperature are listed separately in Appendix A3.

(e) Greek alphabet

Small letter	Capital letter	Name	Small letter	Capital letter	Name
α	A	alpha	ν	N	nu
β	B	beta	ξ	Ξ	xi
γ	Γ	gamma	ο	O	omicron
δ	Δ	delta	π	Π	pi
ε	E	epsilon	ρ	P	rho
ζ	Z	zeta	σ	Σ	sigma
η	H	eta	τ	T	tau
θ	Θ	theta	υ	Υ	upsilon
ι	I	iota	φ	Φ	phi
κ	K	kappa	χ	X	chi
λ	Λ	lambda	ψ	Ψ	psi
μ	M	mu	ω	Ω	omega

APPENDIX A5

Glossary of terms†

Absorption – the process in which incident radiant energy is retained by a substance.

Adiabatic process – a thermodynamic change of state of a system in which there is no transfer of heat or mass across the boundaries of the system. Compression always results in warming, expansion in cooling.

Advection – primarily used to describe predominantly *horizontal* motion in the atmosphere.

Albedo – the ratio of the amount of radiation reflected by a body to the amount incident upon it. Usage varies but here 'radiation' is restricted to the short wavelengths (0·15 to 3·0 μm).

Ambient air – the air of the surrounding environment.

Anabatic wind – an upslope wind due to local surface heating.

Anemometer – an instrument for measuring wind speed.

Anticyclone – a large-scale atmospheric circulation system in which (in the northern hemisphere) the winds rotate clockwise. Interchangeable with 'High' pressure system.

Atmospheric 'window' – the relative gap in the absorption spectrum for atmospheric gases (between 8 and 11 μm).

Attenuation – any process in which the flux density of a 'parallel beam' of energy decreases with increasing distance from the energy source.

† Many of the definitions are based on those of Huschke (1959).

Black body – an hypothetical body which absorbs all of the radiation striking it, i.e. allows no reflection or transmission.

Boundary layer – a general term for the layer of air adjacent to a surface (*see also* LAMINAR BOUNDARY LAYER and PLANETARY BOUNDARY LAYER).

Buoyancy – (or buoyant force) the upward force exerted upon a parcel of fluid by virtue of the density difference between itself and the surrounding fluid.

Cascading system – structures within which the output (energy or mass) from one subsystem forms the input for the next and within which a regulator may operate either to divert a part of the input into a store or to create a throughput, producing the subsystem output.

Concentration – amount of a substance contained in unit volume.

Condensation – the process by which vapour becomes a liquid.

Conduction – the transfer of energy in a substance by means of molecular motions without any net *external* motion.

Convection – mass motions within a fluid resulting in transport and mixing of properties (e.g. energy and mass). Usually restricted to predominantly *vertical* motion in the atmosphere.

Cyclone – a large-scale atmospheric circulation system in which (in the northern hemisphere) the winds rotate anti-clockwise. Interchangeable with 'Low' pressure system.

Dewfall – condensation of water from the lower atmosphere onto objects near the ground.

Dew-point – the temperature to which a given parcel of air must be cooled (at constant pressure and constant water vapour content) in order for saturation to occur.

Diffuse-beam short-wave radiation – short-wave radiation reaching the Earth's surface after having been scattered from the DIRECT-BEAM by molecules or other agents in the atmosphere.

Diffusion – the exchange of fluid parcels between regions in space by apparently random motions on a very small (usually molecular) scale.

Diffusivity – rate of DIFFUSION of a property.

Direct-beam short-wave radiation – that portion of short-wave radiation received in a parallel beam 'directly' from the Sun.

Diurnal – daily.

Eddy – (a) by analogy with a molecule a 'glob' of fluid that has a certain life history of its own, (b) circulation in the lee of an obstacle brought about by pressure irregularities.

Eddy diffusion – turbulent diffusion of properties. An extension of the case of pure DIFFUSION wherein eddies are considered to play the role of molecules.

Emissivity – the ratio of the total radiant energy emitted per unit time per unit area of a surface at a specified wavelength and temperature to that of a BLACK BODY under the same conditions.

Evaporation – (or vaporization) the process by which a liquid is transformed into a gas, in the atmosphere usually water changing to water vapour.

Evapotranspiration – the combined loss of water to the atmosphere by EVAPORATION and TRANSPIRATION.

Extra-terrestrial radiation – solar radiation received at the 'top' of the Earth's atmosphere.

Fetch – distance, measured in the upwind direction.

Flux – rate of flow of some quantity.

Flux density – the FLUX of any quantity through unit surface area.

Free convection – motion caused only by density differences in a fluid.

Forced convection – motion induced by mechanical forces such as deflection or friction.

Front – the interface or transition zone between two air masses of different density.

Guttation – the process by which plants expel water from uninjured leaves in excess of TRANSPIRATION.

Heat capacity – the ratio of the heat absorbed (or released) by a system to the corresponding temperature rise (or fall).

Hygrometer – an instrument which measures the water vapour content of the atmosphere.

Hygroscopic – having a marked ability to accelerate the condensation of water vapour.

Inversion – a departure from the usual decrease or increase with height of an atmospheric property. Most commonly refers to a temperature inversion when temperatures increase, rather than LAPSE, with height, but can also be a moisture inversion.

Irradiation – total radiant flux received by unit area of a given surface.

Katabatic wind – any wind blowing down an incline, often due to cold air drainage.

Kinetic energy – the energy a body possesses as a consequence of its motion, defined as one-half the product of its mass and the square of its speed.

Laminar flow – a flow in which the fluid moves smoothly in parallel STREAMLINES; non-turbulent.

Laminar boundary layer – the layer immediately next to a fixed boundary in which LAMINAR FLOW prevails.

Lapse rate – the decrease of an atmospheric variable with height. Usually refers to temperature unless otherwise specified.

Latent heat – the heat released or absorbed per unit mass by a system in changing phase.

Lysimeter – an instrument for measuring evaporation by monitoring the weight changes of a representative soil plus vegetation monolith.

Mixed layer – the layer of air (usually sub-inversion) within which pollutants are mixed by TURBULENCE.

Mixing height – (or mixing depth) the thickness of the layer measured from the surface upward, through which pollutants are presumed to mix due to CONVECTION caused by daytime heating of the surface.

Momentum – that property of a particle which is given by the product of its mass with its velocity.

Oxidant – any oxidizing agent (i.e. a substance that acquires electrons in a chemical reaction). Ozone (O_3) and atomic oxygen (O) are very effective oxidants.

Perturbation – any departure introduced into an assumed steady state of a system.

Planetary boundary layer – the atmospheric boundary layer or Ekman layer, or the layer of the atmosphere from the surface to the level where the frictional influence is absent.

Potential energy – the energy which a body possesses as a consequence of its position in a field of gravity. Numerically equal to the work required to bring the body from a reference level (usually mean sea level) to its present position.

Potential temperature – the temperature a parcel of dry air would have if brought adiabatically from its present position to a standard pressure of 100 kPa.

Process-response system – the linkage of at least one morphological system with a cascading system thereby demonstrating how form (effect) is related to process (cause).

Profile – a graph of an atmospheric quantity versus a horizontal, vertical or time scale. Usually refers to vertical representation.

Radiant energy – the energy of any type of electromagnetic radiation.

Radiation – the process by which electromagnetic radiation is propagated through free space by virtue of joint undulatory variations in the electric and magnetic fields in space.

Runoff – the water, derived from precipitation, that ultimately reaches stream channels.

Scattering – the process by which small particles suspended in a medium of a different index of refraction diffuse a portion of the incident radiation in all directions.

Schlieren method – an experimental (photographic) technique for detecting the presence of slight density (and hence temperature/pressure) variations in gases and liquids by virtue of refraction effects.

Scavenging – the sweeping out of airborne particulates by rain or snow.

Sensible heat – that heat energy able to be sensed (e.g. with a thermometer). Used in contrast to LATENT HEAT.

Sky view factor – the ratio of the amount of the sky 'seen' from a given point on a surface to that potentially available (i.e. the sky hemisphere subtended by a horizontal surface).

Smog – a mixture of smoke and fog.

Solar altitude – vertical direction of the Sun above the horizon expressed in degrees.

Solar azimuth – horizontal direction of the Sun relative to a reference direction (usually true north) expressed in degrees.

Solar zenith angle – vertical direction of the Sun relative to the ZENITH expressed in degrees. The reciprocal of SOLAR ALTITUDE.

Specific heat – the HEAT CAPACITY of a system per unit mass.

Streamline – a line whose tangent at any point in a fluid is parallel to the instantaneous velocity of the fluid.

Sublimation – the transition of a substance directly from the solid to the vapour phase or vice versa.

Synoptic – referring to the use of meteorological data obtained simultaneously over a wide area for the purpose of presenting a comprehensive and nearly instantaneous picture of the state of the atmosphere.

Thermal conductivity – a physical property of a substance describing its ability to conduct heat by molecular motion.

Thermal diffusivity – the ratio of the THERMAL CONDUCTIVITY to the HEAT CAPACITY of a substance. It determines the rate of heating due to a given temperature distribution in a given substance.

Thermocline – a vertical temperature gradient in a water body, which is appreciably greater than gradients above or below.

Trajectory – a curve in space tracing the points successively occupied by a particle in motion.

Transpiration – the process by which water in plants is transferred as water vapour to the atmosphere.

Troposphere – the lowest 10–20 km of the Atmosphere, characterized by decreasing temperature with height, appreciable water vapour and vertical motion, and weather.

Turbidity – any condition of the atmosphere which reduces its transparency to solar radiation.

Turbulence – a state of fluid flow in which the instantaneous velocities exhibit irregular and apparently random fluctuations so that in practice only statistical properties can be recognized. These fluctuations are capable of transporting atmospheric properties (e.g. heat, water vapour, momentum etc.) at rates far in excess of molecular processes.

Zenith – that point in the sphere surrounding an observer that lies directly above him.

References

ACS, 1969, *Cleaning Our Environment, The Chemical Basis for Action.* American Chemical Society, Washington, D.C.

AISENSHTAT, B. A., 1966, Investigations on the heat budget of Central Asia. In BUDYKO, M. I. (ed.) *Sowremennye Problemy Klimatologii,* Meteorol. Gidrol, Leningrad, 83–129.

ALEXANDER, G., 1974, Heat loss from sheep. In MONTEITH, J. L. and MOUNT, L. E. (eds) *Heat Loss from Animals and Man,* Butterworth, London, 173–203.

ANON, 1962, *US Standard Atmosphere,* U.S. Govt Printing Office, Washington, D.C.

AVERY, D. J., 1966, The supply of air to leaves in assimilation chambers. *J. Experiment. Bot.,* 17, 655–77.

BACH, W., 1972, *Atmospheric Pollution.* McGraw-Hill, New York.

BARTHOLOMEW, G. A., 1964, The roles of physiology and behaviour in the maintenance of homeostasis in the desert environment. *Symposia of the Society for Experimental Biology, No. 18,* Academic Press, New York, 7–29.

BARTHOLOMEW, G. A., 1968, Body temperature and energy metabolism. In GORDON, M. S. (ed.) *Animal Function: Principles and Adaptations,* Macmillan, New York, 290–354.

BATES, D. V., 1972, *A Citizen's Guide to Air Pollution.* McGill-Queen's Univ. Press, Montréal.

BAYLISS, P., 1976, Photograph in *Weather*, 31, 346.

BIERLY, E. W. and HEWSON, E. W., 1962, Some restrictive meteorological conditions to be considered in the design of stacks. *J. Appl. Meteorol.*, 1, 383–90.

BLACK, T. A. and GOLDSTEIN, M., 1977, Effect of ditch water level on root zone water regime and crop water use. *Report to B.C. Dept. Agric.*, Dept. Soil Sci., Univ. B.C., Vancouver.

BROWN, K. W. and ROSENBERG, N. J., 1970, Effects of windbreaks and soil water potential on stomatal diffusion resistance and photosynthetic rate of sugar beets (*Beta vulgaris*). *Agron. J.*, 62, 4–8.

BROWN, K. W. and ROSENBERG, N. J., 1971, Energy and CO_2 balance of an irrigated sugar beet (*Beta vulgaris*) field in the Great Plains. *Agron. J.*, 63, 207–13.

BRUNT, D., 1932, Notes on radiation in the atmosphere. *Quart. J. Royal Meteorol. Soc.*, 58, 389–420.

BUCKMAN, H. O. and BRADY, N. C., 1960, *The Nature and Properties of Soils*. Macmillan, New York.

BUDYKO, M. I., 1958, *The Heat Balance of the Earth's Surface* (transl. Stepanova, N.). Office of Tech. Services, US Dept Commerce, Washington.

BUDYKO, M. I. (ed.), 1963, *Atlas Teplovogo Balansa*, Gidrometeorologicheskoe Izdatel'skoe, Leningrad.

BUETTNER, K. J. K., 1967, Valley wind, sea breeze, and mass fire: Three cases of quasi-stationary airflow, *Proc. Symp. Mountain Meteorol., Atmos. Sci. Pap. No. 122*, Dept. Atmos. Sci., Colorado State Univ., Ft Collins, 104–29.

BYERS, H. R., 1965, *Elements of Cloud Physics*. Univ. Chicago Press, Chicago.

CAREY, F. G., 1973, Fishes with warm bodies, *Scientific American*, 228, 36–44.

CHANDLER, T. J., 1967, Absolute and relative humidities in towns. *Bull. Amer. Meteorol. Soc.*, 48, 394–9.

CHANGNON, S. A. Jr, 1972, Urban effects on thunderstorm and hailstorm frequencies. *Preprints Conf. Urban Environ. and Second Conf. Biometeorol.*, Amer. Meteorol. Soc., Boston, 177–84.

CHANGNON, S. A. Jr, HUFF, F. A. and SEMONIN, R. G., 1971, METROMEX: An investigation of inadvertent weather modification. *Bull. Amer. Meteorol. Soc.*, 52, 958–67.

CHOW, V. T., 1975, Hydrologic cycle. *Encyclopaedia Brittanica*, 15th edn, 9, 102–25.

CLARKE, J. F., 1969a, A meteorological analysis of carbon dioxide concentrations measured at a rural location. *Atmos. Environ.*, 3, 375–83.

CLARKE, J. F., 1969b, The nocturnal urban boundary layer over Cincinnati, Ohio. *Monthly Weather Rev.*, 97, 582–9.

DAVENPORT, A. G., 1965, The relationship of wind structure to wind loading. *Proc. Conf. Wind Effects on Struct.*, Sympos. 16, Vol. 1, HMSO, London, 53–102.

DAVIES, J. A., ROBINSON, P. J. and NUNEZ, M., 1970, Radiation measurements over Lake Ontario and the determination of emissivity. *First Report, Contract No. HO 81276*, Dept. Geog., McMaster Univ., Hamilton, Ont.

DAVIES, J. A., SCHERTZER, W. and NUNEZ., M., 1975, Estimating global solar radiation. *Boundary-Layer Meteorol.*, 9, 33–52.

DEACON, E. L., 1969, Physical processes near the surface of the Earth. In FLOHN, H. (ed.), *World Survey of Climatology*, Vol. 2, *General Climatology*, Elsevier, Amsterdam, 39–104.

DEAN, R. S., SWAIN, R. E., HEWSON, E. W. and GILL, G. C., 1944, Report submitted to the Trail Arbitral Tribunal. *U.S. Bureau of Mines Bull.*, 453.

EAST, C., 1968, Comparaison du rayonnement solaire en ville et à la campagne. *Cahiers de Geog. de Québéc*, 12, 81–9.

ELSASSER, W. M. and CULBERTSON, M. F., 1960, Atmospheric radiation tables. *Meteorol. Monograph, 4*, No. 23, Amer. Meteorol. Soc., Boston.

FANGER, P. O., 1970, *Thermal Comfort: Analysis and Applications in Environmental Engineering*. McGraw-Hill, New York.

FEDEROV, S. F., 1965, Evaporation from forest and field in years with different amounts of moisture. *Soviet Hydrol.*, 4, 337–48.

FLEAGLE, R. and BUSINGER, J., 1963, *An Introduction to Atmospheric Physics*. Academic Press, New York.

FLOHN, H., 1971, Saharization: Natural causes or management? *WMO Spec. Environmental Rep. No. 2*, WMO No. 312, World Meteorol. Organization, Geneva, 101–6.

FORSGATE, J. A., HOSEGOOD, P. H. and MCCULLOCH, J. S. G., 1965, Design and installation of semi-enclosed hydraulic lysimeters. *Agric. Meteorol.*, 2, 43–52.

GARNIER, B. J. and OHMURA, A., 1968, A method of calculating the direct short wave radiation income of slopes. *J. Appl. Meteorol.*, 7, 796–800.

GATES, D. M., 1965, Radiant energy, its receipt and disposal. In WAGGONER, P. E. (ed.), *Agricultural Meteorology*, *Meteorol. Monog.*, *6*, No. 28, Amer. Meteorol. Soc., Boston, 1–26.

GATES, D. M., 1972, *Man and his Environment: Climate*. Harper and Row, New York.

GAY, L. W., 1970, Energy balance estimates of evapotranspiration. In *Water Studies in Oregon*, Oregon Water Resources Instit., Corvallis.

GAY, L. W. and KNOERR, K. R., 1970, The radiation budget of a forest canopy. *Archiv. Meteorol. Geophys. Biokl.*, Ser. B., 18, 187–96.

GAY, L. W., KNOERR, K. R. and BRAATEN, M. O., 1971, Solar radiation variability on the floor of a pine plantation. *Agric. Meteorol.*, 8, 39–50.

GAY, L. W. and STEWART, J. B., 1974, Energy balance studies in coniferous forests, *Report No. 23*, Instit. Hydrol., Natural Environ. Res. Council, Wallingford, Berks.

GEIGER, R., 1965, *The Climate Near the Ground*. Harvard Univ. Press, Harvard, Mass.

GRIFFITHS, J. F., 1966, *Applied Climatology: An Introduction*. Oxford Univ. Press, Oxford.

HAGE, K. D., 1975, Urban-rural humidity differences. *J. Appl. Meteorol.*, 14, 1277–83.

HALITSKY, J., 1963, Gas diffusion near buildings. *Trans. Amer. Soc. Heating, Refrig., Air-Condit. Engr.*, 69, 464–85.

HANLON, J., 1972, Taming man-made winds. *New Scientist*, 54, 732–4.

HAYWARD, J. S., COLLIS, M. and ECKERSON, J. D., 1973, Thermographic evaluation of relative heat loss areas of man during cold water immersion. *Aerospace Medicine*, 44, 708–11.

HAYWARD, J. S., ECKERSON, J. D. and COLLIS, M. L., 1975, Thermal balance and survival time prediction of man in cold water. *Can. J. Physiol. Pharmacol.*, 53, 21–32.

HEINRICH, B., 1974, Thermoregulation in endothermic insects. *Science*, 185, 747–56.

HEMMINGSEN, A. M., 1960, Energy metabolism as related to body size and respiratory surfaces. *Rep. Steno Meml. Hosp., Copenhagen*, 9, Pt. 2, Copenhagen.

HICKS, B. B. and MARTIN, H. C., 1972, Atmospheric turbulent fluxes over snow. *Boundary-Layer Meteorol.*, 2, 496–502.

HOCEVAR, A. and MARTSOLF, J. D., 1971, Temperature distribution under radiation frost conditions in a central Pennsylvania Valley. *Agric. Meteorol.*, 8, 371–83.

HOLLAND, J. Z., 1971, Interim report on results from the BOMEX core experiment. *BOMEX Bull.* No. 10, NOAA, U.S. Dept. Commerce, 31–43.

HOLLAND, J. Z., 1972, The BOMEX Sea-Air interaction program: background and results to date. *NOAA Tech. Memo.* ERL BOMAP-9, NOAA, U.S. Dept. Commerce, 34 pp.

HOLMES, R. M., 1969, Airborne measurements of thermal discontinuities in the lowest layers of the atmosphere. Paper presented at Ninth Conf. Agric. Meteorol., Seattle, 18 pp.

HOLMGREN, B., 1971, Climate and energy exchange on a sub-Polar ice cap in summer. Pt F. On the energy exchange of the snow surface at Ice Cap station. Meteorol. Instit., Uppsala Univ., Uppsala.

HOUGHTON, H. G., 1954, On the annual heat balance of the Northern Hemisphere. *J. Meteorol.*, 11, 1–9.

HUSCHKE, R. E. (ed.), 1959, *Glossary of Meteorology*. Amer. Meteorol. Society, Boston.

IDSO, S. B. and JACKSON, R. D., 1969, Thermal radiation from the atmosphere. *J. Geophys. Res.*, 74, 5397–403.

IRVING, L., KROG, H. and MONSON, M., 1955, The metabolism of some Alaskan animals in winter and summer. *Physiol. Zool.*, 28, 173–85.

JARVIS, P. G., JAMES, G. B. and LANDSBERG, J. J., 1976, Coniferous forest. In MONTEITH, J. L. (ed.), *Vegetation and the Atmosphere, Vol. 2, Case Studies.* Academic Press, London, 171–240.

KALMA, J. D., 1970, Some aspects of the water balance of an irrigated orange plantation. Ph.D. thesis publ. Volcani Institut. Agric. Res., Bet Dagan, Israel.

KALMA, J. D. and BYRNE, G. F., 1975, Energy use and the urban environment: some implications for planning. *Proc. Symp. Meteorol. Related to Urban, Regional Land-Use Planning, Ashville, N.C.*, World Meteorol. Organiz., Geneva.

KEPNER, R. A., 1951, Effectiveness of orchard heaters. *Bull. No. 723*, Calif. Agric. Expt. Stn., Calif.

KOPEC, R. J., 1973, Daily spatial and secular variations of atmospheric humidity in a small city. *J. Appl. Meteorol.*, 12, 639–48.

KRAUS, E. B., 1972, *Atmosphere-Ocean Interaction.* Clarendon Press, Oxford.

LANDSBERG, H. E., 1954, Bioclimatology of housing. *Meteorol. Monog.*, 2, No. 8, Amer. Meteorol. Soc., 81–98.

LATIMER, J. R., 1972, Radiation measurement. *Int. Field Year for the Gt Lakes Tech. Manual Ser.*, No. 2, NRC/USNAS/IHD, Ottawa.

LEGGET, R. F. and CRAWFORD, C. B., 1952, Soil temperatures in water works practice. *J. Amer. Water Works Assoc.*, 44, 923–39.

LEMON, E. R., GLASER, A. H. and SATTERWHITE, L. E., 1957, Some aspects of the relationship of soil, plant, and meteorological factors to evapotranspiration. *Proc. Soil Sci. Soc. Amer.*, 21, 464–8.

LETTAU, H. H., 1970, Physical and meteorological basis for mathematical models of urban diffusion processes. *Proc. Symp. on Multiple-Source Urban Diffusion Models*, U.S. Environ. Protect. Agency, Publ. AP-86 Research Triangle Park, N.C., 2.1–2.26.

LEWIS, H. E., FOSTER, A. R., MULLAN, B. J., COX, R. N. and CLARK, R. P., 1969, Aerodynamics of the human microenvironment. *The Lancet*, 1273–7.

LINAWEAVER, F. P. Jr, 1965, *Residential Water Use Project.* Report II, Phase 2, Johns Hopkins Univ., Baltimore.

LIST, R. J., 1966, *Smithsonian Meteorological Tables*, 6th edn. Smithsonian Instit., Washington, D.C.

LONG, I. F., MONTEITH, J. L., PENMAN, H. L. and SZEICZ, G., 1964. The plant and its environment. *Meteorol. Rundsch.*, 17, 97–101.

LOWRY, W. P., 1967, *Weather and Life: An Introduction to Biometeorology.* Academic Press, New York.

LYONS, W. A. and OLSSON, L. E., 1973, Detailed mesometeorological studies of air pollution dispersion in the Chicago lake breeze. *Monthly Weather Rev.*, 101, 387–403.

MATHER, J. R., 1974, *Climatology: Fundamentals and Applications.* McGraw-Hill, New York.

MCNAB, B. K., 1966, The metabolism of fossorial rodents: a study of convergence. *Ecology*, 47, 712–33.

MCNAUGHTON, K. and BLACK, T. A., 1973, A study of evapotranspiration from a Douglas fir forest using the energy balance approach. *Water Resources Res.*, 9, 1579–90.

MILLER, D. R., ROSENBERG, N. J. and BAGLEY, W. T., 1973, Soybean water use in the shelter of a slat-fence windbreak. *Agric. Meteorol.*, 11, 405–18.

MONTEITH, J. L., 1957, Dew. *Quart. J. Royal Meteorol. Soc.*, 83, 322–41.

MONTEITH, J. L., 1959, The reflection of short-wave radiation by vegetation. *Quart. J. Roy. Meteorol. Soc.*, 85, 386–92.

MONTEITH, J. L., 1962, Attenuation of solar radiation: a climatological study. *Quart. J. Royal Meteorol. Soc.*, 88, 508–21.

MONTEITH, J. L., 1965a, Radiation and crops. *Exp. Agric. Rev.*, 1, 241–51.

MONTEITH, J. L., 1965b, Evaporation and environment. *Symp. Soc. Exptl. Biol.*, 19, 205–34.

MONTEITH, J. L., 1973, *Principles of Environmental Physics.* Edward Arnold, London.

MONTEITH, J. L. and SZEICZ, G., 1961, The radiation balance of bare soil and vegetation. *Quart. J. Royal Meteorol. Soc.*, 87, 159–70.

MONTEITH, J. L., SZEICZ, G. and WAGGONER, P. E., 1965, The measurement and control of stomatal resistance in the field. *J. Appl. Ecol.*, 2, 345–55.

MOORE, W. L. and MORGAN, C. W. (eds), 1969, *Effects of Watershed Changes on Streamflow.* Univ. of Texas Press, Austin.

MOUNT, L. E., 1968, *The Climatic Physiology of the Pig.* Edward Arnold, London.

MOUNT, L. E., 1974, The concept of thermal neutrality. In MONTEITH, J. L. and MOUNT, L. E. (eds), *Heat Loss from Animals and Man.* Butterworth, London, 425–39.

MUKAMMAL, E. I., 1965, Ozone as a cause of tobacco injury. *Agric. Meteorol.*, 2, 145–65.

MUNN, R. E., 1970, *Biometeorological Methods.* Academic Press, New York.

MUNN, R. E. and BOLIN, B., 1971, Global air pollution – meteorological aspects. A survey. *Atmos. Environ.*, 5, 363–402.

MUNRO, D. S., 1975, Energy exchange on a melting glacier. Unpubl. Ph.D. thesis, McMaster Univ., Hamilton, Ont.

NÄGELI, W., 1946, Weitere Untersuchungen uber die Windverhaltnisse im Bereich von Windschutzanlagen. *Mitteil. Schweiz. Anstalt Forstl. Versuchswesen*, Zurich, 24, 659–737.

NEIBURGER, M., 1969, The role of meteorology in the study and control of air pollution. *Bull. Amer. Meteorol. Soc.*, 50, 957–65.

NUNEZ, M., DAVIES, J. A. and ROBINSON, P. J., 1972, Surface albedo at a tower site in Lake Ontario. *Boundary-Layer Meteorol.*, 3, 77–86.

NUNEZ, M. and OKE, T. R., 1977, The energy balance of an urban canyon. *J. Appl. Meteorol.*, 16, 11–19.

OKE, T. R., 1973, City size and the urban heat island. *Atmos. Environ.*, 7, 769–79.

OKE, T. R., 1974, *Review of Urban Climatology 1968–1973*. W.M.O. Tech. Note No. 134, World Meteorol. Organiz., Geneva.

OKE, T. R., 1976a, The distinction between canopy and boundary layer urban heat islands. *Atmosphere*, 14, 268–77.

OKE, T. R., 1976b, Inadvertent modification of the city atmosphere and the prospects for planned urban climates. *Proc. Symp. Meteorol. Related to Urban and Regional Land-Use Planning, Asheville, N.C.*, World Meteorol. Organiz., Geneva, 151–75.

OKE, T. R. and EAST, C., 1971. The urban boundary layer in Montréal. *Boundary-Layer Meteorol.*, 1, 411–37.

OKE, T. R. and HANNELL, F. G., 1966, Variations of temperature within a soil. *Weather*, 21, 21–8.

OKE, T. R. and MAXWELL, G. B., 1975, Urban heat island dynamics in Montréal and Vancouver. *Atmos. Environ.*, 9, 191–200.

OLIVER, H. R., 1974, Wind-speed modification by a very rough surface. *Meteorol. Mag.*, 103, 141–5.

PATERSON, W. S. B., 1969, *The Physics of Glaciers*. Pergamon, Oxford.

PENWARDEN, A. D. and WISE, A. F. E., 1975, Wind environment around buildings. *Build. Res. Establ. Report*, Dept. Environ., HMSO, London.

PEREIRA, H. C., 1973, *Land use and water resources*. Cambridge Univ. Press, Cambridge.

PHILLIPS, D. W., 1972, Patterns of monthly turbulent heat flux over Lake Ontario. In ADAMS, W. P. and HELLEINER, F. M. (eds), *International Geography 1972*, Univ. Toronto Press, 180–84.

PITTS, J. N. Jr, 1969, Environmental appraisal: Oxidants, hydrocarbons and oxides of nitrogen. *J. Air Poll. Control Assoc.*, 19, 658–67.

PRIESTLEY, C. H. B., 1959, *Turbulent Transfer in the Lower Atmosphere*. Univ. Chicago Press, Chicago.

RAUNER, Ju. L., 1976, Deciduous forests. In MONTEITH, J. L. (ed.), *Vegetation and the Atmosphere*, Vol. 2, *Case Studies*, Academic Press, London, 241–64.

REIFSNYDER, W. E., 1967, Forest meteorology: The forest energy balance. *Int. Reviews Forest. Res.*, 2, 127–79.

RIPLEY, E. A. and REDMANN, R. E., 1976, Grassland. In MONTEITH, J. L. (ed.), *Vegetation and the Atmosphere*, Vol. 2, *Case Studies*, Academic Press, London, 349–98.

RIPLEY, E. and SAUGIER, B., 1974, Energy and mass exchange of a native grassland in Saskatchewan. In DE VRIES, D. A. and AFGAN, N. H. (eds), *Heat and Mass Transfer in the Biosphere.* Vol. I, *Transfer Processes in the Plant Environment*, Hemispheric Publ. Corp., New York, 311–25.

ROSS, J., 1975, Radiative transfer in plant communities. In MONTEITH, J. L. (ed.), *Vegetation and the Atmosphere, Vol. 1, Principles*, Academic Press, London, 13–55.

ROTTY, R. M. and MITCHELL, J. M. Jr, 1974, Man's energy and the World's climate. Paper to 67th Annual Meeting Amer. Instit. Chem. Engineers, Washington, D.C.

ROUSE, W. R. and WILSON, R., 1972, A test of the potential accuracy of the water-budget approach to estimating evapotranspiration. *Agric. Meteorol.*, 9, 421–46.

RUTTER, A. J., 1967, An analysis of evaporation from a stand of Scots pine. In SOPPER, W. E. and LULL, H. W. (eds), *Forest Hydrology*, Pergamon, Oxford, 403–17.

RUTTER, A. J., 1972, Evaporation from forests. In *Research Papers in Forest Meteorology*, Cambrian News (Aberystwyth) Ltd, Aberystwyth, 75–90.

RUTTER, A. J., 1975, The hydrological cycle in vegetation. In MONTEITH, J. L. (ed.), *Vegetation and the Atmosphere*, Vol. 1, *Principles*, Academic Press, London, 111–54.

SCHMIDT-NIELSEN, K., 1970, *Animal Physiology*, 3rd edn. Prentice-Hall, Inglewood Cliffs, N.J.

SCHOLANDER, P. F., WALTERS, V., HOCK, R. and IRVING, L., 1950, Body insulation of some arctic and tropical mammals and birds. *Biol. Bull.*, 99, 225–36.

SCHWERDTFEGER, P. and WELLER, G., 1967, The measurement of radiative and conductive heat transfer in ice and snow, *Archiv. Meteorol. Geophys. und Bioklim.*, Ser. B, 15, 24–38.

SELLERS, W. D., 1965, *Physical Climatology*. Univ. Chicago Press, Chicago.

SHAW, R. W. and MUNN, R. E., 1971, Air pollution meteorology. In MCCORMAC, B. M. (ed.), *Introduction to the Scientific Study of Atmospheric Pollution*, Reidel, Dordrecht-Holland, 53–96.

SLAGER, U. T., 1962, *Space Medicine*. Prentice-Hall, Englewood Cliffs, N.J.

SMAGORINSKY, J., 1974, Global atmospheric modeling and the numerical simulation of climate. In HESS, W. N. (ed.), *Weather and Climate Modification*, Wiley, New York.

STANHILL, G., 1965, Observations of the reduction of soil temperatures. *Agric. Meteorol.*, 2, 197–203.

STANHILL, G., 1970, Some results of helicopter measurements of albedo. *Solar Energy*, 13, 59–66.

STEADMAN, R. G., 1971, Indices of windchill of clothed persons. *J. Appl. Meteorol.*, 10, 674–83.

STEWART, J. B., 1971, The albedo of a pine forest. *Quart. J. Royal Meteorol. Soc.*, 97, 561–4.

SUMMERS, P. W., 1962, Smoke concentrations in Montréal related to local meteorological factors. *Symposium, Air over Cities*, U.S. Public Health Service, S.E.C. Tech. Report, A62-5, 89–112.

SUMMERS, P. W., 1964, An urban ventilation model applied to Montréal. Unpubl. Ph.D. thesis, McGill Univ., Montréal.

SUTTON, O. G., 1953, *Micrometeorology*. McGraw-Hill, New York.

SWINBANK, W. C., 1963, Long-wave radiation from clear skies. *Quart. J. Royal Meteorol. Soc.*, 89, 339–48.

SZEICZ, G., 1974, Gaseous wastes and vegetation. *Preprint Symp. on Waste Recycl. and the Environ.*, Royal Soc. (Canada), Ottawa.

SZEICZ, G., ENRODI, G. and TAJCHMAN, S., 1969, Aerodynamic and surface factors in evaporation. *Water Resources Res.*, 5, 380–94.

TENNEKES, H., 1974, The atmospheric boundary layer. *Physics Today*, 27, 52–63.

THEKAEKARA, M. P. and DRUMMOND, A. J., 1971, Standard values for the solar constant and its spectral components. *Nature*, Phys. Sci., 229, 6–9.

THOM, A. S., 1975, Momentum, mass and heat exchange of plant communities. In MONTEITH, J. L. (ed.), *Vegetation and the Atmosphere, Vol. 1, Principles*, Academic Press, London, 57–109.

THRELKELD, J. L., 1962, *Thermal Environmental Engineering*. Prentice-Hall International, London.

TURNER, D. B., 1969, *Workbook of Atmospheric Dispersion Estimates*. U.S. Dept. Health Educ. and Welfare, Public Health Service Publication No. 999-AP-26.

UNDERWOOD, C. R. and WARD, E. J., 1966, The solar radiation area of man. *Ergonomics*, 9, 155–68.

URBACH, F., 1969, Geographic pathology of skin cancer. In *The Biologic Effects of Ultraviolet Radiation*, Pergamon Press, Oxford, 635–61.

U.S. AEC, 1968, *Meteorology and Atomic Energy*. SLADE, D. H. (ed.), U.S. Atomic Energy Commission, Div. Techn. Inform., Oak Ridge, Tenn.

U.S. DHEW, 1970a, *Air Quality Criteria for Particulate Matter*. National Air Poll. Control Admin., U.S. Public Health Service, Publication No. AP-49.

U.S. DHEW, 1970b, *Air Quality Criteria for Photochemical Oxidants.* National Air Poll. Control Admin., U.S. Public Health Service, Publication No. AP-63.

VAN ARSDEL, E. P., 1965, Micrometeorology and plant disease epidemiology. *Phytopathology*, 55, 945–50.

VAN BAVEL, C. H. M., 1967, Changes in canopy resistance to water loss from alfalfa induced by soil water depletion. *Agric. Meteorol.*, 4, 165–76.

VAN STRAATEN, J. F., 1967, *Thermal Performance of Buildings.* Elsevier, Amsterdam.

VAN WIJK, W. R. (ed.), 1963, *Physics of Plant Environment.* North-Holland Publishing Co., Amsterdam.

VAN WIJK, W. R. and DE VRIES, D. A., 1963, Periodic temperature variations. In VAN WIJK, W. R. (ed.), *Physics of Plant Environment*, North-Holland Publishing Co., Amsterdam.

VARNEY, R. and MCCORMAC, B. M., 1971, Atmospheric pollutants. In MCCORMAC, B. M. (ed.), *Introduction to the scientific study of atmospheric pollution*, Reidel, Dordrecht, Holland, 8–52.

VEHRENCAMP, J. E., 1953, Experimental investigation of heat transfer at an air-earth interface, *Trans. Amer. Geophys. Union*, 34, 22–30.

WAGGONER, P. E., MILLER, P. M., and DE ROO, H. C., 1960, Plastic mulching-principles and benefits. *Bull. No. 634*, Conn. Agric. Expt. Stn, New Haven.

WHITE, W. H., ANDERSON, J. A., BLUMENTHAL, D. L., HUSAR, R. B., GILLANI, N. V., HUSAR, J. D. and WILSON, W. E. Jr, 1976, Formation and transport of secondary air pollutants: Ozone and aerosols in the St Louis urban plume. *Science*, 194, 187–9.

WILKINS, E. T., 1954, Air pollution aspects of the London fog of December 1952. *Quart. J. Royal Meteorol. Soc.*, 80, 267–71.

WILLIAMSON, S. J., 1973, *Fundamentals of Air Pollution.* Addison-Wesley, Reading, Mass.

YABUKI, K., 1957, Studies on temperatures of water layer in paddy fields. *Bull. Univ. Osaka Pref.*, Ser. B-7, 113–45.

YABUKI, K. and IMAZU, T., 1961, Studies on the temperature control of glasshouses I. On the temperature and heat balance in an empty unventilated glasshouse. *J. Japan Soc. Hortic. Sci.*, 30, No. 2.

YAMAMOTO, G., 1952, On a radiation chart. Tohoku Univ., Science Reports, Ser. 5, *Geophys.*, 4, 9–23.

YAP, D. and OKE, T. R., 1974, Sensible heat fluxes over an urban area – Vancouver, B.C. *J. Appl. Meteorol.*, 13, 880–90.

Supplementary reading

Part I

ANTHES, R. A., PANOFSKY, H. A., CAHIR, J. J. and RANGO, A., 1975, *The Atmosphere.* Charles Merrill, Columbus, Ohio.

BARRY, R. G. and CHORLEY, R. J., 1976, *Atmosphere, Weather and Climate*, 3rd edn. Methuen, London.

CHORLEY, R. J. and KENNEDY, B. A., 1971, *Physical Geography: A Systems Approach.* Prentice-Hall International, London.

DEACON, E. L., 1969, Physical processes near the surface of the Earth. In FLOHN, H. (ed.), *World Survey of Climatology, Vol. 2, General Climatology*, Elsevier, Amsterdam.

GEIGER, R., 1965, *The Climate Near the Ground.* Harvard Univ. Press, Cambridge, Mass.

MCINTOSH, D. H. and THOM, A. S., 1969, *Essentials of Meteorology.* Wykeham, Andover.

NEIBURGER, M., EDINGER, J. D. and BONNER, W. D., 1973, *Understanding our Atmospheric Environment.* Freeman, San Francisco.

PETTERSSEN, S., 1969, *Introduction to Meteorology*, 3rd edn. McGraw-Hill, New York.

ROSE, C. W., 1966, *Agricultural Physics.* Pergamon, London.

SELLERS, W. D., 1965, *Physical Climatology.* Univ. Chicago Press, Chicago.

Part II

BARTHOLOMEW, G. A., 1968, Body temperature and energy metabolism. In GORDON, M. S. (ed.), *Animal Function: Principles and Adaptations*, Macmillan, New York, 290–354.

FLOHN, H., 1969, Local wind systems. In FLOHN, H. (ed.), *World Survey of Climatology*, Vol. 2, *General Climatology*, Elsevier, Amsterdam, 139–71.

GATES, D. M., 1962, *Energy Exchange in the Biosphere.* Harper and Row, New York.

GATES, D. M., 1972, *Man and his Environment: Climate*. Harper and Row, New York.

GEIGER, R., 1965, see Pt I.

GEIGER, R., 1969, Topoclimates. In FLOHN, H. (ed.), *World Survey of Climatology*, Vol. 2, *General Climatology*. Elsevier, Amsterdam, 105–38.

HARDY, R. N., 1972, *Temperature and Animal Life*, Studies in Biology No. 35, Edward Arnold, London.

LOWRY, W. P., 1967, *Weather and Life: An Introduction to Biometeorology*. Academic Press, New York.

MONTEITH, J. L., 1973, *Principles of Environmental Physics*. Edward Arnold, London.

MONTEITH, J. L., 1975, 1976, *Vegetation and the Atmosphere*, Vols 1 and 2. Academic Press, London.

MUNN, R. E., 1966, *Descriptive Micrometeorology*. Academic Press, New York.

REIFSNYDER, W. E. and LULL, H. W., 1965, Radiant energy in relation to forests. *Tech. Bull. No. 1344*, USDA Forest Service, U.S. Govt Printing Office, Washington, D.C.

Part III

BACH, W., 1972, *Atmospheric Pollution*. McGraw-Hill, New York.

BATES, D. V., 1972, *A Citizen's Guide to Air Pollution*. McGill-Queen's Univ. Press, Montréal.

DETWYLER, T. R. and MARCUS, M. G., 1972, *Urbanization and Environment*. Duxbury Press, Belmont, Calif.

HESS, W. N. (ed.), 1974, *Weather and Climate Modification*. Wiley, New York.

LANDSBERG, H. E., 1970, Man-made climatic changes. *Science*, 170, 1265–74.

MCCORMAC, B. M. (ed.), 1971, *Introduction to the Scientific Study of Atmospheric Pollution*. Reidel, Dordrecht, Holland.

OKE, T. R., 1974, *Review of Urban Climatology, 1968–1973*. W.M.O. Tech. Note No. 134, World Meteorol. Organiz., Geneva.

PETERSON, J. T., 1969, *The Climate of Cities*. U.S. Dept. Public Health and Welfare, AP-59, Washington, D.C.

SCORER, R. S., 1968, *Air Pollution*. Pergamon, London.

SMIC, 1971, *Inadvertent Climate Modification*. MIT Press, Cambridge, Mass.

VAN EIMERN, J., KARSCHON, R., RAZUMOVA, L. A. and ROBERTSON, G. W., 1964, *Windbreaks and Shelterbelts*. W.M.O. Tech. Note No. 59, World Meteorol. Organiz., Geneva.

WILLIAMSON, S. J., 1973, *Fundamentals of Air Pollution*. Addison-Wesley, Reading, Mass.

Appendices

LATIMER, J. R., 1972, Radiation measurement. *Int. Field Year for the Gt. Lakes, Tech. Manual Ser.*, No. 2, NRC/USNAS/IHD, Ottawa.

NATIONAL PHYSICAL LABORATORY, 1970, *The International System of Units*. HMSO, London.

SZEICZ, G., 1976, Instruments and their exposure. In MONTEITH, J. L. (ed.), *Vegetation and the Atmosphere*, Vol. 1, *Principles*, Academic Press, London.

WADSWORTH, R. M. (ed.), 1968, *The Measurement of Environmental Factors in Terrestrial Ecology*. Blackwell, Oxford.

Author Index

Subject Index

Bold *type indicates main entries for multi-listed subjects*